Water Resource Management and the Law

Water Resource Management and the Law

Edited by

Erkki J. Hollo

Professor Emeritus, University of Helsinki, Finland

Edward Elgar
PUBLISHING

Cheltenham, UK • Northampton, MA, USA

Published by
Edward Elgar Publishing Limited
The Lypiatts
15 Lansdown Road
Cheltenham
Glos GL50 2JA
UK

Edward Elgar Publishing, Inc.
William Pratt House
9 Dewey Court
Northampton
Massachusetts 01060
USA

A catalogue record for this book
is available from the British Library

Library of Congress Control Number: 2017947095

This book is available electronically in the **Elgar**online
Law subject collection
DOI 10.4337/9781785369834

FSC
www.fsc.org
MIX
Paper from
responsible sources
FSC® C013056

ISBN 978 1 78536 982 7 (cased)
ISBN 978 1 78536 983 4 (eBook)

Typeset by Columns Design XML Ltd, Reading
Printed and bound in Great Britain by TJ International Ltd, Padstow, Cornwall

Contents

Contributors

Bankes, Nigel; Professor and Chair of Natural Resources Law, Faculty of Law, University of Calgary, and Adjunct Professor, University of Tromsø, Faculty of Law and KG Jebsen Centre for the Law of Sea. nd.bankes@ucalgary.ca

Belinskij, Antti; Professor of Environmental Law, Law School, University of Eastern Finland, and Research Professor, Finnish Environment Institute SYKE. Specialised in water, natural resources and nature conservation law, he heads legal research in the Winland project funded by the Finnish Strategic Research Council. He is Editor-in-chief of the leading environmental law journal in Finland (Finnish Environmental Law Review) and a member of the Board of the Finnish Society for Environmental Law. antti.belinskij@uef.fi

Coetzee, Hennie; Lecturer in the Department of Public, Constitutional and International Law at the University of South Africa (UNISA). B.Iuris, LLB (PU for CHE), PGDip Compliance and Risk Management (University of Johannesburg), LLM (NWU). His research focuses on aspects of environmental regulation and governance within sustainable development, with a specific focus on the regulatory and governance framework(s) of the nexus between energy security, water security and climate change. coetzjh@unisa.ac.za

Couzens, Ed; Associate Professor, University of Sydney Law School, Australia. BA Hons LLB (Wits) LLM Environmental Law (Natal & Nottingham) PhD (KZN). Attorney of the High Court, South Africa. He has participated in the research project 'Legal Framework to Promote Water Security' (WATSEC), financed by the Academy of Finland (268151). Main area of study and research: international and national environmental law: biodiversity and wildlife law specifically. Author of *Whales and Elephants in International Conservation Law and Politics*. couzens.ed@gmail.com

Couzens, Meda; Honorary Research Fellow, School of Law, University of KwaZulu-Natal, Durban, South Africa. Law (Babeş-Bolyai) MA (Bucharest) LLM (London) MChPr (UKZN). Doctoral candidate:

University of Leiden. Main area of study and research: international law relating to the protection of children. meda.couzens@gmail.com

Curran, Deborah; Associate Professor, University of Victoria Faculty of Law and School of Environmental Studies, Acting Executive Director Environmental Law Centre. Her research focuses on subjects of water law, regional sustainability and the pedagogy of field courses and clinical learning. She supervises Environmental Law Centre students working on environmental law projects for community organization and First Nation clients. Professor Curran was named the 2016 Land Champion by the Real Estate Foundation of British Columbia. dlc@uvic.ca

Dai, Liping; Associate professor (Hubei Water Affairs Research Centre, Hubei University of Economics), postdoctoral researcher (Utrecht Centre for Water, Oceans and Sustainability Law, Utrecht University). Doctoral degree in Water Law (Utrecht University 2015). Visiting researcher at the China International Water Law Programme of Xiamen University (China) in 2013, and at the Research Institute of Environment Law of Wuhan University (China) in 2016. Her research interests include water governance, urban resilience, and climate change adaptation. L.Dai@uu.nl

Fisher, Douglas; Emeritus Professor of Law at the Queensland University of Technology in Australia. He holds the degrees of MA, LLB and PhD of the University of Edinburgh. His involvement in natural resources law began in 1970 and he has pursued both academic and professional careers since then in Scotland, New Zealand and Australia. fishers74@icloud.com

Hollo, Erkki J.; Professor Emeritus of Environmental Law at the University of Helsinki. LLD (Helsinki Law Faculty), dr. iur. (University of Tübingen). Formerly Justice at the Finnish Supreme Administrative Court. Member of the Board of the Finnish Academy of Sciences and Humanities. Founder and chair of the Finnish Society for Environmental Law. Président d'honneur of the European Council for Agricultural Law (CEDR). About 300 publications on environmental and property law, legal sciences and comparative law. erkki.hollo@helsinki.fi

Kornfeld, Itzchak E.; Researcher (PhD) Faculty of Law, Hebrew University of Jerusalem. kornfeld.itzchak@mail.huji.ac.il

Kotzé, Louis J.; Research Professor of Law at the Faculty of Law, North-West University (Potchefstroom Campus), South Africa and Visiting Professor of Environmental Law at the University of Lincoln (UK). B.Com, LLB, LLM (cum laude), LLD (North-West University), PhD

(Tilburg University). Author, co-author, and co-editor of over 100 publications on South African and global environmental law. His research focuses on the Anthropocene, environmental constitutionalism, human rights, and global environmental governance. Louis.Kotze@nwu.ac.za

Kuokkanen, Tuomas; Docent of International Environmental Law, University of Eastern Finland, Docent of International Law, University of Helsinki. Doctor of Laws (University of Helsinki). Currently, Justice at the Supreme Administrative Court of Finland. Previously, he has worked as a Ministerial Adviser with the Ministry of Environment of Finland and as a part-time Professor of International Law at the University of Eastern Finland. He has published a number of articles on international environmental law. tuomas.kuokkanen@oikeus.fi

Mascher, Sharon; University of Calgary Faculty of Law, Professor. Professor Mascher's research focuses on the subjects of climate change law, environmental law and indigenous land governance. She teaches and supervises graduate students working in the areas of climate change regulation, environmental law and property law. Professor Mascher is the co-editor of the Journal of Environmental Law and Practice and an Honorary Fellow in the University of Western Australia Faculty of Law. smascher@ucalgary.ca

Nyanchaga, Ezekiel Nyangeri; Professor, Department of Civil and Construction Engineering, University of Nairobi. He is a qualified licensed practicing consulting engineer with over 37 years of professional experience in planning, design and supervision of water supply and wastewater engineering projects, irrigation and drainage, water resources engineering, contract management, technical audit, monitoring and evaluation. He has authored several books, e.g. *History of Water Supply and Governance in Kenya* (1895–2005). enyangeri@uanbi.ac.ke

Onestini, Maria; Director, Centro de Estudios Ambientales (CEDEA)/ Centre for Environmental Studies in Buenos Aires, Argentina. She has been an invited lecturer at several universities in different countries and is a consultant to various agencies of the United Nations and of other international and regional institutions. She has published widely in environment and development issues. rponesti@criba.edu.ar

Paloniitty, Tiina; LLM (University of Helsinki), doctoral candidate. Paloniitty's doctoral thesis deals with agricultural runoff governance and the regulation of adaptive water management in the EU. Previously Paloniitty's work has been published in, for example, the Journal of

Environmental Law and Journal of Human Rights and the Environment. tiina.paloniitty@helsinki.fi

Reese, Moritz; Senior researcher (Dr. iur.) at the Department for Environmental and Planning Law of the Helmholtz Centre for Environmental Research in Leipzig, Germany. Core fields of research are water law and governance as well as climate adaptation. Chair of the Governance Division of the Center for Advanced Water Research (CAWR) and German rapporteur to the European Water Law Network. He is also chairman of the European Environmental Law Forum (EELF) and editor in chief of the Journal for European Environmental & Planning Law. moritz.reese@ufz.de

Schmidt, Bram; Environmental lawyer at Stibbe, the Netherlands, with research experience in environmental law, sustainability and law and water law. BramSchmidt@live.nl

van Rijswick, Marleen; Professor (Dr./PhD) of European and Dutch Water Law at Utrecht University The Netherlands) and Visiting Professor of European Environmental Law at Université Panthéon-Assas, Paris (France). She leads the Utrecht Centre for Water, Oceans and Sustainability law. In 2015 she was Visiting Professor of International and European Environmental Law at Wuhan University (China). h.f.m.w.vanrijswick @uu.nl

Vihervuori, Pekka; LL.D. President of the Supreme Administrative Court of Finland from 2012. Previously justice of the said court, professor of environmental law or land law in four Finnish universities, and ministerial legislative counsellor. About 180 publications on the various fields of environmental law in a broad sense, and on procedural law. pekka.vihervuori@oikeus.fi

Acknowledgements

The present publication covers various aspects of water management, a topic of increasing importance worldwide. The chapters deal with legal aspects at the international as well as, randomly, at the national level. The first part of the book presents selected water right systems, and the following chapters introduce the main principles that are applied in water management strategies and projects. Such an approach should facilitate the comparison of models for solving conflict situations and for adopting sustainable and democratic strategies. The authors are renowned experts in their field of water and environmental law.

My thanks go to the Publisher and its editorial staff for the valuable support in the realization of this book project. I also express my special gratitude to Mr Otso Myrsky, editorial assistant at the University of Eastern Finland, for his thorough work on the technical finalization of the publication.

Erkki J. Hollo
Editor
Helsinki, February 2017

Introduction: the concept and traditions of water management law

Erkki J. Hollo

1 BACKGROUND REMARKS

As the basic natural source of life, water resources play a decisive role in the development and welfare of societies. Due to the economic and social value of the resource and related interests, conflicts between different categories of users as well as states have gradually led to the adoption of rules and guiding principles for the distribution and protection of water resources.[1] At the same time, legally supported priority positions concerning the use of waters have emerged. Such positions and priorities within groups of interested owners and users or actors enjoy at least occasional legal protection on a constitutional basis. This is a reality where some regions suffer from water scarcity,[2] public interests tend to restrict the full enjoyment of water priorities under the control of a national water policy, for instance concerning the right to establish water-consuming plants. Also, where states have no problems with water quantity, they are, for different reasons, interested in adopting national or regional water management plans as the basis for legal decision-making in permitting cases and for the organization of public water services. Partners to international water regimes and member states of regional alliances are obliged to follow jointly adopted requirements concerning environmental quality, planning procedures, international cooperation,

[1] See e.g. the Water Conflict Chronology, maintained by the Pacific Institute, <http://worldwater.org/water-conflict/> accessed 12 December 2016.

[2] The UN 'Water for Life Decade' states: 'Around 1.2 billion people, or almost one-fifth of the world's population, live in areas of physical scarcity, and 500 million people are approaching this situation. Another 1.6 billion people, or almost one quarter of the world's population, face economic water shortage (where countries lack the necessary infrastructure to take water from rivers and aquifers).' <http://www.un.org/waterforlifedecade/scarcity.shtml/> accessed 10 January 2017.

civil rights and more. This legal trend works in favour of international harmonization, which again enables a better approach to manage large water areas also beyond state borders.

Environmental philosophy and ethics emphasize the need to respect nature.[3] In order to build a foundation for such actions, legal theory has developed instruments, based on political programmes and environmental principles (soft law).[4] To some extent those principles are sector-specific and, in case of water law, they serve as a basis for further development of better and more dynamic practices (the basin principle). A functioning legal system for water management includes models, techniques and standards for solving imminent threats and meeting challenges, not only nationally but also internationally. The origin of soft law instruments is partly a consequence of the open structure of international agreements, as these are incomplete in wording, and enforcement allows for the use of guidelines, standards and recommendations. In environmental law, interpretative principles are developed and sometimes incorporated in hard, i.e. 'clear', law elements. Typical principles of this kind are the causation or polluter pays principle and the precautionary principle.[5]

One problem with soft law arguments is the denouncement of the categorical interpretation of hard law sources, since legal uncertainty in decision-making affects the management of projects as long as soft law instruments are not 'hardened' and tend to be political rather than legal. But the distinction between hard and soft law is not clear: environmental (hard) law consists of a large number of dynamic or discretionary elements, and the margin of 'valid' alternative interpretations and solutions is broad. Within this margin, soft law instruments tend to play a significant role by complementing traditional 'command and control' instruments with voluntary measures, for instance agreements by the project manager with the state or the municipality. One observes hardening of soft law instruments especially in the fields of environmentally based human rights and nature conservation. Here for instance the vague term 'interest of future generations' may be used as a soft argument with

---

[3] Henryk Skolimowski, *Eco-Philosophy* (Boston M Boyars 1981), 36: 'Eco-philosophy, perceived as global and comprehensive, is a process philosophy which is integrative, hierarchical and normative – self-actualizing with regard to the individuals, and symbiotic with regard to the cosmos.'

[4] For an overview of the topic see Gregory Shaffer and Mark A. Pollack, 'Hard vs. Soft Law: Alternatives, Complements and Antagonists in International Governance' (2010) Minnesota Law Review, Vol. 94, 706–89.

[5] Nicolas de Sadeleer (ed.), *Implementing the precautionary principle. Approaches from the Nordic countries, EU and USA* (Earthscan 2007), 10–35.

varying interpretations in different situations or, following Alan Boyle: 'An important advantage of the procedural interpretation of environmental right is that it avoids problems of anthropocentricity to the extent that such rights can be exercised on behalf of the environment ... They can also be employed in the interests of future generations.'[6]

A democratic concept of national water stewardship should function in line with the population's needs and expectations concerning health, food safety and private economy.[7] However, most kinds of water utilization require technical installations, operating institutions and, primarily, capacity to finance works, pumps, dams, purification plants and pipelines. A state or a municipality would hardly be capable of satisfying all needs in a growing community without participation of specialized actors and financing institutions. Such actors again would not be interested in investing without expectation of an economic profit or a legally protected status in the water management system. The involvement of investors in projects that are carried out in a globalized world by enterprises, without responsibility to investors, is a challenge to ecological and also to social sustainability.

2 THE ORIGIN OF WESTERN WATER LAW

The concept of water law as a modern legal discipline has its origin in the eighteenth and nineteenth centuries, when industrialization multiplied the modes of profitable utilization of waters for various purposes. Diversified rules were adopted for different categories of water management. Already at the beginning of the eighteenth century, environmental concerns drew the attention of legislators. Two main lines of the

---

[6] Alan Boyle, 'The Role of International Human Rights Law in the Protection of the Environment' in Alan Boyle and Michael Anderson (eds), *Human Rights Approaches to Environmental Protection* (OUP 1996), 62.

[7] The following definition of water stewardship is adopted by EWS (European Water Stewardship: 'Water Stewardship: Use of freshwater that is socially beneficial, environmentally responsible and economically sustainable. (Source: AWS Glossary):

(a) Socially beneficial: water use recognizes basic human needs and ensures long term benefits (including economic benefits) for local people and society at large.
(b) Environmentally responsible: water use maintains or improves biodiversity and ecological processes at the watershed level.
(c) Economically sustainable: water use is secure, reliable and financially viable in the long term.'

development of water legislation have been dominant. The first one is sectoral and manifests certain areas of water law as closed systems. This is the case in states where specific or sectoral laws are enacted for the establishment and maintenance of waterways, for water supply and sanitation, in some states also for the installation of hydropower plants. In addition, in several cases there is a general set of rulings (water acts) concerning all categories of water utilizations (Germany, Austria, Finland). Such general water law provides for valid principles of conduct, rules for compensation and protection of civil rights. The other line incorporates traditional water law in a broader environmental context: environmental 'codes' minimize the particularities of water law (Sweden). The integration of water regulations may also take place in the context of land-use planning (France).

It is common knowledge that documented rules concerning the utilization of watercourses already existed in ancient times, for instance in the Middle East.[8] The best known, but not oldest example is the Code of Hammurabi, dating from the seventeenth century B.C.[9] In the preamble to the Code one finds a vision of good governance, which in the region of Mesopotamia even today means well-organized water conditions: 'They have prospered under my protection. I have made the people lie down in well-watered pastures. I am the shepherd who brings peace.' Some examples of rules concerning the use of waters (Harper's translation):

> 53. If a man neglect to strengthen his dyke and do not strengthen it, and a break be made in his dyke and the water carry away the farm-land, the man in whose dyke the break has been made shall restore the grain which he has damaged …
>
> 55. If a man opens his canal for irrigation and neglect it and the water carry away an adjacent field, he shall measure out grain on the basis of the adjacent fields.
>
> 56. If a man opened up the water and the water carry away the improvements of an adjacent field, he shall measure out ten GUR of grain per GAN.

Classical Roman law dealt with water issues on a practical basis. According to Roman law, waters were classified as public domain (*res*

---

[8] For the history of water resources management see Thomas V. Cech, *Principles of Water Resources: History, Development, Management, and Policy* (3rd edn, John Wiley & Sons 2010), 9–14.

[9] The text, including the account of its historical origin is quoted from the Harper translation: Robert Francis Harper: The Code of Hammurabi, King of Babylon: About 2250 B.C. (The Lawbook Exchange, Ltd 1904).

publicae), an object belonging to everyone (*res communis omnium*). Private use of minor watercourses (*aquae privatae*) or access to major watercourses was possible as long as the utilization of public waters was not negatively affected. Private ownership of minor waters was therefore limited to small waterbodies.[10] The use of water resources was also under some circumstances submitted to authorization. Private disputes concerning modes of water utilization were dealt with on the basis of *actio in rem* (claim based on possession). The classification of waters as *res publicae* was consolidated by later codes, especially *Corpus Iuris Civilis*, the most famous legal code of the Roman Empire, enacted by the Byzantine Emperor Justinian (AD 535). The *Institutes* of this Code is considered fundamental for the further development of western legal systems, since it reflected early ideals of humanism[11] as natural law and introduced the separation of public (law of nations) and private law.[12]

The Roman law tradition has prevailed in most successor states of the Roman Empire. French water law in the thirteenth century developed and through centuries maintained the principle that sovereigns were empowered to control the use of watercourses. Presently there are three categories of waters, those under public domain (*cours d'eaux domaniaux*), the waters under private regime (*cours d'eaux non-domaniaux*) and the 'mixed' waters (*cours d'eaux mixtes*).[13] In modern times the state domain has been enlarged to cover not only the bed of navigable rivers but also the flow from tributaries and land-based sources (section 29 of the Act of 16 December 1964).[14] This corresponds with the basin principle, which in that period became a guiding concept for the planning of new water enterprises and for the overall management of

---

[10] The distinction or categorization of waters, especially private ones, was in the end not clear. Sea shores were not included in the system of watercourses; public waterways were characterized as being open for the use of everyone (*usus publicus*). For details see Alfred Ossig, *Römisches Wasserrecht* (Leipzig 1898), 47–140.

[11] Book I, Title I, '(1) Jurisprudence is the knowledge of matters divine and human, and the comprehension of what is just and what is unjust.' Translation by S. P. Scott (1912) <http://www.constitution.org/sps/sps02.htm/last> accessed 9 January 2017.

[12] Book I, Title II, 1–2.

[13] Jean-Louis Gazzaniga, Jean-Paul Ourliac and Xavier Larrouy-Castera, *Léau: usage et gestion* (Litec 1998), 116–42.

[14] Michel Prieur, *Droit de l'environnement* (4th edn, Dalloz 2001), 674–6.

water areas (so it is also present in the European Water Framework Directive 2000/60/EC[15]).

In other parts of the western legal orders the legal development was different. The distinction between public waters and private waters was not fully followed in the countries of the common law tradition. Common law maintained the Roman principle that flowing waters are *res publicae*, and flowing water is considered *publici iuris* in the sense that none has an exclusive title to the water itself.[16] The access principle of Roman law developed into the doctrine of riparian rights in the nineteenth century and still is characteristic for the water regimes in England and North America.

In old Germanic, including Scandinavian law, the title to use waters was connected to the acquisition of land for new agricultural settlement areas. Within settlement areas water resources were used or allocated as joint units (*Allmende*), especially for traffic and fisheries.[17] During the later development, specific rights of utilization were recognized, in particular for the construction of mills, water works and more. During the feudalist period in continental Europe the influence of landlords and crowns increased and led towards public governance of major watercourses. In most parts of Scandinavia that development did not take place and the principle of joint private property rights within communities prevailed.[18] Watercourses beyond the borders of communities were open to the public as joint property or *res communes omnium*, later declared as state dominium. According to the present federal system of Germany[19] watercourses are public goods (*öffentliche Sachen*). The legislative competence is shared between the federal state and the states (Länder); each state has its own water act. The states define the extension of private ownership to waters but, in principle, the uniform rules concerning planning and utilization of waters of the Water Management Act apply to

---

15 Directive 2000/60/EC of the European Parliament and of the Council of 23 October 2000 establishing a framework for Community action in the field of water policy [2000] OJ L327/1.

16 A.S. Wisdom, *The Law of Rivers and Watercourses* (3rd edn, Shaw and Sons 1976), 86.

17 Michael Kloepfer, *Umweltrecht* (2. Auflage, C.H. Beck 1998), 68–70.

18 The principle of private ownership to adjacent waters of a community had already been declared in Sweden in provincial laws of the thirteenth century. Annette Hoff, *Recht und Landschaft: Der Beitrag der Landschaftsrechte zum Verständnis der Landwirtschafts- und Landschaftsentwicklung in Dänemark ca. 900–1250* (Walter de Gruyter 2006), 18–22.

19 Water Management Act 1957 (Wasserhaushaltsgesetz) with amendments.

all water resources.[20] The federal legislation also transposes European water law.[21]

Comparative studies of national laws show that despite the differences in legislative structures concerning property rights and administration, the goal-setting in terms of water management projects is comparable. In states with centralized state control, national planning interests have a stronger guiding impact on the possibility of starting new installations than is the case in states with a private-initiative approach because, here, owners' participation may be required. Since global environmental regimes (UN water conventions) and supranational water guidance (European Union) influence national laws in a harmonized manner, differences in national legislative structures tend to decrease continuously.

3 CATEGORIES OF WATER RIGHTS

Water rights and interests may be structured into two groups, the original ones (owners, shareholders of water resources) and derived ones (operators and owners of licenced water installations and water services). A peculiarity particularly in developing states and within areas of indigenous populations is that the utilization of natural resources has historically been organized by customary law only – and with respect for the belief that Nature cannot be owned![22] In this context the later legislative and administrative measures of modern states may neglect such traditional positions and overrule them by legal decision-making in favour of, for instance, internationally financed water management projects. Also, in western traditions customary law serves historically as the foundation for water regimes but the legislators have incorporated those principles into written law (e.g. public use of private waters or the protection of minority peoples).

In addition to these individual property rights, including riparian rights, other 'rights', i.e. legally admitted modes of public use, have long traditions, based on the principle of accessibility to waters or openness of watercourses. Access rights are typical for systems based on the rule of public water ownership or public stewardship (Roman law). No less in

---

20 Rüdiger Breuer, *Öffentliches und privates Wasserrecht* (2. Auflage, München 1987), 50–56.

21 Ibid., 244.

22 James L. Merrill, 'Aboriginal water rights' (1980) Natural Resources Journal Vol. 20, 45–70.

states with private water rights or mixed systems (Germanic and Anglo-Saxon law), there is a strong public interest in allowing access to private waters for transportation and water catchment as well as for fisheries. Such public and neighbourhood civil rights are implemented by means of restrictions on owners' rights to forbid access to private waters; owners and possessors have to tolerate a reasonable use of their waters by the public.[23]

The human rights approach extends access rights to all citizens and obliges society to organize necessary services unless the water resource is naturally available otherwise. This category of the enlarged public of citizens and other individuals, when depending on access to water, should, in a democratic state, be entitled to have free access to water resources at a reasonable cost, notwithstanding the property-rights context.[24] These access rights are internationally acknowledged and in some states also constitutionally manifested.[25] In a modern society access to drinking water is organized by institutions, which in practice means that the public-user has to pay for the services, since financers of public services expect to be compensated for offering free access to the population, either by the user or the state administration.

4 MAINSTREAM OF INTERNATIONAL WATER LAW

Most large waterbodies straddle the borders of one or more states. National water rights and licences do not normally prevail over the interests of other states, and national authorities should not grant allocation permits that have negative influence on citizens' rights in neighbouring states. However, without agreements between states, an indirect impact on over-the-border rights is difficult to prevent. The international principle of proper conduct in inter-state relations would be suitable for such situations, but the use of the principle would be problematic due to the principle of state sovereignty.[26]

---

[23] In Scandinavian states usus publicus is called 'allemansrätt', in German law 'Gemeingebrauch'.

[24] See e.g. Article 27 of the South African Constitution: 'Everyone has the right to have access to … sufficient food and water …'

[25] Jonathan Verschuuren, 'Climate change adaptation and water law' in Jonathan Verschuuren (ed.), *Research Handbook on Climate Change Adaptation* (Edward Elgar Publishing 2013), 253–7.

[26] See e.g. the Finnish-Russian Agreement on the Utilisation of Transboundary Watercourses (1964), Article 4: 'Without a procedure, provided for in chapter

Water law originated as an international doctrine in the 1950s, when the United Nations started to adopt rules for cooperation between states in cases of water conflicts and crises. Before that, international case law had cemented certain leading principles to be applied especially in water conflicts.[27] In cases where there is no bilateral agreement between states, customary international law still applies today.[28] Such conflicts originated to some extent from the state sovereignty principle, and case law started to set out rules concerning the respect for justified needs of other states and in fact to some extent overruled the sovereignty principle.[29] Cooperation between states and regions presupposes consideration of the impacts on waters beyond state borders. Case-law concerning projects in transboundary watercourses proves that the involvement of neighbouring states or jurisdictions in planning and decision-making procedures facilitates the setting of mutually acceptable conditions for the project.[30] The state sovereignty argument must not be used to the extent that states or actors beyond the border suffer losses: '... the "Harmon Doctrine" of absolute territorial sovereignty should, one hundred years after it was enunciated, be laid to a richly-deserved rest.'[31] Unfortunately, under

---

II of this agreement, no measure may be executed in a transboundary watercourse or on the banks thereof where it is likely to cause damage or harm to the water area, fisheries, land, buildings within the territory of another Party or other property, or it may cause a risk of flooding or significant water scarcity, alter the streamlet, impede the use of traffic or floating ways, or in any other manner jeopardizes the public interest.'

[27] The Corfu Channel case, the Lake Lanoux case and others, see Tuomas Kuokkanen (ed.), *Seminal cases of international environmental law* (Helsinki 1999) passim.

[28] In the famous Trail Smelter case (United States v. Canada) the Tribunal Decision 18 April 1938 confirmed (Part three) that states shall refrain from causing damage in the other state when deciding on harmful activities; Patricia Birnie and Alan Boyle, *International Law and the Environment* (OUP 1992), 139–52.

[29] Charles B. Bourne, 'Contribution to international water resources law' in Slavko Bogdanović (ed.), *International Law of Water Resources* (Kluwer Law International 2001), 18–20.

[30] International Court of Justice (ICJ) Case concerning Pulp Mills on the River Uruguay (Argentina v. Uruguay), Judgment of 20 April 2010: '101. The Court points out that the principle of prevention, as a customary rule, has its origins in the due diligence that is required of a State in its territory. It is every State's obligation not to allow knowingly its territory to be used for acts contrary to the rights of other States (Corfu Channel (United Kingdom v. Albania).'

[31] Stephen C. McCaffrey, 'The Harmon Doctrine One Hundred Years Later: Buried, Not Praised' (1996) Natural Resources Journal Vol. 36, 965. 'The theory

difficult political or diplomatic conditions, conflicts may emerge into water crises, and eventually into 'water wars' between states. One reason for conflicts is that downstream states do not receive the necessary quantity of drinking water while upstream states are using water abundantly for irrigation and water installations.

International water law consists of two main sets of rulings, the law on the international sea and the law on transboundary watercourses or so-called international lakes and rivers. Within these regimes specified subcategories emerged, mainly in order to regulate water quality, oil spills, and waste dumping and, later, to meet challenges caused by climate change, including flood control, threats to food safety and social welfare as a whole. The work of the United Nations has resulted in several important international conventions, to start with the global 'Code' UNCLOS (UN Convention on the Law of the Sea), which was signed in 1982 and entered into force in 1994.[32] One particularity of this Convention is the establishment of exclusive economic zones (EEZ), that extend legislative competences of shore states beyond their territorial waters up to 370 kilometres from the shore line. This instrument enables states to control impacts from seabed-based activities and also facilitates the establishment of national installations in the close international sea. The regional Convention of the UNECE on the Protection and Use of Transboundary Watercourses and International Lakes, was adopted in Helsinki (1992) and entered into force in 1996. This Convention provides for detailed rules concerning activities affecting water resources in transboundary basins. Almost all states in the region of the United Nations Economic Commission for Europe (ECE) are parties to the Convention. A third international water convention of the UN is the Convention on the Law of the Non-navigational Uses of International Watercourses (1997, entered into force in 2014). This global Convention is the main treaty governing shared freshwater resources internationally. According to this Convention (Article 3),

---

that states have exclusive or sovereign rights over the waters flowing through their territory which they can use regardless of their infringement of the rights of other states.' (Definition in *The Manual of Human Rights* by the Council of Europe, 2012, 136), <http://www.echr.coe.int/LibraryDocs/DH_DEV_Manual_Environment_Eng.pdf/> accessed 9 January 2017.

[32] Within the UN the International Maritime Organization (IMO) is the specialized agency with responsibility for the safety and security of shipping and the prevention of marine pollution by ships.

> Watercourse States shall in their respective territories utilise an international watercourse in an equitable and reasonable manner. In particular, an international watercourse shall be used and developed by watercourse States with a view to attaining optimal and sustainable utilisation thereof and benefits therefrom, taking into account the interests of the watercourse States concerned, consistent with adequate protection of the watercourse.

In addition to setting up general and, to some extent, detailed provisions for international respect and cooperation in matters concerning the sea and transboundary waters, the Conventions also commit to modern environmental principles, in particular the basin principle and the polluter pays principle.[33] In addition, general international environmental agreements and principles apply for the most part also to water resources. This is in line with the modern environmental integration principle[34] and concerns in particular the environmental quality control and the regulation of civil rights concerning participation. The integration principle is repeatedly pronounced in the international context; for instance, the UNECE Convention on the Protection and Use of Transboundary Watercourses and International Lakes (1992) declares (Article 3.i) that '[S]ustainable water-resources management, including the application of the ecosystems approach, is promoted ...' For states without former national legislation in this area the development of international water law favoured and intensified the elaboration of national provisions. This kind of interaction between national and international regimes has been substantially fruitful for the adoption of a globally acceptable concept of water management, concerning especially the responsibility of states and authorities for the sustainable maintenance of water resources and to safeguard the fundamental civil rights of the population.

Along with international water conventions there are numerous bilateral agreements for transboundary waters between the states concerned. Such agreements often have their background in the principles of

---

[33] The polluter pays principle, originally developed by the OECD in the 1960s, is based on the calculation that environmental costs for prevention are lower than the costs for reparation of a polluted site. Also, in case of reparation the lesson had been that in the end the state often had to take the costs because the original polluters could not be found or made liable.

[34] In European law, one good example is the IPPC (Integrated Pollution Prevention and Control) Directive 96/61/EC: '8. Whereas the objective of an integrated approach to pollution control is to prevent emissions into air, water or soil wherever this is practicable, taking into account waste management, and, where it is not, to minimize them in order to achieve a high level of protection for the environment as a whole.'

national laws but they may be enriched or motivated by the cited UN Conventions. An important source for the development of rules for transboundary waters is the set of rules elaborated by the International Law Association (ILA). The work of the ILA has functioned as a support for the preparation of the UN Conventions on transboundary waters (the Helsinki Rules 1966). Despite lack of legal validity the ILA Rules enjoy considerable recognition worldwide. Today the Rules have enlarged their focus to cover environmental threats as well (the Berlin Rules 2004).[35]

5 CATEGORIES OF WATER MANAGEMENT PROJECTS

Water governance is a term used for the organization and management of waters and services. According to Rogers and Hall 'water governance refers to the range of political, social, economic and administrative systems that are in place to develop and manage water resources, at different levels of society, and the delivery of water services (Global Water Partnership 2002)'.[36] It is not reasonable to study water-related interests without consideration of the type or mode of water use. Groundwater plays an important role in terms of water supply; surface watercourses serve different purposes. Also, the technical realization of water-related projects varies from case to case, from region to region and from state to state. At a very general level water management, water strategies and water legislation are composed of two modern approaches and there is also a strong link between them, namely (1) the profitable or economic use of water resources and (2) the protection of the aquatic environment. Modern law to a high extent combines these modes of water management in one common principle, the principle of sustainable and environmentally friendly use of all water resources.

What are, globally, the guiding principles when initiating new water management projects? There are varying perspectives, some of them are purely financial by nature, and others are more social or development-oriented. Generally, all water management projects are based on the principle of economic viability. No water management initiative is successful without the involvement of economic interests. This does not

---

[35] Salman M.A. Salman, 'The Helsinki Rules, the UN Watercourses Convention and the Berlin Rules: Perspectives on International Water Law' (2007) Water Resources Development, Vol. 23, No. 4, 625–40.

[36] Peter Rogers and Alan W. Hall, *Effective Water Governance* (GWP Novum 2003), 7.

mean that water management has a balanced or equitable ground of interest considerations. In reality, the situation may be the opposite. In particular, mega size water management projects may locally affect existing human societies, local economies and environmental conditions in a hard and negative way. In such situations it is reasonable to argue also in favour of open participation, public hearings, human rights, minority protection and more. But there is also another, strongly opposing reality in addition to that of preservation, namely the reality and the needs of growing and developing economies. Often, economic sustainability requires rarely internationally supported and subsidized large-scale infrastructures, traffic and industrial areas. Satisfactory access to water resources is not naturally available in all such cases. Therefore internationally funded developing projects in the land-use sector are supported by or even based on water management projects of remarkable extension. In addition to growth, other aspects also demand improvement in water conditions, especially health and food safety. Dellapenna and Gupta state in a study concerning the evolution of law and politics of water that

> Global water governance is diffuse and the related international law, while binding, suffers from the shortcomings of international law generally. Furthermore, global water goals are defined but scarcely adequate or achievable … Global water meetings and discussions provide only an elusive arena of global governance and global water 'ideology' is marketed through a variety of instruments worldwide.[37]

Water management projects may be classified into different categories according to their purpose. Enforcement strategies and financing systems vary depending on the impact and the extent of the project. Examples of such categories of water management are projects aimed at the improvement of water supply, health structures, traffic networks, energy supply, food production and agriculture. In addition to the operative task of a project, the infrastructure, including installations or the creation of industrial and urban sites, is also significant. This kind of management project will here be classified as 'growth-oriented water management projects'.[38]

---

[37] Joseph W. Dellapenna and Joyeeta Gupta (eds), *The Evolution of the Law and Politics of Water* (Springer Science 2009), 5.

[38] The World Bank project 'The first component is investment in integrated water resources development and management. This component includes three sub-components. The first sub-component is identification, preparation and implementation of selected priority investments through a participatory planning

Other types of water management projects are initiated in particular to ensure maintenance or balance of resources in a changing environment. These projects are here characterized as 'adaptive water management'. Examples are found in modern land-use strategies that are adopted in order to combat negative impacts caused by natural or anthropocentrically induced erosion or deforestation. In relation to climate change the term 'adaptation' or 'adaption' is used to cover measures or strategies aimed at the limitation of further destruction. The intention may also be to adapt human societies to cope with new agricultural technologies and services. In that sense adaptive management projects may be growth-oriented but the policy basis is different. Behind adaptive measures there is often a direct national or regional strategy, not a direct economic incentive. It seems that social losses caused by adaptive water management projects are of minor intensity compared to those caused by growth-based water management projects. On the other hand, it seems that the differences between the two types of water management projects are vanishing because development projects often include adaptive aspects as well.

6 INVESTMENT PRINCIPLES FOR WATER MANAGEMENT PROJECTS

Today the regulations and the technological solutions tend to be softer, more sophisticated than was the case before the 1980s. Financing institutions have started to stand up for human and environmentally friendly values in developing countries and they require that measures, while still being economically profitable, are sufficiently adapted to local and regional conditions and support.[39] In addition to development-oriented projects, targeted nature conservation is one category of water

---

process in the Kyoga and Upper Nile Water Management Zone (WMZs) … The second component is infrastructure investments in urban water supply and sanitation/sewerage and catchment/source protection. This component includes two sub-components. The first sub-component is town water supply and sanitation under National Water and Sewerage Corporation (NWSC) …' <http://projects.worldbank.org/P123204/water-management-development-project?lang=en/> accessed 9 January 2017.

[39] See e.g. the Uttarakhand Watershed Sector Project in India, financed by the World Bank. The objective of the Sustainable Land, Water and Biodiversity Conservation, and Management for Improved Livelihoods Project is to improve the productive potential of natural resources and increase incomes of rural inhabitants in selected watersheds through socially inclusive, institutionally

management project, in particular in the framework of biodiversity strategies.[40] A large number of endangered sites and ecological systems are situated in aquatic areas, mainly wetlands.[41]

Some developing countries and their governments may welcome even mega size water management projects if there is an international financial basis and the outcome would be a benefit for the majority of the population. But how does the national system, often without proper legislation, ensure that the rights of citizens, rights to homes and life, rights of minorities, human rights and serious environmental considerations are taken into account when projects are to be carried out by a foreign project manager? In the history of large water management projects it has not been unusual for foreign or international financing to be granted without any realistic conditions concerning respect for human rights and environmental values. Today it seems that most international banking institutions expect from project managers a solid analysis-based plan that takes into account internationally practiced rules concerning social and environmental protection. This should be the conduct code for all international investors. The World Bank is the leading representative for modern project investment practices. Environmental and other conditions are determined by agreement with the borrower; the investor still has no risk or legal liability, other than suspension of the loan, for negative impacts caused by the financed project. Mohammed Abdelwahah Bekhechi states:

---

and environmentally sustainable approaches, <http://documents.worldbank.org/curated/en/887531468051261974/India-Sustainable-Land-Water-and-Biodiversity-Conservation-and-Management-for-Improved-Livelihoods-in-Uttarakhand-Watershed-Sector-Project-additional-financing/> accessed 9 January 2017.

[40] Most of the large-scale nature protection programmes are initiated by the Convention on Biological Diversity (1992), Article 6: 'Each Contracting Party shall, in accordance with its particular conditions and capabilities: (a) Develop national strategies, plans or programmes for the conservation and sustainable use of biological diversity or adapt for this purpose existing strategies, plans or programmes …' At the international level, the main conventions under the UN are seen as tools for the enforcement of biodiversity strategies (Secretariat of the Convention on Biological Diversity, Technical Series 40, Transboundary water resources management (2008, 12) <https://www.cbd.int/doc/publications/cbd-ts-40-en.pdf/> accessed 14 January 2017.

[41] The Ramsar Convention on protection of wetlands of international importance was adopted in 1971. For the future the task of its strategies is to achieve reliance in ecosystems, to contribute to mitigation of climate change and to combat desertification, <http://www.ramsar.org/document/the-fourth-ramsar-strategic-plan-2016-2024/> accessed 10 January 2017.

> … projects are the borrowers' own projects where the Bank plays the role of a supportive financier … These general covenants requiring the borrower to exercise due regard to the environment are viewed as no more than non-binding guidelines for project execution and operation, especially when they do not refer to specific standards to be applied to project activities. These covenants are based on a vague standard ('due regard'), and do not provide for remedies.[42]

One problem with international financing is corruption, which will just be mentioned here. If states and state authorities act as initiators for internationally financed development projects they should also control compliance with legislative conditions in their territory. Here a conflict of interests cannot be excluded because the state may be keen to get a politically supported water management project financed on profitable terms and therefore turn a blind eye when environmental risks occur. It may also be that the developing state has no enforceable environmental legislation and the project manager makes use of the lack of control.

A report by Aled Williams and others presents the example of Kyrgyzstan in relation to corruption in local water management. The authors report the following findings: 'Poor water infrastructure in Kyrgyzstan from the Soviet era led international donors to support investments in agricultural irrigation and potable freshwater systems. The financial investments made, however, did not always underpin improvements in local water delivery and the Kyrgyz Vice President once noted that "the lion's share of the credit was stolen".'[43] Also this case shows that international financing institutions play an important role in control of the enforcement of a financed project. The conditions of financing agreements should, in terms of project enforcement, be as strict as contractual provisions legally can be in this respect. One solution might be that the project manager would have to apply in the project state the same rules, practices and principles concerning environmental liability and human rights as are required in the home state of the project manager. One could, theoretically, even envisage a system where the

---

42 Mohammed Abdelwahah Bekhechi, 'Some Observations Regarding Environmental Covenants and Conditionalities in World Bank Lending Activities' (1999) Max Planck UNYB 3, 302.

43 Aled Williams, Gulnaz Isabekova, Kubanychbek Ormushev, Toktobek Omokeev and Natalia Zakharchenko, 'Leaking projects: Corruption and local water management in Kyrgyzstan' (U4 Practice Insight 2013:3), 1(-8). The authors mention that 20 per cent of the World Bank investment was lost to corruption.

home state authority of the project manager had to decide upon the minimum of environmental conditions to be respected for the project in a third state.

7 FINAL REMARKS

Due to the geological and climatic characteristics of water resources as a basin or a flow, the legal settlement of rights to use water turns out in most cases to be artificial but also loaded with changing demands and safeguards. Comparative studies report different ways of addressing the use of and rights to water resources legally but, basically, law requires uniformity, equality and universality. The diversity of water resources in location and importance constitutes problems when striving for uniform legal rules for the protection of individual needs, as is the case for instance in the European Union and also in the international conventions on transboundary waters. As a result, legal history has developed certain, still rather general, rules or principles to serve as guidance for the utilization of rights to waters. One additional difficulty is related to the fact that different types of large water management projects are composed of several measures on land and on water. In addition, the often far-reaching impacts vary from local nuisance to erosion, flood and even climatic problems. An integrated and balanced legal and financial model for large-scale water management projects should be grounded on political, preferably international, strategies, adaptive land use planning, settlement of legal disputes and more. Modern and future law on water management law should at the same time be practice-oriented, dynamically precautionary and predictive for future risks and changes.

PART I

The role of public and private law in the context of water allocation

1. Common law and public domain approaches to water governance – an Australian perspective[1]

Douglas Fisher

INTRODUCTION

Access to water and use of water are critical issues for individuals, corporate and institutional entities, all levels of government and nation states. The legal arrangements for water resources governance are equally critical: in particular, who controls access and use and according to what criteria. Is it the nation state in the exercise of rights of sovereignty; a public sector entity in the exercise of civil law, common law, codified or statutory rights; a group of individuals who are part of an interested community in the exercise of rights of common; or an individual person or a juridical entity in the exercise of a right of property? Is the exercise of these rights limited in any way? What is the nature of these rules?

There are perhaps three ways of responding to these questions. These responses reflect the characteristics of the relevant legal arrangements: those based on practice and custom; a rules based system according to which the content of these rules is public knowledge; and an administration based system according to which the content of each person's rights and duties is determined by a public sector agency exercising a discretion or in accordance with a set of criteria of which there is public knowledge. These are complex and intellectually demanding questions that invite a discussion of a range of normative and legal issues. This chapter seeks to address these issues in their wider context and more

[1] This chapter builds upon material discussed in Fisher, Douglas *Water Law* (LBC Information Services 2000) (hereafter *Water Law*) and Fisher, Douglas *The Law and Governance of Water Resources* (Edward Elgar Publishing 2009) (hereafter *Water Resources*) and analyses it from the perspective relevant to this chapter.

specifically from the perspective of the way the legal arrangements in Australia have developed in different ways over the years.

1 THE DOCTRINAL FOUNDATIONS OF WATER AS A COMMON RESOURCE – A COMPARATIVE PERSPECTIVE

1.1 Water as a Common Resource in International Law[2]

The normative and legal issues mentioned in the abstract have arisen not fortuitously but rather as a result of a range of hydrological, geographical, ecological, economic, social, cultural and political factors. Although international law in its current form is a relatively recent phenomenon, its acknowledgement of water as a common resource is a current reflection of the doctrinal debate that has occurred over thousands of years. A brief review of water as a common resource in international law is a useful point of commencement of this wider review of the doctrinal foundations of water as a common resource.

An international watercourse may flow through two or more nation states whose need for and use of the water may differ considerably. The use by the upstream state may impact upon the needs of the downstream state and vice versa. Similarly states adjacent to an inland lake may have competing interests in the water. Out of these circumstances there emerged in international law the doctrine of a community of interest. As early as 1911 the Madrid Declaration, an unauthoritative instrument of the Institute of International Law, recognised that 'riparian states with a common stream are in a position of permanent physical dependence on each other which precludes the idea of the complete autonomy of each state in the section of the natural watercourse under its sovereignty'.[3]

States have always enjoyed the capacity to negotiate a consensual arrangement about the use of these shared water resources. In the absence of such an agreement between states, the international judiciary was originally hesitant to impose restrictions on the exercise of their sovereignty. However, as early as 1985 the approach of the international judiciary was summarised in this way:

---

[2] Fisher, *Water Resources* (n 1), 114–22, 192–3.

[3] Food and Agriculture Organisation of the United Nations, 'International Regulation regarding the Use of International Watercourses for purposes other than Navigation' in *Sources of International Water Law* (Food and Agriculture Organisation 1998), 269–70.

> The common import of all these decisions is that where a river flows between two or more states or similar entities, each of those states or entities has a vital interest in the waters of the river and none has an absolute right to them; consequently, every state must, in the exercise of its sovereignty, consider the needs of the neighbouring state.[4]

The idea of water as a shared or common resource had become the basis of the approach of international law to the governance of water resources.

But according to what criteria were the needs of neighbouring states to be determined? In 1958, the International Law Association introduced the doctrine of reasonable and equitable use in these terms:

> Except as otherwise provided by treaty or other instruments or customs binding upon the parties, each co-riparian state is entitled to a reasonable and equitable share in the beneficial uses of the waters of the drainage basin. What amounts to a reasonable and equitable share is a question to be determined in the light of all the relevant factors in each particular case.[5]

It was not until 1997 that this approach was formalised by the Convention on the Law of the Non-navigational Uses of International Watercourses[6] which entered into effect in 2014. The fundamental principle stated in article 5(1) provides:

> Watercourse states shall in their respective territories utilize an international watercourse in an equitable and reasonable manner. In particular, an international watercourse shall be used and developed by watercourse states with a view to obtaining optimal and sustainable utilization thereof and benefits therefrom, taking into account the interests of the watercourse states concerned, consistent with adequate protection of the watercourse.

Structurally, this limitation on the exercise of sovereign rights comprises four duties: a duty to use in an equitable and reasonable manner; a duty to use and develop with a view to obtaining optimal and sustainable utilisation; a duty to take into account the interests of watercourse states;

---

[4] Godana, B.A. *Africa's Shared Water Resources* (Francis Pinter Publishers 1985), 47–8.

[5] Food and Agriculture Organisation of the United Nations, 'Resolution on the Use of the Waters of International Rivers' in *Sources of International Water Law* (Food and Agriculture Organisation 1998), 285–6.

[6] Convention on the Law of the Non-Navigational Uses of International Watercourses 1997 (1998) *International Legal Materials* 36, 700–718.

and a duty for the utilisation of a watercourse to be consistent with its adequate protection.[7]

What emerges is a rules-based system for the governance of international watercourses. There is no involvement of international institutional decision makers. The states in question decide how to comply with these rules subject only to a judicial determination of any dispute. Water in these circumstances has become a shared or common resource capable of use in accordance with a set of rules formulated in general but objective terms and able in accordance with the supporting provisions in the Convention to be applied in specific circumstances.

1.2 The Range of Approaches to Water as a Common Resource

These rules of international law are of recent origin. Do they represent a new doctrinal approach or are they reflective of ancient doctrine? Individuals as such or as members of a group have been using water to which they have access since time immemorial. The source of these arrangements has variably been practice, custom, judicial recognition or a code. In modern parlance this has been the outcome: 'most natural resource systems used by multiple individuals can be classified as common-pool resources.'[8]

What is the legal position when the individual accesses the common pool and uses the resource? It has been described in this way in relation specifically to water:

> While water in a stream channel, a lake or sea, or an underground aquifer appears as an undifferentiated mass resource – the quintessential 'common pool' – water extracted from the surface or underground body is an alienable quantity. It can be stored, measured, transported, even bought and sold as a commodity, and consumed. A quantity of water captured for consumption by one user is unavailable to another user, at least, until it is returned to the body of water from which it was taken.[9]

The critical question is the set of rules according to which the water can be extracted. Who has the right to extract: any member of a group, a

---

[7] Fisher, *Water Resources* (n 1), 121.

[8] Ostrom, E. 'Reformulating the commons' in Joanna Burger, Elinor Ostrom, Richard Norgaard, David Policansky and Bernard D. Goldstein (eds), *Protecting the Commons* (Island Press 2001), 18.

[9] Schalger, Edella and Blomquist, W. 'Water resources: the southwestern United States' in Joanna Burger *et al.* (eds), *Protecting the Commons* (Island Press 2001), 134–5.

particular member of a group, the riparian landowner, the surface landowner, or a person authorised by a regulatory entity to do so? The reference to 'owner' immediately acknowledges the relevance of the complex notion of property.[10] And are there any limitations on the exercise of the right or on the grant of authority? The answer depends upon the specific rules of the legal system in question. A number of examples illustrate the range of approaches.

1.3 The Structure of Chinese Law[11]

The approach according to early Chinese law was a combination of custom and central administration. A 'sophisticated administrative system linked the Emperor with the local community of users'.[12] But if this system provided no solution, then 'ancient usages and precedents'[13] filled the gap. Thus individuals had no rights – only duties. In this way, 'everyone seems to have had duties, with rights reduced to the positive manifestation of other people's duties'.[14] The governing principle of Chinese law has been 'water equalization'.[15] This was realised by the imposition of a range of duties on regulators, distributors and users alike:[16] for example, the duty to provide one's labour for one month during the year at the disposal of the administration.[17] The sanctions for breach of the duties were penal. This approach has remained in place. Thus 'the basic principles of water equalization, punishment for offences relating to waters and responsibility of local officials, persist up to modern times'.[18]

1.4 The Focus of Islamic Law[19]

Although Islamic water law has been codified, 'the arrangements for managing water resources in Islamic countries have traditionally been

---

10 Fisher, D. 'Rights of property in water: confusion or clarity' (2004) *Environmental and Planning Law Journal* 21(3), 200–226.

11 See generally Fisher, *Water Resources* (n 1), 78, 80–81, 163–4.

12 Caponera, D.A. 'Water law principles in the Chinese legal system', (1960) *Indian Journal of International Law*, 263.

13 Ibid.

14 Ibid., 248.

15 Ibid., 250, 252.

16 Ibid., 250, 255.

17 Ibid., 249.

18 Ibid., 252.

19 See generally Fisher, *Water Resources* (n 1), 78, 102–3, 165–8.

very much influenced by local practice and custom and by the values stated in the Qur'an'.[20] Water, it has been stated, 'lay at the heart of the legal system'.[21] Indeed, the word *shari'a* meant water law rather than law in general. Out of this there developed a system of rights and duties. More particularly:

> The connection between *shari'a* as a generic term for Islamic law, and *shari'a* as the path as well as the law of water, is not a coincidence, and the centrality of water in Islam is obvious in the economic as well as ritualistic sense. What is more important however, is that the jurists – the exponents and expounders of the *shari'a* – did not fail to develop, in answer to this centrality, a highly sophisticated system of rules.[22]

Thus the 'ritualistic' and moral underpinnings of this system were transformed into rules which largely comprised a set of duties based upon a duty of care owed generally to the public. This required a person 'to exercise a higher standard of care where his action adversely affects the interest of the public or may cause injury to it'.[23] The duties included obligations for human kind to care for the earth; not to damage the environment; not to overuse water; and to keep water free from pollution.[24] It was on these foundations that Islamic law was codified.

1.5 The Foundations of Roman Law[25]

The doctrinal foundations of Roman law acknowledge the complex notion of property. Fundamentally the holder of a right of property has exclusive access to, use of and control of the thing or *res* in question. Roman law classified *res* in four ways:

- *res nullius* – things belonging to no person,
- *res publicae* – things belonging to the state,

---

20 Fisher, *Water Resources* (n 1), 165.

21 Mallat, C. 'The quest for water use principles: reflections on Shari'a and custom in the Middle East', in J.A. Allan and Chibli Mallat (eds), *Water in the Middle East: Legal, Political and Commercial Implications* (Tauris Academic Studies 1995), 128.

22 Ibid.

23 Ahmad, A. 'Islamic water law as a comparative model for maintaining water quality' (2000) *The Journal of Islamic Law and Culture* 5, 173.

24 Ibid., 174–8; and Roughton, G.E. 'The ancient and the modern: environmental law and governance in Islam' (2007) *Columbia Journal of Environmental* Law 32, 6–8.

25 See generally Fisher, *Water Resources* (n 1), 97–8, 164–5.

- *res communes* – things in which all members of the relevant group have an interest,
- *res propriae* or *res privatae* – things belonging to an identifiable person.

According to the Institutes of Justinian, *res communes* were the air, running water, the sea and the seashore.[26] Classical Roman law during the republic recognised a close relationship between land and water.[27] 'Thus water physically connected with public land was *aqua publica* and water physically connected with private land was *aqua privata* or *aqua propria*.'[28]

It was, however, access rather than ownership that lay at the foundations of this system. Accordingly:

> Rivers were regarded as public things if they were perennial, i.e. normally flowing all the year round. Although such a river could not be owned, its beds and banks could be; but it seems that the riparian owners (i.e. the owners of the river banks) could not prevent the public from using the banks to enjoy the river. Remedies were normally available to protect the enjoyment or use of public things.[29]

If so, rivers as *res publicae* disclosed some of the features of *res communes*. In any event, it may be concluded that rivers – running water according to Justinian – were regarded as common or shared resources. But this governance system changed.

During the later years of the Roman empire, this system of water resources governance based upon public waters controlled by the state and private waters controlled by enforceable legal rules was increasingly overtaken by a system of regulation based upon the administrative grant of rights in relation to water.[30] In other words, an administration based system was substituted for a rules based system of public and private rights in relation to water. Although the system may have continued to recognise water as a common or shared resource, the instruments for its management evolved over time.

---

[26] Sanders, T.C. *Institutes of Justinian* (2nd edn, Parkes & Son 1859), 167–70.

[27] Caponera, D.A. *Principles of Water Law and Administration* (A.A. Balkema Publishers 1992), 31.

[28] Fisher, *Water Resources* (n 1), 97.

[29] Borkowski, A. and du Plessis, P. *Textbook on Roman Law* (3rd edn, Oxford University Press 2005), 172.

[30] Caponera (n 27), 40–43.

1.6 The Evolution of Public Domain Regimes[31]

If, as noted in the previous paragraph, waters increasingly became a matter of public administration towards the end of the Roman empire, the rules developed as a system based upon the concept of the public domain. Consistently with this approach, in the eighteenth century a Scottish court based its decision on what it perceived to be the public interest.[32] Underlying these paradigms is the idea of water as a common resource. But how is a common resource to be managed? The outcome of a management system based on this idea is likely to be a set of specific rights and duties attached to a range of individuals and institutions, both public and private, and based upon what is perceived to constitute the public interest. What is the public interest may be left to the person or institution making the decision or determined in accordance with a set of publicly available criteria. It is the legal system which creates the point of commencement of such a system. It may be a set of common law rules or an administrative system managed by a public sector agency. The outcome is a set of rights and duties. Such a system is often described as a public domain regime.

The doctrine of the public domain takes water in its natural location out of any trading regime. It thus becomes a *res extra commercium.*[33] The instruments for its management are variable and include: state ownership; public ownership; public trusteeship; and exclusive public control.[34]

Water resources in France are recognised as an element of *le patrimoine commun de la nation.*[35] More specifically, article 538 of the French Civil Code of 1804 declared that 'rivers or streams which will carry floats ... which are not susceptible of private ownership are considered as dependences on public domain'.[36] The owners of land through, under or across which water flowed had rights conferred and controlled by the civil code. In Italy navigable and non-navigable streams were declared to be public waters but subject to the right of riparian

---

31 See generally Fisher, *Water Resources* (n 1), 235–6, 243–50.

32 For details about this decision see the discussion in later paragraphs about the influence of the civil law on the development of the common law.

33 Fisher, *Water Resources* (n 1), 243.

34 Ibid.

35 Salman, S.M.A. and Bradlow, D.D. *Regulatory Frameworks for Water Resources Management: a Comparative Study* (World Bank 2006), 57.

36 Hu, D. *Water Rights: an International and Comparative Study* (IWA Publishing 2006), 24; and Planiol, M. *Treatise on the Civil Law: Volume 1, Part 2* (Louisiana State Law Institute with the authority of Libraire Generale de Droit et de Jurisprudence Paris tr, 12th edn, Louisiana State Law Institute 1939), 437.

landowners to use non-navigable streams for irrigation and industry provided certain conditions were satisfied.[37] In Spain all water resources are owned and controlled by the state with two exceptions: groundwater under private land and surface waters other than rivers and streams originating on the land of the landowner.[38]

An alternative to state or public ownership is public trusteeship, as in South Africa, where the national government is by statute the 'public trustee of the nation's water resources'.[39] In Germany, the regulatory framework for water resources governance gives to the state – at either federal or *Lander* level – exclusive control over them.[40] The public domain thus provides for the management of water resources at their source in various ways that fundamentally reflect what is perceived to be the public rather than an individual interest.

These public domain regimes exist alongside the civil codes which create a set of liability rules to compensate for the unlawful consequences of the use of water. In much the same way, common law regimes incorporate sets of rules granting rights of access and of use to individuals and distributing liability in the event of injury or damage to the interests of affected members of the community. The principal difference between the two systems is that in the public domain system rights are granted administratively, while in the common law system rights exist in accordance with a set of rules which are a matter of public knowledge. This is an important difference in practice. Under the common law it is the responsibility of the individual to decide what are the relevant rights and duties in any particular set of circumstances and to ensure compliance with them. Under the public domain system, it is the responsibility of a public sector agency to determine what are the relevant rights, to grant these rights formally and to inform the individual accordingly.

---

37 Food and Agriculture Organization, *Water Laws in Selected European Countries* (Food and Agriculture Organization of the United Nations 1975), 119.

38 Dzurik, A.A. *Water Resources Planning* (Rowman and Littlefield Publishers Inc. 1990), 41.

39 Fisher, *Water Resources* (n 1), 245–6.

40 Salman and Bradlow (n 35), 61.

2 THE DEVELOPMENT OF DOCTRINAL FOUNDATIONS OF THE COMMON LAW IN RELATION TO WATER RIGHTS[41]

2.1 Water Rights as Natural Rights

The relationship between land and water – jointly or separately – has influenced the development of English law. In 1625 the acquisition of water rights was judicially determined not to be a right incidental to the possession of land.[42] It was commented that 'the question here is of *aqua profluens*'[43] and two consequences followed:

- 'water hath its natural course, and *natura sua descendit*, and it may be called *usu captio* or usage',[44]
- 'where the thing hath its being by prescription, unity will extinguish it; but where the thing hath its being *ex jure naturae*, it shall not be extinguished.'[45]

Rights in relation to water were accordingly natural rights and, by implication, could be acquired by usage. This led to an acknowledgement of the appropriation doctrine favoured by the philosopher John Locke.[46] According to Blackstone the consequence in terms of legal doctrine was that 'If a stream be unoccupied, I may erect a mill thereon, and detain the water; yet not so as to injure my neighbour's prior mill, or his meadow; for he hath by the first occupancy acquired a property in the current.'[47] Thus there emerged a set of rules comprising a combination of protectable rights and enforceable duties. This approach was judicially endorsed in 1831. 'By the law in England, the person who first appropriates any part of the water flowing through his land to his own use, has the right to the use of so much as he thus appropriates, against any other.'[48] The

---

[41] See generally Fisher, *Water Law* (n 1), 64–74; Fisher, *Water Resources* (n 1), 98–101, 168–72.

[42] *Sury v Pigot* (1625) English Reports, 79, 1263.

[43] Ibid., 1266.

[44] Ibid.

[45] Ibid., 1267.

[46] Fisher, *Water Resources* (n 1) 99.

[47] Blackstone, W. (1765 to 1769), *Commentaries on the Laws of England* (edn of 1809, Professional Texts Ltd) 403.

[48] *Liggins v Inge* (1831) English Reports, 131, 268.

doctrine of appropriation was thus confirmed but it was soon to give way to the riparian doctrine.

2.2 The Influence of the Civil Law

In 1851 the Court of Exchequer[49] was presented with a choice: to continue with the notion of water as a right *ex jure naturae* and capable of acquisition by appropriation – an individual rights approach – or to adopt the approach embraced by the law of Scotland which was based on the principles of the codified Roman law and which recognised water as a shared or common resource – a community rights approach. The reasoning of the Court of Session in the case of 1768[50] was considered in detail by the Court of Exchequer. This was the reasoning of the Scottish court:

> A river, which is in perpetual motion, is not naturally susceptible of appropriation; and were it susceptible, it would be greatly against the public interest that it should be suffered to be brought under private property. In general, by the laws of all polished nations, appropriation is authorised with respect to every subject that is best enjoyed separately; but barred with respect to every subject that is best enjoyed in common. Water is scattered over the face of the earth in rivers, in lakes etc. and for the use of animals and vegetables. Water drawn from a river into vessels or into ponds becomes private property; but to admit of such property with respect to the river itself, considered as a complex body, would be inconsistent with the public interest, by putting it in the power of one man to lay waste a whole country.[51]

This was the conclusion of the Scottish court:

> A river may be considered as the common property of the whole nation; but the law declares against separate property of the whole or part. '*Et quidem naturali jure communia sunt haec, aer, aqua profluens et mare.*' A river is a subject composed of a trunk and branches. No individual can appropriate a river, or any branch of it; but every individual of the nation, especially those who have land adjoining, are entitled to use the water for their private purposes.[52]

[49] *Embrey v Owen* (1851) Exchequer, 6, 353.
[50] *Magistrates of Linlithgow v Elphinstone* (1768) Morison, 15, 12805-case no 28.
[51] Ibid., para. 3.
[52] Ibid., para. 4.

This consists of these elements: flowing water as a common resource; flowing water as common property; every individual has a right to use the water; riparian land owners in particular have a right to use the water; and water extracted in accordance with these rules becomes private property. It is not surprising that, given the civil law context of the law of Scotland, this approach reflected the law of imperial Rome.

2.3 The Emergence of the Riparian Doctrine

Which approach did the Court of Exchequer adopt? In effect a modified version of the Scottish approach out of which emerged the riparian doctrine. It was accepted that flowing water was not the subject of a right of property. There was, however, a link between water and land. Thus 'the right to have the stream to flow in its natural state without diminution or alteration is an incident to the property in the land through which it passes.'[53] And consequently 'each proprietor of the adjacent land has the right to the usufruct of the stream which flows through it.'[54]

How was this usufructuary right characterised in the absence of a right of property in flowing water? Here is the answer. 'Flowing water is *publici juris*, not in the sense that it is *bonum vacans*, to which the first occupant may acquire an exclusive right, but that it is public and common in this sense only, that all may reasonably use it who have a right of access to it.'[55]

This is a rejection of a comprehensive appropriation doctrine because the usufructuary right is limited in two ways:

- to those who have access to the water,
- to use it reasonably.

It will be recalled that the Scottish court extended the usufructuary right to every individual – including the riparian occupiers. The English court effectively limited it to riparian occupiers, a subtle but important difference. The flowing water may therefore be regarded as a common resource or possibly even common property – a *res communis* – but whose usufructuary right is restricted to the group of individuals with a lawful right of access – the riparian landowners.

Water not only flows through a watercourse but also percolates underground and flows across the surface of land. The owner or the

---

[53] *Embrey v Owen* (n 49), 369.

[54] Ibid.

[55] Ibid.

lawful occupier of the land in each of these two sets of circumstances has a similar usufructuary right in relation to the water – underground or surface. It was explained judicially in 1885 in these words:

> No one has at any time any property in water percolating below the surface of the earth, even when it is under his own land, but it is equally clear that everybody has a right to appropriate that percolating water, at least whilst it is under his own land, to the extent that he may take it all so as to prevent any of it going on the land of his neighbour.[56]

The neighbour has a similar right. Why? 'This percolating water below the surface of the earth is therefore a common reservoir or source in which nobody has any property, but of which everybody has, as far as he can, the right of appropriating the whole.'[57]

2.4 The Challenges for the Common Law of Public Water Supply

2.4.1 Introduction

It will be recalled that the Court of Exchequer in 1851 had limited the exercise of the riparian right in two ways: first, to those with lawful access to the water and, second, to use the water reasonably.[58] The second became more problematic. It became an issue for the House of Lords in 1859.[59] Large quantities of underground water were extracted to provide a public water supply to a small township. In the circumstances this was held to be lawful in exercise of the right of appropriation available to the owner of the surface of the land to extract water moving indiscriminately and in an unknown and undefined way below the surface of the land. But Lord Wensleydale had reservations. There was no doubt about the right to sink the relevant well. But:

> According to the rule of reason and law, '*sic utere tuo ut alienum non laedas*', it seems right to hold that he ought to exercise his right in a reasonable manner, with as little injury to his neighbour's rights as may be.[60]

A supply of water for the landowner's own operations would have been reasonable.

---

56 *Ballard v Tomlinson* (1885) Chancery Division, 29, 120.
57 Ibid.
58 *Embrey v Owen* (n 49), 369.
59 *Chasemore v Richards* (1859) House of Lords Cases, 7, 349.
60 Ibid., 388.

> But I doubt very greatly the legality of the defendant's acts in abstracting water for the use of a large district in the neighbourhood, unconnected with his own estate, for the use of those who would have no right to take it directly themselves, and to the injury of those neighbouring proprietors who have an equal right with themselves.[61]

According to the approach of Lord Wensleydale, two conditions for the application of the riparian doctrine under the common law were required to be satisfied:

- the existence of a right of access to the water,
- the requirement for the use of the water to be reasonable.

The approach of the common law was to a limited extent modified by the arrangements for public water supply introduced in England during the late eighteenth century and enduring to varying extents for two hundred or so years. Individual entrepreneurs – personal or more likely corporate – presented a private bill to Parliament inviting it to enact a provisional order approving a particular project: in effect legislation authorising a particular project sponsored by the private or in some cases the public sector.[62] The projects were frequently major infrastructure developments – some involving water resources. The relationship between these statutory rules and the traditional common law rules was an issue for the House of Lords in 1875 in the Swindon Case.[63]

2.4.2 The background to the Swindon Case

One of the litigants in this case was a canal company which had, since 1795, statutory authorisation to construct and maintain a canal and to supply the canal with water taken from sources within 2000 yards of any part of the canal. These arrangements were confirmed by complementary legislation in 1813 and 1821. In 1807 the company bought a mill on the banks of one of these watercourses and so became a riparian proprietor. The company accordingly had statutory and common law rights to water to support its activities in relation to the canal. The other litigant was a waterworks company registered under the Joint Stock Companies Act 1862 and established to supply water to the town of Swindon. This

---

61 Ibid., 389.

62 Hood Phillips, O., Jackson, P. and Leopold, P. *O. Hood Phillips & Jackson: Constitutional and Administrative Law* (8th edn, Sweet & Maxwell 2001), 249–50.

63 *Swindon Waterworks Company v Wilts and Berks Canal Navigation Company* (1875) Law Reports Hose of Lords, 7, 697.

company bought a mill on the banks of the same watercourse as the canal company but upstream of the location of the mill bought by the canal company. This company enjoyed common law riparian water rights but not statutory rights. The waterworks company was thus an upstream riparian proprietor and the canal company a downstream riparian proprietor.

The circumstances giving rise to the litigation are indicative of the relationship between the environment and economic development in nineteenth-century England. The waterworks company extracted water from the river to provide a public supply to Swindon. The canal company extracted water to ensure enough water in the canal to enable successful navigation. 1870 was an eventful year:

> The year 1870 was, no doubt, a year of unusual drought; but in that year … the supply of water to the canal was undoubtedly defective. Heavy traffic – traffic of barges laden heavily, requiring a large draught of water – had to be refused on account of the insufficient supply of water in the canal … How did that deficiency of supply arise? No doubt the evaporation and absence of influx owing to the dryness of the season had something to do with that. But it stands upon the evidence … that a considerable reason for the deficiency of the supply was, that at that very time the [waterworks company] had taken possession of the upper part of the waters of the stream and diverted those waters for the supply of Swindon.[64]

The conclusion was that '[t]he supply of water for the canal, arising out of the stream, was deficient, and was made deficient by, or at all events partly by, the acts of the [waterworks company]'.[65] But were the acts of the waterworks company illegal?

2.4.3 The rights of the waterworks company in the Swindon Case

The waterworks company relied upon their rights as a riparian proprietor under the common law to justify their actions. It will be recalled that in the case of Chasemore[66] the source of the water supplied to the town for public use was an underground source and the extraction determined to be legal. But Lord Wensleydale had reservations and introduced a test of reasonability. The right of a riparian proprietor to extract and use water was not absolute. In the Swindon Case[67] Lord Cairns noted that 'the uses

---

64 Ibid., 707.

65 Ibid., 707.

66 *Chasemore v Richards* (n 59).

67 *Swindon Waterworks Company v Wilts and Berks Canal Navigation* (n 63).

to which the water may be put by the upper owner'[68] were required to be 'uses connected with the tenement of that upper owner',[69] for example, for irrigation or manufacturing. What about public supply? The lengthy response of Lord Cairns is worthy of note:

> But the use which here has been made by the [waterworks company] of the water, and the use which they claim the right to make of it, is not for the purpose of their tenements at all, but is a use which virtually amounts to a complete diversion of the stream – as great a diversion as if they had changed the entire water-shed of the country, and in place of allowing the stream to flow towards the south, had altered it near its source, so as to make it flow towards the north.[70]

On the basis of this factual analysis, Lord Cairns placed the issue in its legal context:

> That is not a user of the stream which could be called a reasonable user by the upper owner; it is a confiscation of the rights of the lower owner; it is an annihilation, so far as he is concerned, of that portion of the stream which is used for those purposes, and that is done, not for the sake of the tenement of the upper owner, but that the upper owner may make gains by alienating the water to other parties, who have no connection whatever with any part of the stream.[71]

The relevant test was thus reasonability. Not a surprising conclusion given its pivotal function in the common law.

2.4.4 The relevance of the statutory rights of the canal company in the Swindon Case

The canal company, whose rights under the common law had been infringed, was vested with statutory rights in relation to the water in question. Were they relevant? The legislation establishing the canal company, it will be recalled, authorised the company to 'supply the canal with water' from the sources stated in the legislation and situated within 2000 yards of the canal. This statutory power was exercised by the company. Lord Cairns concluded that 'the canal proprietors had the right then and there to take that water for the supply of their canal.'[72]

---

[68] Ibid., 704.
[69] Ibid., 704.
[70] Ibid., 704–705.
[71] Ibid., 705.
[72] Ibid., 706.

The critical words were 'for the supply of their canal'. The use of the water for this purpose may not have been a use recognised by the common law riparian doctrine – a point not decided by the House of Lords. It was clearly a use authorised by the legislation. But if the taking of the water for such a use 'stopped it from flowing in its accustomed course to any tenement lower down upon its course, [the canal company] would be obliged to compensate the owner of that inferior tenement under their Act of Parliament.'[73] In this respect common law liability and statutory liability were much the same. But this was not the relationship between the litigants in this case. The waterworks company was the upper riparian proprietor and the canal company the lower riparian proprietor. Hence liability was determined in accordance with the common law simply because both companies were riparian proprietors. The analysis by Lord Cairns of the canal company's statutory rights – strictly *obiter dicta* – is nevertheless particularly illuminating.

Lord Hatherley developed this point in a rather significant way. Although both companies were riparian proprietors, the existence of the canal company's statutory authority to take and use the water to supply the canal gave 'to the canal proprietors a particular right'.[74] What was the substance of this particular right? Here is the answer:

> For the purpose in question, namely, the making of the canal and the keeping it filled from time to time, either in its then state or in its enlarged state ... and either for the existing navigation, or for any navigation which might come on it, they were entitled to have this canal adequately and properly supplied with water.[75]

The amount of water to be available for this purpose was not, as it had been argued under the common law riparian doctrine, restricted in any way. Why?

> The answer is that the Legislature has not so limited the supply [to so many thousand gallons a day], and therefore the Legislature not having so limited it, we must take the larger purpose for which it was granted, namely, the continuous use of the water for the purposes of navigation. In that case the canal company would have the right which every riparian proprietor would have, namely, a right to the flow of the stream for the purpose of satisfying their requirements.[76]

---

[73] Ibid., 706–707.
[74] Ibid., 711.
[75] Ibid.
[76] Ibid.

This is a subtle combination of the statutory and common law rules. As a riparian proprietor, the canal company had the right to extract and use the water for the purposes recognised by the common law. As the holder of the rights conferred by the legislation, the canal company had the common law right extended to include the purposes stated in the legislation. In this case the waterworks company's activities were governed only by the common law which did not recognise the public supply of water as a lawful use. Although the activities of the canal company were not directly an issue, the existence of statutory rights might well in appropriate circumstances have proved critical. This case demonstrates that the legal arrangements in England to support the achievement of public purposes in managing water resources tended to be project based and ad hoc rather than based upon appropriate doctrinal foundations – other than the common law which did not directly address these issues. The response in other jurisdictions, already noted in previous paragraphs, has been quite different – the recognition of public domain regimes.

3 WATER GOVERNANCE IN AUSTRALIA BEFORE THE TWENTIETH CENTURY[77]

3.1 Water Governance by Indigenous Practices and Customs

Before the arrival of British colonists in 1788, access to and use of water in Australia were governed by the practices and customs of the indigenous community in accordance with their ancient traditions.[78] These were not uniform across the indigenous community. Individual tribes engaged in their own practices and customs. These practices disclosed a close relationship not only between water and land but also between water, land and culture. Accordingly, these rights to water could 'simultaneously contain ecological, spiritual, cultural, economic and social dimensions'.[79] When native title was judicially recognised in 1992 as part of the common law and implemented by legislation in 1993, it was stated to mean quite specifically:

---

[77] See generally Fisher, *Water Law* (n 1), 64–89; Gardner, A., Bartlett, R. and Gray, J. *Water Resources Law* (Lexis Nexis Butterworths 2009) 151–77; and Stoeckel, K., Webb, R., Woodward, L. and Hankinson, A. *Australian Water Law* (Thomson Reuters, Lawbook Co 2012), 13–17.

[78] The information in this paragraph is derived from Gardner *et al.* (n 77), 534–46 and the sources therein referenced.

[79] Gardner *et al.* (n 77), 535.

- the communal, group or individual rights and interests in relation to land or waters:
- possessed under the traditional laws acknowledged and traditional customs observed
- where, by those laws or customs, there is a connection with the land or waters and
- that are recognised by the common law.[80]

This is no doubt a reflection of how indigenous rights in relation to water were traditionally perceived. The nature of indigenous rights may now be seen as essentially communal and to some extent individual: a perception far removed from the doctrines underlying the common law of England. The reception of the common law of England in 1778 rendered irrelevant indigenous water rights until their subsequent recognition by the High Court of Australia in 1992.

3.2 The Environmental Context of Water Governance During the Nineteenth Century

During the nineteenth century, water resources in Australia were governed in the first instance by the common law and then to the very limited extent to which the common law was varied by local legislation. It became increasingly obvious to the colonial administrators that the sources and quantity of water resources and the nature of the environment in Australia – in particular in the first colony of New South Wales – were significantly different from those in the British Isles. How were the rules inherited from the English system applied in Australia during the nineteenth century – bearing in mind the climatic differences between a relatively humid climate and a relatively dry climate?

It is useful to begin with this historical perspective:

> On 15 May 1788 Governor Phillip chose Sydney Cove as the landing place for the First Fleet because it was the only satisfactory site thus far encountered which appeared to have a reliable stream of fresh water. Baron Augustus Theodore Henry Alt, a well-seasoned military engineer, designed and supervised the essential buildings and public works and was largely responsible for the deepening of the river and the excavation of tanks in its sandstone bed, thereby ensuring the capacity of the 'Tank Stream' to provide Sydney's only public water supply for about 40 years.[81]

---

[80] Native Title Act 1993 (Commonwealth), s. 223(1).

[81] Powell, J.M. *Environmental Management in Australia 1788–1914* (Oxford University Press 1976), 18.

Significantly this project was designed to provide a supply of water to the public. The institutional developer of the project and the supplier of the water were the government, represented formally by the Governor.

Equally illustrative is this judicial perspective. Sources of water for private use were as important as those for public supply when the economy began to develop – particularly in relation to agriculture, cattle and sheep farming, and mining. These activities were largely those of the private sector and were based upon the rights and duties of individuals and private sector institutions. In the context of litigation between two individuals in 1867 these judicial comments illustrate the economic and social importance of water resources governance:

> The subject-matter of this action is obviously of very great importance to the people of this province, where, by reason of the peculiar aridity of the climate, running water possesses special value to the cultivators of gardens and vineyards situate on the banks of creeks, as well as to the proprietors of mills worked by water-power. Hereafter, also, the question may assume a special interest to the proprietors of mines in cases where subterranean water channels may happen to be cut in the progress of mining operations. For these reasons, it has been considered fitting that on delivering judgment on this rule the law relating to watercourses should be declared by the Court. Fortunately, we have for our guidance many cases decided in the English Courts in this branch of the law.[82]

The Supreme Court of South Australia proceeded to do precisely that and, in doing so, reviewed many of the relevant judgments. In this case there was no relevant legislation. It was decided purely on the basis of the common law of England that applied to the rights of individuals. Would it have been different if it had involved the acquisition of water rights to provide a supply of water to the public? To answer this question requires an examination of a different set of legal rules: those relating to the reception of English law in the Australian colonies, how it was applied and how it could be modified in these colonies.

3.3 The Rules Providing for the Reception of English Law

The Governor of a colony as the representative of the British Crown in the colony exercised the proprietorial powers of the Crown in relation to the use of land, water and related resources. These powers were sourced in the common law. The Governor had the executive power to proclaim

---

[82] *Dunn v Collins* (1867) South Australian Law Reports (Pelham), 1, 126, 135.

ordinances effective as enforceable rules of law. Although these arrangements originated much earlier, they were clarified judicially as late as 1889 in relation to the colony of New South Wales in a judgment of the Privy Council which indirectly, as it will be later noted, raised issues about the law relating to water governance. Lord Watson explained that 'the extent to which English law is introduced into a British Colony, and the manner of its introduction, must necessarily vary according to circumstances.'[83]

The reference to 'English law' includes both the common law and statutory law. Are they treated differently or the same? Much depends on the status of the colony in question. Having determined the status of the colony of New South Wales at that time, Lord Watson went on to explain the rules for New South Wales and thus declared that 'the Crown may by ordinance, and the Imperial Parliament, or its own legislature when it comes to possess one, may by statute declare what parts of the common law and statute law of England shall have effect within its limits.'[84] There are accordingly potentially three forms of legislation: (1) statutes enacted by the Parliament of the United Kingdom; (2) ordinances promulgated by the representative of the Crown in the colony – the Governor; and (3) statutes enacted by the legislature of the colony once one has been established.

In the absence of such legislation declaring what parts of the common law and statute law of England apply in the colony, the 'law of England must (subject to well-established exceptions) become from the outset the law of the Colony, and be administered by its tribunals.'[85] Lord Watson summarised the position very succinctly in this conclusion: 'In so far as it is reasonably applicable to the circumstances of the Colony, the law of England must prevail, until it is abrogated or modified, either by ordinance or statute.'[86]

The language used by Lord Watson – especially the expression 'reasonably applicable' – gave to the judiciary a degree of discretion in determining what law applied in the colony. In the context of water resources governance in the colony of New South Wales the relationship between the common law of England and any relevant local legislation – either ordinances or colonial statutes – became a critical issue for the courts.

---

[83] *Cooper v Stuart* (1889) Appeal Cases, 14, 286, 291.
[84] Ibid.
[85] Ibid.
[86] Ibid.

3.4 The Relationship Between Common Law Rules and Statutory Rules

3.4.1 Introduction

The development and construction of infrastructure facilities for the benefit of the public and the commercial sector in the United Kingdom during the nineteenth century – for example, railways, canals and to a lesser extent water supply – were for the most part authorised by legislation sponsored by the private sector or by a public sector agency in relation to a specific – sometimes a major – project. Public general legislation prescribed the procedures for such legislation but not its content. The common law appropriation and riparian doctrines enabled individual entrepreneurs to take and use the waters of rivers for their cotton and wool mills built on the banks of these rivers. In the colony of New South Wales, infrastructure facilities – especially those related to public water supply – were structured differently – no doubt because of the powers available to the Governor under the colonial system for managing land and promulgating legislation. There were traditionally three: (1) the exercise by the Crown of water rights incidental to riparian land ownership; (2) the reservation to the Crown of water rights in grants by the Crown of interests in land; and (3) the promulgation of local legislation. The doctrinal issues associated with these mechanisms are revealed by a consideration of a small number of nineteenth-century judicial decisions.

3.4.2 The 1853 Cooper Case

The Supreme Court of New South Wales was faced in 1853 with a dispute involving water that flowed naturally on to land in private ownership.[87] The local council – Sydney Corporation, hereafter 'the council' – proposed to construct a trench on its water reserve in order to obtain water from a swamp located in the water reserve. An adjacent landowner sought an injunction to restrain the council from excavating a trench on the ground that the trench would intercept the water flowing naturally on to the adjacent land. One of the issues was whether the flow of water was temporary and casual or whether the water flowed continually and habitually between banks. If the former, the landowner had no right to any water flowing on to his land. If the latter, he had a right that there should be no interference with the flow of water and that the water

---

[87] *Cooper v Sydney Corporation* (1853) Legge, 1 and 2, 765.

should not be abstracted. The evidence on this was contradictory and the court ordered a trial of these issues.

What rules of law did the court apply? The court turned to the doctrines of the common law. Did the Crown enjoy any privileges? Were its rights the same as any other person's? The response to the first question was negative and to the second it was positive. 'The Crown, or the Corporation of the City, representing the Crown in this matter, irrespective of any powers conferred by the Legislature, can have no right, which an individual in such a case would not have.'[88] There had not been any grant of the water reserve but the council had been in possession of it for many years 'by the authority of the Crown'.[89] It followed that 'they are entitled to exercise, therefore, all the rights of the Crown in respect of that land'[90] and also that 'whether the Crown granted, in addition, any casual or accidental advantage, such as that of the enjoyment of water, temporarily overflowing (or which at times might overflow) from underground sources on land remaining in its possession, is to be shown.'[91]

In other words, the reciprocal rights and duties of the landowner as owner and occupier of land and of the council, representing the Crown as owner and occupier of land, were the same. The landowner enjoyed a right 'to the reasonable use of the water of any perennial stream … flowing through that land'[92] but not to any temporary or casual overflow. These were the matters of fact sent for trial. The court thus applied the rules of the common law in relation to land and water by relying specifically upon the appropriation and riparian doctrines. In doing so, it reviewed many of the authoritative judgements of the first half of the nineteenth century.

There was, however, a second complicating issue: legislation authorising the council to undertake a range of works associated with the public supply of water. The language of the legislation proved critical. The council was authorised to engage in an extensive range of activities and to construct and maintain infrastructure in the broad context of what may be described generically as waterworks. This part of the judicial analysis of the legislation proved critical:

88 Ibid., 771–2.
89 Ibid., 772.
90 Ibid.
91 Ibid.
92 Ibid.

> They may enter upon and pass over any river, stream, road, waste lands, or street, in the City *or its vicinity*; and construct, and do, all such matters and things as may be *deemed necessary*, for making or improving such waterworks, and *for bringing a sufficient supply of water for the city*, provided that they shall make full satisfaction to the owners and occupiers of all lands *or other hereditaments*, taken, used, or injured, for all damage sustained by them by the execution of any of those powers.[93]

The court was impressed by the words 'bringing a supply of water to the city'. These words, it is here suggested, represent a statement of the object or outcome to be achieved as a result of the exercise of these powers. In an oblique sense it is a power to bring a supply of water to the city. Although the court did not express it in this way, a power to supply water is not a power to abstract water. The court recognised this distinction. It stated unambiguously that 'so, they may enter on and over streams; but nothing is said respecting abstracting the water from them. And waters, not being streams, it would appear, are excluded.'[94]

The court thus preferred a limited rather than an unlimited interpretation of the content of the powers conferred on the council. The legislation did not authorise the abstraction of water: only the construction of water infrastructure. The legality of the council's activities in relation to the water depended upon the application of the relevant rules of the common law

3.4.3 The 1860 Hood Case

Seven years later the Supreme Court addressed a similar set of issues.[95] A riparian landowner – hereafter 'the landowner' – carried on a wool-washing establishment on the banks of a tributary to the watercourse in question. In carrying out these activities, 'dirt and filth' were thrown into the tributary from which it went into the watercourse and then into the reservoir managed by Sydney Corporation – hereafter 'the council'. This was the response of the council:

> To prevent the watercourse and reservoir being so defiled by the dirt and filth thrown into the tributary ... the [council] prevented the [upper riparian landowner] from using the tributary and watercourse for his wool-washing. By such act of the [council] almost every kind of use to which the [upper riparian landowner's] land could be put has been prevented.[96]

---

93 Ibid., 772–3.
94 Ibid., 773.
95 *Hood v Sydney Corporation* (1860) Legge, 1 and 2, 1294.
96 Ibid., 1295.

The landowner sought compensation from the council for the loss he had sustained. On the face of it, the landowner's riparian activities had interfered with the right of lower riparian landowners to receive water of the same quantity and quality as enjoyed by upper riparian landowners. But was the council a lower riparian landowner or a public agency exercising statutory powers, or perhaps both? Significantly the powers of the council were derived from three sources: (1) as the holder of water rights reserved from a grant by the Crown; (2) as a downstream or lower riparian landowner; and (3) as the recipient of powers conferred by legislation.

Consider each in turn. In relation to the first, the land of the landowner 'was part of a large portion granted by the Crown ... by a deed, wherein the Crown reserved to itself any quantity of water and any quantity of land not exceeding ten acres on any part of the grant as might be required for public purposes'.[97] The council – as representative of the Crown – had taken from time to time water and portions of the land for public purposes. The council thus had an interest in the water – no doubt including its quality – recognised by the law.

Second, it was noted that, if a polluting activity had been carried on for twenty years, then 'the neighbouring riparian proprietors could be bound to endure the nuisance'.[98] However, the wool-washing had not been carried on for twenty years. So the council could, according to the common law, have abated the nuisance. Significantly, in this context, the court inferred that the council had 'become possessed of land below that occupied by the [landowner in question] and so in effect have become sub-riparian proprietors'.[99] Apart from any liability of the upper riparian landowner in nuisance, the landowner could be liable under the riparian doctrine because the council 'had become sub-riparian proprietors'. Accordingly, the council in that capacity 'would have been entitled to sue the[landowner] for, and restrain him by injunction from, fouling the water which they would have been entitled to receive in its natural state'.[100] From the landowner's perspective, he had received no injury 'for which he could have obtained redress at common law'.[101] His loss was a *damnum sine injuria.*

Third, what was the function of the legislation? It will be recalled that the council in this case derived its rights in relation to the water from the

---

97 Ibid.
98 Ibid., 1296.
99 Ibid.
100 Ibid., 1297.
101 Ibid.

reservation in the Crown grant and from its status as a riparian proprietor of the land through which the watercourse flowed. The council's waterworks were constructed and managed under the Sydney Water Act 1853. The upper riparian landowner in this case sought compensation under the Act for his loss. Having decided that the loss was *damnum sine injuria* under the common law, the court noted that 'the question that remains is whether he is entitled to the compensation which he seeks under the Act'.[102] The court once again adopted a limited rather than an expansive view of the legislation. Relying upon earlier decisions of courts in England and Scotland, here is the answer:

> The compensation clauses in statutes authorising public works apply only where the act complained of is such that it would have afforded a cause of action before the act; and further that such act is authorised by the statute. Now, as the acts committed by the [council] clearly ... would not before the act have conferred a right of action to the [landowner], and as those acts are clearly authorised by the Sydney Water Act 1853, we are of opinion that the[landowner] is not entitled to succeed in this action.[103]

Significantly, it will be recalled, the legislation authorised the undertaking of public works. It was silent on the issue of water rights. During the nineteenth century, the colonial legislation was doctrinally secondary to the clear rules of the common law.

3.4.4 The 1859 Lord Case

In the 1853 Cooper Case the issue, it will be recalled, was the nature of the water rights of two adjacent landowners at common law irrespective of the status of the council as a landowner and irrespective of any statutory provisions. In the Hood Case in 1860 the statutory provisions for compensation and their relationship with the cognate rules of the common became relevant. In the Lord Case[104] before the Privy Council in 1859, the question was the relationship between two related Crown grants, one of which involved water rights resumed by the council under the Sydney Water Act 1853. This Act authorised the council to 'divert, take and use' not only the waters arising from or flowing through the lands in possession of the Crown but also the waters flowing into or from these waters.[105] The Crown grant of 1810 contained no reservation of

---

[102] Ibid.
[103] Ibid.
[104] *Lord v Commissioners for the City of Sydney* (1859) English Reports, 14, 991.
[105] Ibid., 992.

water rights but the land was bounded by a small creek. Presumably the grantee as a riparian proprietor was entitled to any common law water rights. The Crown grant of 1823 of land above the land granted by the 1810 grant – above in terms of the flow of any water – contained a reservation of 'any quantity of water ... as might be required for public purposes' provided there was no interference with, injury to or prevention of the working of related water-mills.[106] The owner of the land which was the subject of the 1823 grant – the 'above' land – had become the owner of the land which was the subject of the 1810 grant – the 'below' land. In 1855 parts of the 1810 and 1823 land grants were resumed by the Crown and the water of the creek associated with the 'above' land appropriated by the Crown.[107] Was the owner of the 'below' land who was also the owner of the 'above' land entitled to compensation for the removal of water rights associated with the 'above' land?

The argument presented to the Privy Council seems to have been along these lines.[108] The 1823 grant contained the provision reserving water to the Crown for use for public purposes. This had been acted upon and the water was no longer available to the holder of the 1823 grant. Was the holder of the 1810 grant similarly deprived of any water right dependent upon the flow of water from the upper riparian land? The answer depended upon the extent of the original grant. According to Sir John Coleridge:

> If you grant anything, you are presumed to grant, to the extent of your power, that also without which the thing cannot be enjoyed. But what is the effect of this reservation? Ten acres of land in any part of the grant are reserved, at the election of the Crown, if required for public purposes; these are not to be granted to the Crown by Lord [the 1823 grantee], but are provisionally saved out of the grant to him. As to the water, Lord, as owner of the land, had no power to grant it to the Crown; the Crown could not grant any property in it to him, nor he to the Crown; the effect of the saving is only that he waives his own rights as riparian owner to the use of it as it flowed.[109]

To a considerable extent this personalises the saving in relation to the original grantee irrespective of subsequent changes to ownership of the land. The reporter has summarised neatly the essence of this complicated decision:

---

106 Ibid., 993.
107 Ibid., 993–4.
108 Ibid., 1001.
109 Ibid.

> That the right to the use of the flowing water of the creek in respect of the land below, originally granted to Redmond [in 1810], was not lost by the acceptance of Lord of the land above [in 1823], although in the latter grant the Crown had reserved the right to take the water, the only effect of the reservation being that Lord waived his own rights as riparian owner to the use of the water as it flowed past his land.[110]

In 1889 in the case of Cooper v Stuart[111] Lord Watson said this about the Lord Case:

> But though the decision is not directly in point, its circumstances throw some light upon the present question. It was an action for compensation under the Sydney Water Act of 1853. The compensation sought and awarded was in respect of putting in force a reservation under a grant of 1810, made by Governor Macquarie in terms identical with the grant of 1823, and the Water Act seems fully to recognise the validity of such reservations.[112]

The person successfully seeking compensation in the Lord Case was not the original grantee of the 1823 grant but rather the devisee of that grant, also named Lord.[113] This reinforces the individuality of these water rights – entirely consistent with the doctrinal foundations of the common law.

4 TOWARDS A PUBLIC DOMAIN REGIME IN AUSTRALIA

4.1 Policy Developments Towards the End of the Nineteenth Century[114]

During the nineteenth century water resources governance in Australia was based upon three different approaches: (1) the application of the rules of the common law; (2) the reservation of water rights to the Crown in grants by the Crown of interests in land to individuals; and (3) the

---

110 Ibid., 992.

111 (1889) Appeal Cases, 14, 286. This case is discussed in Section 3.3.

112 Ibid., 294.

113 *Lord v Commissioners for the City of Sydney* (1859) English Reports, 14, 991, 1001.

114 See generally Fisher, *Water Law* (n 1), 4–5; Lucy, J. *Water Regulation – The Laws of Australia* (Thomson Reuters 2008), 100; Gardner *et al.* (n 77), 183–9; and Stoeckel *et al.* (n 77), 17–18.

enactment of legislation authorising on an ad hoc project basis the construction and use of public works related to the supply of water for public use.

The governance of water resources was thus largely unplanned, uncoordinated and unregulated. Were arrangements such as these impeding the economic and social development of Australia? The three particular areas of political concern were the availability of water for public supply, mining and irrigation. The response was a series of inquiries prompted by a severe drought during the years 1877 to 1881.[115] A report of 1871 for the colony of South Australia recognised the problems of 'semi-aridity, rainfall variability and inadequate water storages'.[116] The solutions to these problems included 'the proper assertion of the public right to all streams throughout the country, and the desirability of at once, if possible, taking the most effectual measures to stop anything approaching to their permanent alienation.'[117]

A similar approach was recommended by the Royal Commission on Water Supply undertaken in 1884 for the colony of Victoria. In these terms: 'It is essential the State should exercise the supreme control of ownership over all the rivers, streams and sources of water supply, except springs arising upon private lands.'[118] The several colonial legislatures began to respond to these recommendations. The first response emerged as section 4 of the Irrigation Act 1886 of the colony of Victoria in these terms: 'The right to the use of all water at any time in any river stream watercourse lake lagoon swamp or marsh shall for the purposes of this Act in every case be deemed to be vested in the Crown until the contrary is proved.'[119] This, it has been suggested, 'prevented the establishment of further riparian rights and effectively nationalised Victoria's surface waters.'[120] The other colonies followed with their own legislation but in terms that were not consistent.[121] The result was – and remains – a somewhat diverse and variable set of arrangements pointing generally in the direction of the public control and regulation of water resources.

---

115 Gardner *et al.* (n 77), 184.
116 Powell (n 81), 66.
117 Ibid., 68.
118 Gardner *et al.* (n 77), 184.
119 Ibid., 185.
120 Powell (n 81), 132.
121 Gardner *et al.* (n 77), 185–9.

4.2 The Relationship Between the Early Public Control Legislation and the Common Law[122]

4.2.1 Introduction

The legislation was enacted in the context of the approaches adopted by the common law. There emerged, not surprisingly, tensions between perceptions of how water resources should be managed in the public interest and how they should be managed by the individual holders of water rights in their own interest. It is again no surprise that the courts themselves have been challenged by these tensions. The physical, biological and ecological relationships between land and water and the reciprocal consequences of their uses have in most – perhaps in all – legal systems presented a range of doctrinal challenges for the courts, certainly in Australia. While the rights of surface landowners are protected by the common law rules of nuisance and to some extent of negligence and the rights of riparian landowners are protected by the common law rules of the riparian doctrine, the creation of a system of public rights in relation to water which impact upon the use of land has compounded the tensions within the common law. This can be demonstrated by a brief review[123] of how the courts in Australia have responded.

4.2.2 The 1900 Hanson Case

Such an issue emerged in 1900 in the Hanson Case.[124] The litigants were engaged in mining and occupied adjacent land through which a stream flowed intermittently. The lower tenement holder constructed a dam across the stream. The effect was that the water which normally flowed through and away from the land of the upper tenement holder was diverted back and it flowed into mines and shafts of the upper tenement holder. If the common law riparian doctrine applied, the rights of the upper tenement holder would appear to have been infringed. What was the effect of section 1(1) of the Water Rights Act 1896 of New South Wales which stated that '[t]he right to the use and flow and to the control of the water in all rivers and lakes which flow through or past or are situate within the land of two or more occupiers, and of the water contained in or conserved by any works to which this Act extends, shall,

---

122 See generally Fisher, *Water Law* (n 1), ch. 5; and Gardner *et al.* (n 77), 193–7.

123 For a detailed review, see Fisher, *Water Law* (n 1), 101–122.

124 *Hanson v Grassy Gully Gold Mining Co.* (1900), New South Wales Reports, 21, 271.

subject only to the restrictions hereinafter mentioned, vest in the Crown'? Section 1(2) provided for the exercise of this right in operational terms: namely the construction and management of the relevant infrastructure. What did section 1(1) mean? According to Justice Stephen: 'I do not think the language of the Act could be clearer, and plainly the rights of the riparian owners were divested and vested in the Crown.'[125] The rights were undoubtedly vested in the Crown but the Act said nothing about the divestment of the rights of the riparian owners. This remained an unresolved issue for decades.

4.2.3 The 1950 Jones Case

The issue was debated by the High Court of Australia twice during the 1950s. In the Jones Case[126] the litigants were not two individuals as in the Hanson Case but a riparian landowner and a local council exercising statutory powers to provide a public water supply. Section 209 of the Local Government Act 1906 of Tasmania vested in each council every river, creek or watercourse within the limits of every water district within the area of the council together with the 'absolute control and regulation of the supply of water' from these sources but subject to 'the previously existing rights of any riparian proprietors to the use of the water flowing into' these sources. In this case the council had taken water from a river at a point located above the land of the riparian landowner to the disadvantage of the riparian landowner. The Chief Justice of Tasmania decided that the riparian rights were diminished to the extent necessary to enable the council to supply water from the stream.[127] The Chief Justice of the High Court of Australia presented a different analysis:

> With much respect to his Honour, it appears to me that this interpretation of section 209 makes riparian rights subject to the exercise of the power of the council to supply water, whereas the section provides that the exercise of the power given to the council is to be subject to the riparian rights of persons who own the land on the banks of the river.[128]

Riparian rights existed side by side with statutory rights. Which rights would have priority, however, was not entirely clear. Much would depend upon the statutory provisions.

---

[125] Ibid., 275.

[126] *H Jones & Co Pty Ltd v Municipality of Kingborough* (1950), Commonwealth Law Reports, 82, 282.

[127] Ibid., 302; See also Fisher, *Water Law* (n 1), 104–105.

[128] Ibid.

4.2.4 The 1955 Thorpes Case

The litigants in the Thorpes Case[129] in 1955 were private institutions. Levee banks had been constructed along a creek which flowed into a river that adjoined the land of the upper landowner. This altered the direction of the flow of the water and on occasions caused damage to the lands of the lower landowner for which the affected lower landowner sought a remedy. The upper landowner argued that the action was an action for damages for the infringement of a riparian right and that this riparian right had been removed by the Water Rights Act 1896 of New South Wales. This argument was rejected by the High Court of Australia. The right in question was not a riparian right but a right held by any owner of land. As Justice Fullagar pointed out, the action was 'an ordinary action of nuisance'.[130]

His Honour went on to comment on the effect of the Water Rights Act 1896 of New South Wales in a way different from the approach of Justice Stephen in the Hanson Case, in these words:

> The real object of the *Water Rights Act* 1896, as revealed by the latter part of section 1, was to enable the Crown, in a country in which water is a comparatively scarce and important commodity, to exercise full dominion over the water of rivers and lakes and to undertake generally the conservation and distribution of water. For the attainment of that object it was not necessary to destroy anybody's rights, but it was necessary to give to the Crown, or to some other statutory authority, overriding rights to which private rights must, if need arise, give way.[131]

There are four points critical to this analysis: (1) the importance of the object of the Act; (2) the relevance not only of section 1(1) but also of section 1(2); (3) the use of the expression 'full dominion'; and (4) statutory rights as overriding rights. His Honour expanded upon this analysis in these words:

> The view which I am disposed to take is that the Act does not directly affect any private rights, but gives to the Crown new rights – not riparian rights – which are superior to, and may be exercised in derogation of, private riparian rights, but that, until those new and superior rights are exercised, private rights can and do co-exist with them.[132]

---

[129] *Thorpes Ltd v Grant Pastoral Co Pty Ltd* (1955), Commonwealth Law Reports, 92, 317.

[130] Ibid., 329.

[131] Ibid., 331.

[132] Ibid.

Bearing in mind that this case was decided on the basis of liability in nuisance, a further set of four points emerges: (1) common law rules in relation to riparian rights and liabilities relate only to water and its uses; (2) common law rules in relation to the use of land can relate to associated uses of water; (3) the strategic nature of statutory rights in relation to the use, flow and control of water; and (4) the operational nature of statutory rights in relation to the construction, use and management of infrastructure for water supply

A determination of the relationship between common law and statutory rules is no easy task. It is a reflection of the jurisdiction in question; the precise issue before the court – land based or riparian based; the detail of the common law rules relevant to the action; the terms of the particular statutory provision in question; and the broad structure and object of the legislation in question.

4.2.5 The 1962 Gartner Case

The courts in Australia have grappled with these challenges not only at a federal level but also at a state and territory level.[133] The most authoritative analysis emerged in 1962 with the decision of the High Court of Australia in the Gartner Case[134] which essentially endorsed the earlier analysis of the court. In this case the litigants were individuals whose interests were private. Their lands were contiguous at certain points. Water gathered on the surface of these lands and flowed across it. A drain had been constructed to assist with the flow of water away from a sand pit operated by the defendant. Barriers were then constructed by the defendant and these barriers impeded the flow of water so that water remained on the plaintiff's land during wet seasons. The plaintiff sought a remedy for an infringement of common law rights. It was decided that the flow of water in the circumstances of this case was not through a watercourse but across the surface of the land. The riparian doctrine did not apply. Liability was determined on the basis of the law of nuisance.

A review of the approach adopted by Justice Windeyer illuminates the range of doctrinal issues underlying the governance of water resources in Australia. His Honour prefaced his comments by referring to the rules of the common law in this way:

> These rules are very old. They have been applied in England in innumerable cases to determine and define rights and obligations in relation to streams

---

[133] See Fisher, *Water Law* (n 1), 107–119.

[134] *Gartner v Kidman* (1962), Commonwealth Law Reports, 108, 12.

> along the banks of which men have been settled for centuries, using the waters and enjoying the benefits of their flow.[135]

After noting the very different climatic and geographical conditions in Australia, this was the conclusion:

> And this is to be borne in mind when particular decisions of English courts are brought forward as analogies. But it is beyond doubt that these rules are a part, and an important part, of the common law that Australia has inherited.[136]

What rules of the common law were relevant in this case? Not riparian rules but those relating to nuisance because of the particular circumstances of this case. It was necessary to identify a right which had potentially been infringed. As his Honour explained:

> In considering whether there has been an actionable interference with the beneficial enjoyment of land, it is necessary, in every case, to know whether some particular right in or in relation to that land is said to have been invaded, and if so to ascertain the limits of that right.[137]

The relevant question was whether the common law recognised a right to obstruct the natural flow of water and the circumstances in which such a right could be exercised. Justice Windeyer reviewed earlier judicial discussion and stated:

> The fundamental rule is, I think, the reverse of what Chief Justice Madden in Vinnicombe v MacGregor[138] stated it to be. He said that the lower owner cannot obstruct the natural flow of surface water unless he can and does so in a manner which will not injure the upper land. On the contrary the law is, I think, that he may block it by any works on his own land, so far as they are reasonably necessary to protect his land for his reasonable use and enjoyment; but that in doing so he must not act recklessly of his neighbour so as to cause wanton damage to him. That, in my opinion, accords with the broad principles of the law of nuisance today.[139]

Justice Windeyer responded to the Chief Justice's analysis by translating a negative obligation coupled with a positive but conditional right into a positive right coupled with a negative but conditional obligation. These rights and obligations relate to the use of land and the impact of such a

---

135 Ibid., 23.
136 Ibid.
137 Ibid., 38–9.
138 (1902) Victorian Law Reports, 28, 144.
139 *Gartner v Kidman* (1962) Commonwealth Law Reports, 108, 12, 46.

use of land. The circumstances in which they can arise include activities involving water and by association such activities impacting on land: namely, a doctrinal interface between water and land – a concept thousands of years old. Perhaps in this way:

> There is, in other words, a fundamental recognition of the mutuality of interests of adjacent landowners and occupiers, protected by the law of nuisance, just as the law protects the mutual rights and interests of riparian proprietors by the common law riparian doctrine. There is no absolute right to deal with the land in an unrestricted way. This is so, not only in relation to the land itself, but also in relation to any interference with the flow of water across the land as a result of activities on the land.[140]

4.2.6 Conclusion

There can be little doubt that the rules of the common law had not been automatically extinguished by the creation of a Crown right to the flow, use and control of water. Each set of circumstances, each rule of the common law and each statutory provision would need to be analysed, interpreted and applied both separately and together so that the relationship between all of the relevant elements of the legal system could be viewed consistently and coherently. Not an easy task. This is no doubt the reason why some jurisdictions later in the twentieth century specifically abolished common law rights in relation to water. Now is the time to turn to a more detailed assessment of the current legislation in Australia.

4.3 An Analysis of Current Australian Legislation

4.3.1 The constitutional framework[141]

It is important to note that, after the establishment of the federal Commonwealth of Australia in 1900 until 2007, the management of water resources had been undertaken in accordance with the legislation of the states and territories – that is the former colonies. In a negative sense this principle of state legislative exclusivity was recognised by the inclusion of section 100 in the Commonwealth Constitution which states that 'The Commonwealth shall not, by any law or regulation of trade or commerce, abridge the right of a State or of the residents therein to the reasonable use of the waters of rivers for conservation or irrigation.'

---

[140] Fisher, *Water Law* (n 1), 121.

[141] See generally Fisher, *Water Law* (n 1), ch. 3; Gardner et al. (n 77), ch. 5; and Stoeckel *et al.* (n 77), 5–10, 62–82.

The right in relation to water recognised in this way – probably in the context of interstate rivers – is protected by the duty imposed on the Commonwealth not to interfere with it in the exercise of the power of the Commonwealth to enact laws concerning trade or commerce. Justice Mason of the High Court of Australia has commented that the relationship between the water and the trade and commerce provisions suggests that '[t]he primary purpose of section 100 was to safeguard the rights of a State and its residents to the use of waters and rivers used for interstate trade and commerce including navigation and shipping, viz the Murray River.'[142]

The Constitution of the Commonwealth otherwise makes no reference to water. If the Commonwealth were minded to become involved in water resources governance, it either seeks a referral of legislative power from the states or exercises somewhat indirectly a potentially relevant power over another matter. For example, powers in relation to:

- trade and commerce with other countries and among the states – water as a commodity,
- trading corporations formed within the limits of the Commonwealth – corporations trading in water resources,
- external affairs – implementing international obligations,
- grants of financial assistance to any state.

The grant of financial assistance and the implementation of international obligations have become important examples of Commonwealth involvement in water resources governance. The Water Act enacted by the Commonwealth in 2007 – the most important legislation enacted by the Commonwealth about water – relies largely upon the referral of powers by the states and to a lesser extent on the other sources of legislative authority. The Act of 2007 creates a planning and management regime for the Murray-Darling Basin – a very large and important interstate water catchment area responsible for a considerable part of Australia's agricultural production.[143] The Murray-Darling Basin Plan is formulated

---

142 *Commonwealth v Tasmania* (1983) Commonwealth Law Reports, 158, 1, 241.

143 For a detailed discussion see Fisher, D. 'A Sustainable Murray-Darling Basin: the Legal Challenges' in D. Connell and R. Quentin Grafton (eds), *Basin Futures – Water Reform in the Murray-Darling Basin* (ANU E Press 2011).

under the Act but it is given effect through the legal arrangements in the states. The legal nature of the legal arrangements in the states remains critical.

4.3.2 The generic elements of the water legislation in Australia

The legislation of the nine jurisdictions in Australia[144] is broadly similar in content, reflective of how legal rules for the governance of natural resources and the environment are evolving to meet the challenges of sustainability, but often different in terms of form and language.[145] Apart from the residual but important functions of the common law, water resources governance is essentially statutory. This is no surprise given that the function of legislation is often to formalise what is perceived to be in the public interest – the essence of a public domain regime. The structure, form and language of the legislation are critical to its interpretation and application.[146] Hence the need to assess the structure and text of a number of examples.

The end of the nineteenth century saw the emergence of the foundations of a public domain regime in Australia with the conferment on a public authority of a statutory right to the flow, use and control of water in watercourses – hereafter the 'foundational' right of the state. This, it has been noted, created a new right alongside existing common law rights without automatically extinguishing them but overriding them in some circumstances. During the twentieth century and, most importantly, at the beginning of the twenty-first century a comprehensive range of legal rules developed further the idea of the public domain. In some jurisdictions this included the complete or partial abolition of common law rights in relation to water. Their ultimate objective has been to achieve the sustainable use and development of water resources.

What has emerged is a range of structures each with some of these elements: (1) the foundational right of the state; (2) a statement of the objectives of the legislation – sometimes supported by a duty to achieve them – hereafter the objects; (3) a statement of the principles for managing water – sometimes supported by a duty to be guided by them – hereafter the principles; (4) the formulation of a plan or a series of plans for managing the water; (5) a duty not to take or use water without a

---

[144] Namely the Commonwealth, the six states and the two territories.

[145] Fisher, D. 'Delivering the National Water Initiative – the emergence of innovative legal doctrine' in K. Hussey and S. Dovers (eds), *Managing Water for Australia – the social and institutional challenges* (CSIRO Publishing 2007), 115.

[146] See generally Fisher, D. *Legal Reasoning in Environmental Law – a Study of Structure, Form and Language* (Edward Elgar Publishing 2013).

licence, an approval or other form of authorisation; (6) an individual right to take or use water or to receive water in accordance with such an authorisation; (7) a statutory right conferred upon individual landholders to take or use water for specific purposes – analogous in some respects to common law riparian rights; (8) arrangements for transferring water rights; and (9) arrangements for environmental water entitlements.[147]

No jurisdiction contains all of these instruments. It has been concluded that:

> these arrangements disclose a set of mosaics or patterns that are structurally linked in different ways, but each lead to the sustainable use and development of water resources in one form or another. While there is no thematic or doctrinal consistency – let alone uniformity – each set of arrangements strive to achieve its objectives in its own way.[148]

4.3.3 The functions of the foundational right, objects, principles and plans

Consider, first, the relationship between the foundational right of the state and its associated objects and duties. Section 19 of the Water Act 2000 of Queensland states that 'all rights to the use, flow and control of all water in Queensland are vested in the State'. This 'right' is in the nature of a power. The exercise of this right or power is not unlimited. It is controlled in two ways:

1. by the purpose stated in section 10(1);
2. by the duty imposed by section 12 upon all entities to perform their functions or to exercise their powers in a way that advances the purpose stated in section 10(1).

According to section 10(1), the purpose is to advance sustainable management and efficient use of water and other resources by establishing a system for the planning, allocation and use of water. The terms 'sustainable management' and 'efficient use' are defined in some detail.[149] Although it is ultimately the responsibility of the state to ensure that the planning, allocation and use of water achieve the purpose of the Act, the implication of section 12 is that everyone shares a common responsibility to do so because everyone is performing statutory functions or exercising statutory powers – if not directly, at least indirectly because

---

[147] For more detail see Fisher (n 145), 116–20.
[148] Ibid., 120.
[149] Water Act 2000 (Queensland), s. 10(2) and (3).

the Act is the source of all authority in relation to water resources governance.

Consider, second, the relationship between the management principles that guide the exercise of statutory powers and their associated objects, duties, plans and specific rules. Section 392 of the Water Management Act 2000 of New South Wales creates the foundational right; section 393 abolishes common law riparian rights; section 3 states the objects of the Act; and section 9 imposes a duty in relation to the management principles set out in section 5. The object of the Act is directed at sustainable development which is achieved by applying the associated management principles. Section 5(3)(a) states as a principle that 'sharing of water from a water source must protect the water source and its dependent ecosystems.' Section 9(1)(b) states:

> [i]t is the duty of all persons exercising functions under this Act:
>
> (b) as between the principles for water sharing set out in section 5(3), to give priority to those principles in the order in which they are set out in that sub-section.

These provisions have attracted judicial comment.

In the Nature Conservation Council Case in 2005,[150] the issue was the validity of a clause in a water management plan which contained environmental rules. These rules were to identify planned environmental water which is water 'committed' for fundamental ecosystem health. The clause in question prohibited the taking of water from the source in question in excess of the long term average annual extraction. The Court of Appeal in New South Wales took the view that 'virtual' rather than 'real' water had been 'committed' to protect 'fundamental ecosystem health'.[151] The clause in the rules was formally invalid because priority had not been given to fundamental ecosystem health. The water management plan, however, was valid because it achieved substantively the objects of the Act. Here is the reason:

> The objects of the Act set out in section 3(a) and section 3(b), the water management principles set out in section 5(2)(a) to (d) and section 5(3)(a), together with a priority established by section 9(1)(b) strongly suggest that the

---

150 *Nature Conservation Council of New South Wales Inc v Minister Administering the Water Management Act 2000* (2005) New South Wales Court of Appeal 9.

151 Ibid., paras. 65–6.

> Parliament was concerned with matters of substance rather than form when it required the establishment of environmental water rules.[152]

The importance of this decision is the way in which the court interpreted and applied the language in the Act which expressed the relationship between the objects, the management principles, the duty in relation to these principles and the provision in the rules – a complex set of concepts but clearly stated in the legislation.

4.3.4 An integrated model for water resources governance

Consider a different model: namely, a fully integrated, planning and regulatory regime for natural resources governance of which water resources are a critical part. The Natural Resources Management Act 2004 of South Australia does not prescribe a foundational right of the state. It is a combination of objects, principles, general duties, more specific duties, plans, rights, restrictions on the exercise of rights, and supporting administrative procedures. The foundation upon which the Act is constructed is the set of objects and principles declared in section 7. In the context of water resources, section 7(1)(c) provides:

> The objects of this Act include to assist in the achievement of ecologically sustainable development in the State by establishing an integrated scheme to promote the use and management of natural resources in a manner that – provides for the protection and management of catchments and the sustainable use of land and water resources and, insofar as is reasonably practicable, seeks to enhance and restore or rehabilitate land and water resources that have been degraded.

Significantly, this provision incorporates several related objects or outcomes: (1) ecologically sustainable development; (2) promotion of the use of natural resources; (3) protection of catchments; (4) sustainable use of water resources; and (5) restoration of degraded water resources.

Section 7(2) goes on to define ecologically sustainable development in a way that links economic, social and ecological outcomes. Section 7(3) states in great detail the principles that 'should be taken into account in connection with achieving ecologically sustainable development for the purposes of this Act'. The principle in section 7(3)(k) is particularly significant. Namely the responsibility to achieve ecologically sustainable development should be seen as a shared responsibility between the public sector, the private sector, and the community more generally. The notion

---

[152] Ibid., para. 95.

of shared or common responsibility is a reflection of water resources as common resources.

The achievement of the objects of the Act is facilitated by the creation of three duties in descending order of specificity: (1) the overriding duty in section 8 in relation to the objects of the Act; (2) the duty imposed by section 9(1) on all persons to 'act reasonably in the management of natural resources within the State' supported by the criteria in section 9(2) for determining what is reasonable; and (3) the duty imposed by section 133(1) on 'the owner of land' situated on or adjoining 'a watercourse or lake to take reasonable measures to prevent damage' to their beds, banks or shores and 'to the ecosystems that depend on the watercourse or lake'.

The duty in section 8 requires further analysis. It states:

> The Minister, the Court and all other persons or bodies involved in the administration of this Act, or performing, exercising or discharging a function, power or duty under this Act, must have regard to, and seek to further, the objects of this Act.

The duty applies effectively to every person and institution in the state. It is imposed not only on those involved in administration but also on those seeking authorisation for or engaging in an activity to which any provision in the Act applies. It is a duty of comprehensive and probably total application. The duty, however, is not absolute. It is limited to having regard to the objects of the Act and to seeking to further them. The duty to 'have regard' is certainly justiciable. Whether the duty to 'seek to further' is justiciable is less clear. In any event each of these three duties sends a clear message to everyone involved in the management of natural resources – including water resources – particularly when they are applied in conjunction with the principle of common responsibility.

It is against this background that the range of rights and regulatory arrangements needs to be discussed. The range of rights is extensive. Section 124(8) abolishes rights at common law in relation to the taking of naturally occurring water. In substitution, water rights are conferred by section 124. First, a person who has lawful access to a watercourse, lake or well may take water therefrom for any purpose.[153] Second, the occupier of land is entitled to take surface water from the land for any purpose.[154] However, if the watercourse, lake or well is prescribed or the

---

[153] Natural Resources Management Act 2004 (South Australia), s. 124(1).
[154] Ibid., s. 124(2).

area is a prescribed surface water area, then either an authorisation under section 128 or a water allocation is required.[155] If the watercourse, lake or well is not prescribed, water must not be taken if – to paraphrase – the interests of others are detrimentally affected.[156] However, the right of the occupier of land to take water from a watercourse, lake or well or water that is surface water is not restricted by these provisions provided the water is used by the occupier for domestic purposes or for watering stock but not stock subject to intensive farming.[157] These rules are in some respects a statutory reflection of those of the common law.

There are two types of regulatory arrangements: water resource planning and control of water related activities. There are obligations to prepare three types of plan:

- a State Natural Resources Management Plan,[158]
- a regional natural resources management plan for each region,[159]
- a water allocation plan for each of the prescribed water resources in each region.[160]

Unless there is in force an authorisation or a water allocation, water must not be taken contrary to the provisions of a regional or a water allocation plan.[161] There is accordingly a hierarchical relationship between these three types of plan and the statements of objects and principles in section 7 supported by the range of duties imposed by sections 8, 9 and 133.

The range of water and water related activities controlled is very extensive. Activities affecting water include taking water from a prescribed source and those involving wells, dams and similar structures.[162] The range of relevant instruments designed to grant approval to water and water related activities includes: (1) well driller licences;[163] (2) water licences;[164] (3) water allocations;[165] (4) water resource works

---

[155] Ibid., s. 124(3)(a).
[156] Ibid., s. 124(3)(b).
[157] Ibid., s. 124(4).
[158] Ibid., s. 74.
[159] Ibid., s. 75.
[160] Ibid., s. 76.
[161] Ibid., s. 124(7).
[162] Ibid., s. 127(1) and (3).
[163] Ibid., s. 139.
[164] Ibid., s. 146.
[165] Ibid., s. 152.

approvals;[166] (5) site use approvals;[167] and (6) delivery capacity entitlements.[168]

What emerges from these arrangements for water resources governance in South Australia is the integration of water resources management into a system of natural resources governance directed at their ecologically sustainable development. These arrangements include an obligation to take into account the evolving set of principles in support of sustainability. The capacity of the system to ensure compliance – always a major challenge – relies upon the duties imposed upon the community at large to move in the direction of ecologically sustainable development. Equally important are the water rights conferred upon everyone, while the exercise of these rights is controlled by a very comprehensive planning regime and an equally extensive number of detailed sets of circumstances requiring administratively granted approval. Despite the absence of a foundational right conferred on the state, these arrangements reflect most – perhaps all – of the features of a public domain system. There can be little doubt that the legislature has determined that ecologically sustainable development is in the public interest and that these arrangements can go some way to achieve it.

5 CONCLUSION

5.1 From Common Resources to an Institutionalised Public Domain

Water resources have traditionally been regarded as common resources to be used for the benefit of the community at large. From a legal perspective, they may be regarded as *res communes* – common not private property. This was part of the approach stated in the Institutes of Justinian. It was modified towards the end of the Roman empire, when the management of water resources was institutionalised by the introduction of supporting administrative arrangements – the emergence of a public domain regime. Although the common law was influenced by the doctrinal foundations of Roman law, it did not evolve into a public domain regime. The rules of the common law essentially granted rights of access to water to eligible landowners and occupiers and to be used for private rather than public purposes. The reluctance of the common law to

---

166 Ibid., s. 159.
167 Ibid., s. 164A.
168 Ibid., s. 164G.

grant rights of access to water to be used for public purposes was solved by the enactment of legislation authorising the construction of water infrastructure and in some cases access to water and its use for public purposes. This model began to be adopted in England in the late eighteenth century and later in the colony of New South Wales. In New South Wales, in addition, water rights were reserved to the Crown in grants of interests in land to individuals. Finally, in Australia, the Crown – thereafter the state – was invested with the foundational right to water. This led to the emergence of an inchoate public domain regime in Australia.

In these various ways the management of water resources in Australia was institutionalised along the lines of a public domain regime. There developed a set of private rights and a set of public rights. The relationship between them presented doctrinal challenges to the courts. The exercise of common law rights of access to water changes the status of the water abstracted in this way from a *res communis* to a *res propria* or a *res privata.* The institutional arrangements associated with a public domain regime facilitate the management and use of water in the public interest. The relationship between the rules supporting the management of water in the public interest and the rules of the common law protecting the rights of individuals has been discussed in a number of cases in the courts. The most recent analysis undertaken by the High Court of Australia was the Thorpes Case in 1955[169] until the court took the opportunity to consider the doctrinal foundations of water law in the ICM Case in 2009.[170] Did the court acknowledge the emergence of a public domain regime in Australia?

5.2 Judicial Endorsement of Water Resources as Common Resources

The issue in this complicated case was whether variations made to water plans and consequential reductions in the quantity of water available for irrigation under water entitlements granted under the New South Wales legislation in accordance with a scheme partly funded by the Commonwealth constituted an acquisition of property other than on just terms

[169] *Thorpes Ltd v Grant Pastoral Co Pty Ltd* (1995), Commonwealth Law Reports, 92, 317. See the discussion in Section 4.2.4 above.

[170] *ICM Agriculture Pty Ltd v Commonwealth* (2009), Australian Law Journal Reports 84, 87.

contrary to section 51(xxxi) of the Commonwealth Constitution.[171] This was a very complex issue but one which raised questions about the status of water under the common law and under the statutory regimes that had emerged in Australia. In particular:

> The tension in relation to water resources between, on the one hand, the public interest in ensuring that water resources are used and developed sustainably for the future and, on the other hand, the need to protect private interests in the use of water resources in the context of the rights of existing users – in this case, irrigators.[172]

How did the court respond? Given the provision in the Constitution about 'acquisition of property', it is no surprise that water as property was discussed. This was the first response:

> First and foremost, there was then [in the nineteenth century], and still must be, a clear recognition of the difficulty of applying notions of ownership or property to water in the ground or in a flowing stream. What exactly would be the subject of property rights? While still allowed to flow, no part of the water that flows in the stream can be isolated and tagged as the water 'owned' by some person. And water in the ground may move more slowly but there is no less difficulty in identifying what would be the subject of the proprietary rights.[173]

This was the second response:

> Next, even if these difficulties of identifying the object in respect of which proprietary rights were to exist could be overcome, *should* any proprietary right be recognised?[174]

The first response was an acknowledgement of hydrological reality. The second raised a fundamental issue of principle. How was it resolved?

It was accepted that water exists in its natural state for the common benefit of humans. Accordingly, it has been suggested, 'while the common law recognised individual or private rights of access in relation

---

[171] There is no discussion here of the complex constitutional issues that arose in this case. These are discussed in Fisher, D. 'Water law, the High Court and techniques of judicial reasoning' (2010) *Environmental and Planning Law Journal* 27, 85–97.

[172] Fisher (n 171), 89.

[173] *ICM Agriculture Pty Ltd v Commonwealth* (2009), Australian Law Journal Reports 84, 112.

[174] Ibid., 113. Emphasis added.

to water, they were based upon water as common property.'[175] This was certainly not a novel proposition. The court itself referred to the riparian doctrine that emerged in the nineteenth century and concluded:

> The underlying proposition was that water, like light and air, is common property (for the common benefit of man).[176]

The common law, however, was silent about how to ensure that water was collected and used 'for the common benefit of man'. Hence, as noted on several occasions, the need for institutionalised arrangements to promote the public interest in the management of water resources. The court then addressed whether the statutory arrangements introduced since the end of the nineteenth century did so.

It will be recalled that in the Thorpes Case in 1955, Justice Fullagar said that the 'real object' of the Water Rights Act 1896 of New South Wales was to give the Crown 'full dominion' over water. The Oxford English Dictionary describes dominion as sovereignty or lordship. In 2009 the court explained further that:

> The vesting of rights to the 'use' and 'control' of water constituted an exercise of sovereignty in the sense that the rights so vested were based on the political power of the state.[177]

The exercise of political power would normally be taken to be a reference to its exercise for the public benefit. So, in this case, the question was the nature of the issue before the court: protection of private interests or promotion of the public interest. The court responded in these words:

> Finally, it is of the very first importance to recognise that the common law principles established in the nineteenth century were directed to the adjustment of rights between landowners. The issue in this case arises, not because there has been some adjustment of those rights, but because the polity has sought to regulate generally the access allowed to a common resource.[178]

---

[175] Fisher (n 171), 91.

[176] *ICM Agriculture Pty Ltd v Commonwealth* (2009), Australian Law Journal Reports 84, 109.

[177] Ibid., 54.

[178] Ibid., 115.

The institutions exercising power on behalf of the 'polity' represent the means to ensure water resources as common resources are used for the common benefit of humans. Perhaps this synthesises the result:

> The High Court has told us that water resources – either from the point of view of the common law or of the public domain regime created by the legislation – are quintessentially common resources or common property. Any private interests in or in relation to water derive from that source and from no other source.[179]

5.3 A Synoptic Analysis

Perhaps in these terms:

- water is and always has been a common resource;
- as such it is common not private property;
- it is to be used for the common benefit;
- the common law protects the private rights of individuals to water;
- the exercise of common law rights may indirectly benefit the public as well as the individual;
- the foundational right of the state and its associated administrative arrangements are the means by which water as a common resource is used for the common benefit.

---

179 Fisher (n 171), 97.

2. A comparison of constitutional provisions on water-related rights in Southern African states

Ed Couzens and Meda Couzens

1 INTRODUCTION

Southern African states[1] face many environmental constraints, and insufficient water is one of the gravest of these. The region is generally arid or even water-stressed, and is subject also to extreme regional and/or economic class differences in availability of water. In many areas, even where there ought not to be a shortage of water, infrastructural deficiencies will prevent people from accessing the available water optimally. Exacerbating these problems is that the region is one of those with the greatest imperatives for rapid economic growth in order to lift its inhabitants from poverty. According to the Southern African Development Community's (SADC's) own website, Southern Africa's water resources are vital for sustainable economic and social development in the region – in fact, it describes these resources as being 'critical for meeting the basic needs related to water supply for domestic, industrial and agricultural requirements and for sanitation and waste management for over 250 million people'.[2]

In such a context one would assume that the constitutions of such states, as the most important national legal instruments and drivers of policy, would reflect the concerns and realities that affect people in this

1 Here considering the states which are member states to the Southern African Development Community (SADC): Angola, Botswana, DR Congo, Lesotho, Madagascar, Malawi, Mauritius, Mozambique, Namibia, Seychelles, South Africa, Swaziland, Tanzania, Zambia, and Zimbabwe.

2 Southern African Development Community: SADC Water Sector <http://www.sadc.int/sadc-secretariat/directorates/office-deputy-executive-secretary-regional-integration/infrastructure-services/sadc-water-sector/> accessed 31 August 2016.

region of the world, including through the recognition and protection of water-related rights. It might be expected therefore that the issue of access to water and the protection of water so as to secure such access would feature in all of the constitutions of the region. Indeed, Southern Africa represents a context where, perhaps more than in other regions of the world, the interdependence between human rights to water access and the protection of water as a natural resource becomes important. However, as shown in this chapter, a surprisingly high number of SADC states do not include specific rights of access to water nor specific clauses aimed at protecting water resources.

The purpose of this chapter is to isolate water-related rights as well as water protection-relevant provisions in constitutions of the 15 SADC member states (as well as Kenya), to explain each briefly, to group the states according to whether they provide a right at all or not, and to ascertain the type of right if there is a right. This chapter will draw a preliminary map of the constitutional protection of water-related rights in SADC states; emphasise strengths and weaknesses of the different approaches taken; and make suggestions for further research necessary for a better understanding and, arguably, more efficient protection of water-related rights in SADC states.

In this chapter, we use a variety of terms which we consider capture best the essence of the diverse constitutional provisions relating to water (from a human rights perspective or from an environmental protection perspective). We use terms such as 'rights of access to water', 'water rights', 'water-related-rights' or 'water security clauses'. A 'right of access to water' might not be the same thing as a 'water-related right' or a 'water right' or even a 'water security clause' – and all of these might or might not relate to environmental protection as distinct from a pure human right of access. Capturing the essence of the wide variety of constitutional provisions in standard or clear-cut legal language is challenging because a water-related right (or a right of access to water) might be subsumed under other rights, such as 'health' or 'environment'; and because, even where states include an explicit water-related provision, the contents are not uniform. For instance, as will be seen later, the rights in the Kenyan ('Every person has the right to clean and safe water in adequate quantities') and Zimbabwean ('Every person has the right to safe, clean and potable water') Constitutions appear stronger than the right in the South African Constitution ('Everyone has the right to have access to sufficient water') and the relevant provision in the Zambian Constitution ('the Government shall strive to provide clean and safe water').

The complexity of these terms (especially in relation to defining the content of the rights they describe, if any) and their perfectible nature must be acknowledged, but a proper consideration of the nature of each potential water-related right reflected in national constitutions would require a closer consideration of each individual case, and would be far beyond the scope of this necessarily preliminary study.[3]

This chapter is structured as follows: water-related constitutional rights are isolated and considered for all of the Southern African states – after which we divide the states into various groups, according to the approach they have taken, and briefly discuss the opportunities and the challenges of each approach. Tentative conclusions are drawn as to the reasons why particular states might have taken particular approaches, and how different states have approached inclusion of environmental rights[4] in their Constitutions. In our conclusions, we also make suggestions for further research.

Some of the Constitutions appear to be largely political instruments that give less attention to environmental rights than do others which are more 'modern' and seek to entrench environmental rights. All will need to consider the issue in the years to come, as water security-related conflicts are likely to increase in the region.[5] What is also considered is that while most states which have included water-related rights in their constitutions have done so by way of imposing duties on the state, some have gone further and imposed duties on individuals also.

It needs to be acknowledged that this study is extremely limited, being a comparative consideration of water-related rights in constitutions only.

---

[3] This is particularly difficult given that there are both translation difficulties and difficulties with accessing judicial interpretations to be taken into account.

[4] As will be explained later, some states have no direct water-related provisions but these may be implied from more general environmental rights.

[5] Opinion is divided on the likelihood of water security-related international conflicts arising. At least three United Nations Secretary-Generals (Boutrous Ghali, Annan and Ki-moon) have suggested that serious conflicts might arise, as have many commentators; but other analysts have argued that there is waning empirical evidence that competing uses of water give rise to conflict rather than cooperation. For a detailed discussion of these different points of view, see Ed Couzens, 'Water-Related Conflict and Security in Southern Africa: The SADC Protocol on Shared Watercourses' in *International Environmental Law-making and Diplomacy Review* 2014 (University of Eastern Finland – UNEP Course Series 14 2015) 91–126 at 92–100, 124–6.

Only extremely limited attention has been given, and then only by way of an illustration,[6] to legislation other than constitutions and to case law.

2 NATIONAL CONSTITUTIONS AND WATER-RELATED RIGHTS

2.1 Introduction

Probably the high point worldwide for protection of citizens' rights in respect of nature, at least on paper, and of citizens' rights to have nature protected, is today to be found in the Constitution of the Republic of Ecuador, 2008,[7] which is famous for providing nature with inherent rights:[8]

> Nature, or Pacha Mama, where life is reproduced and occurs, has the right to integral respect for its existence and for the maintenance and regeneration of its life cycles, structure, functions and evolutionary processes.[9]

Less well-known is that water is also given specific consideration in Ecuador's Constitution.[10] It is provided that the 'State shall guarantee the conservation, recovery and integral management of water resources, watersheds and ecological flows associated with the water cycle'; that '[a]ll activities that can affect the quality and amount of water and the equilibrium of ecosystems shall be regulated, especially in water replenishment sources and zones'; and that '[t]he sustainability of ecosystems and human consumption shall be priorities in water use and development'.[11]

---

[6] See 2.12 below.

[7] Constitution of the Republic of Ecuador, 2008. For an English translation, see <http://pdba.georgetown.edu/Constitutions/Ecuador/english08.html> accessed 31 August 2016.

[8] Title II: Rights. Chapter 7: Rights of Nature.

[9] Article 71. See also Articles 72–4 for more specificity on the rights of nature and the duties resting on the state in respect thereto.

[10] Title VII: The Good Way of Living System. Chapter 2: Biodiversity and Natural Resources. Sixth Section: Water.

[11] Article 411. It is further provided in Article 412 that 'the authority in charge of managing water shall be responsible for its planning, regulation, and control'; and that 'this authority shall cooperate and coordinate with the authority in charge of environmental management to guarantee water management based on an ecosystemic approach'.

Although few constitutions worldwide even begin to approach this entrenching of the rights of nature, and ergo of the rights of citizens in respect of nature, many contain specific rights accorded to people in regard to nature and – specifically – to water. It would seem a truism, in fact, to assert that environmental and human rights jurisprudence has developed in recent decades to the extent that it would be no great surprise to find water and environment-related rights in *any* national constitution.

Of course, different legal systems will accord different values to certain natural resources and give different weight to legal rights thereto. In respect of water, it would be surprising if Finland and Morocco (to take as examples two states not considered in this chapter) were to accord the same significance to the legal protection of rights of access to water – in Finland there might be concerns over water quality, but less so over quantity; in Morocco the opposite might generally apply. However, it is also common to find constitutions which do not provide rights to the environment – or to water specifically – in any form.

2.2 Angola

The Constitution of the Republic of Angola, 2010[12] contains no express water-related right. It is indicated that the 'solid, liquid and gaseous natural resources existing in the soil and subsoil, in territorial waters, in the exclusive economic zone and in the continental shelf under the jurisdiction of Angola shall be the property of the state'.[13] This could be taken to indicate that water is reserved, in the public trust, into the care of the state. However, the article then indicates that the state is to 'determine the conditions for concessions, surveys and exploitation' of these natural resources, which implies that the article is concerned with commercial exploitation rather than with environmental rights.

In Article 39: Environmental Rights, the Constitution stipulates that '[e]veryone has the right to live in a healthy and unpolluted environment and the duty to defend and preserve it';[14] and that:

> [t]he state shall take the requisite measures to protect the environment and species of flora and fauna throughout national territory, maintain the ecological balance, ensure the correct location of economic activities and the

---

[12] See <http://www.comissaoconstitucional.ao/pdfs/constituicao-da-republica-de-angola-versao-ingles.pdf> accessed 31 August 2016.

[13] Article 16: Natural Resources.

[14] Article 39.1.

> rational development and use of all natural resources, within the context of sustainable development, respect for the rights of future generations and the preservation of species.[15]

It is then stated that '[a]cts that endanger or damage conservation of the environment shall be punishable by law'.[16]

Article 39 could, by virtue of the existence of a duty corresponding to the right, be interpreted to be a strong right, with everyone having not just the right to a 'healthy and unpolluted environment' but also a 'duty to defend and preserve it'. A reading of this article together with Article 95, which provides that inland waters are in the public domain of the state,[17] may lead a court to conclude that water is held in public trust by the state, and that the state needs to ensure that its subjects benefit from this natural resource. Whether a court would conclude that a trusteeship by the state necessarily implies a right to be provided with water, or a right to access water, is difficult to predict. If such an argument were to be made, another article which could be used in support would be Article 94: Property of the State.[18] These arguments[19] could be relied upon by a court in concluding that there is a necessarily implied right to water,[20] which is held in the public trust by the state – it being implied that if there is a right to have water held in public trust, then this is not for the benefit of the state itself. However, interpretation external to the Constitution, by a court, is required; and the lack of specific reference to water is somewhat surprising.

---

[15] Article 39.2.

[16] Article 39.3.

[17] Article 95: '1. The following shall constitute property in the public domain: a) Inland waters, territorial waters and the adjacent sea beds, in addition to lakes, lagoons and watercourses, including their beds; ...'

[18] Article 94: 'The property of the state and the various legal persons governed by public law shall belong to the public or the private domain, in accordance with the Constitution and the law'.

[19] Perhaps coupled with Article 89: General Rights, which reads: '1. The organisation and regulation of economic activities shall be based on a general guarantee of overall economic rights and freedoms in general, and an appreciation of work, human dignity and social justice, in accordance with the following fundamental principles: ... h) Consumer and environmental protection'.

[20] Of course, it must be admitted that a court might not find that such a right is necessarily implied – it being possible for a court to decide that, while the state would not hold water in public trust for its own sake, the state could decide to prioritise other legitimate users rather than to provide it directly to the people.

2.3 Botswana

The Constitution of Botswana, 1966[21] contains no specific mention of water. This is surprising in a country which is highly water-stressed.[22]

2.4 Democratic Republic of the Congo

The Constitution of the Democratic Republic of the Congo, 2005[23] contains a specific right of access to drinking water: '[t]he right to decent housing, the right of access to drinking water and to electric energy are guaranteed'; although it is provided that '[t]he law establishes the conditions for the exercise of these rights'.[24] Water quality is also important, with Article 55 providing for the criminalisation of water pollution or of conduct creating a risk of water pollution.[25] There is also a general environmental right which provides that '[a]ll persons have the right to a healthy environment and (one) propitious for their integral development'; that '[t]hey have the duty to defend it'; and that '[t]he State sees to the protection of the environment and the health of the population'.[26]

---

21 Constitution of Botswana, adopted on 30 September 1966; amended 1969, 1970, 1982 and 1997. See <http://www.commonlii.org/bw/legis/const/1966/> accessed 31 August 2016.

22 The importance of water to Botswana can perhaps be seen in the predominance of blue on its flag, with the blue symbolising water and the life that is sustained by water; and by the name of the Botswana currency, 'pula' (which means rain in Setswana).

23 Constitution of the Democratic Republic of Congo, 2005, as amended in 2011. See <https://www.constituteproject.org/constitution/Democratic_Republic_of_the_Congo_2011.pdf> accessed 31 August 2016.

24 Article 48. (An alternate translation is: '[t]he law establishes the modalities of the exercise of these Rights.') A related right is contained in Article 47 which provides that '[t]he right to health and to (a) secure food supply is guaranteed. The law specifies the fundamental principles and the rules of organisation for public health and (a) secure food supply'.

25 Article 55 provides that: '[t]he transportation, the importation, the storage, the spilling [or] the disposal in the internal waters or maritime spaces under national jurisdiction, [or] the release into the airspace[,] of toxic, polluting or radioactive waste or of any other dangerous product, of foreign origin [provenance] or not, constitutes a crime punishable by the law'.

26 Article 53.

2.5 Lesotho

The Kingdom of Lesotho does not suffer from a lack of water,[27] but does face issues related to water quality.[28] The Constitution of Lesotho, 1993[29] contains no explicit mention of water. There is an environmental section in the Constitution, Section 36: Protection of the Environment, which provides that:

> Lesotho shall adopt policies designed to protect and enhance the natural and cultural environment of Lesotho for the benefit of both present and future generations and shall endeavour to assure to all citizens a sound and safe environment adequate for their health and well-being.

There is also a section on health[30] which would impliedly cover issues related to water quality. However, both sections are included in Chapter III, titled 'Principles of State policy', which, according to Section 25, 'shall not be enforceable by any court but … shall guide the authorities and agencies' in fulfilling their functions.

---

[27] In fact, the country supplies South Africa with considerable water, both through its role as a natural watershed and through human-made water diversion projects such as the Lesotho-Highlands Water Project.

[28] Such as the pollution of rivers by industrial projects.

[29] Constitution of Lesotho, adopted in 1993; amended 1996, 1997, 1998 and 2001. See <http://www.ilo.org/wcmsp5/groups/public/—ed_protect/—protrav/—ilo_aids/documents/legaldocument/wcms_126743.pdf> accessed 31 August 2016.

[30] Section 27: Protection of health: '(1) Lesotho shall adopt policies aimed at ensuring the highest attainable standard of physical and mental health for its citizens, including policies designed to –

(a) provide for the reduction of stillbirth rate and of infant mortality and for the healthy development of the child;
(b) improve environmental and industrial hygiene;
(c) provide for the prevention, treatment and control of epidemic, endemic, occupational and other diseases;
(d) create conditions which would assure to all, medical service and medical attention in the event of sickness; and
(e) improve public health'.

2.6 Madagascar

The Constitution of Madagascar, 2010[31] contains no mention of water – or, surprisingly given the recent adoption of the Constitution, of the environment generally.

2.7 Malawi

The Constitution of the Republic of Malawi, 1994,[32] does not contain a specific water-related right; nor does it contain a general environment-related right. However, the Constitution provides, under the heading 'Principles of National Policy',[33] that '[t]he State shall actively promote the welfare and development of the people of Malawi by progressively adopting and implementing policies and legislation aimed at achieving the following goals' … to manage the environment responsibly in order to:[34]

(i) prevent the degradation of the environment;
(ii) provide a healthy living and working environment for the people of Malawi;
(iii) accord full recognition to the rights of future generations by means of environmental protection and the sustainable development of natural resources; and
(iv) conserve and enhance the biological diversity of Malawi.[35]

2.8 Mauritius

The Constitution of the Republic of Mauritius, 1968,[36] contains no specific provision on either water or the environment. The closest

---

31 Constitution of Madagascar, adopted on 17 November 2010. See <https://www.constituteproject.org/constitution/Madagascar_2010.pdf> accessed 31 August 2016.

32 Constitution of the Republic of Malawi, 1994, adopted 16 May 1994, in force 18 May 1995, as amended in 2010. See <http://www.wipo.int/wipolex/en/text.jsp?file_id=317750> accessed 31 August 2016.

33 Article 13: Principles of National Policy; introduced by amendment in 2010.

34 Article 13(d): The Environment.

35 Article 13(d)(i)–(iv).

36 See <http://mauritiusassembly.govmu.org/English/constitution/Pages/constitution2016.pdf> accessed 31 August 2016.

provisions relate to property[37] and effectively reserve to the state the right to override the right to property in the interests of matters such as 'defence, public safety, public morality, public health, town and country planning'.[38]

2.9 Mozambique

The Constitution of the Republic of Mozambique, 2004[39] does not mention water specifically, but has a dedicated article[40] on the environment which provides that:

> 1. All citizens shall have the right to live in a balanced environment and shall have the duty to defend it.
> 2. The State and the local authorities, with collaboration from associations for environmental protection, shall adopt policies to protect the environment and shall promote the rational use of all natural resources.

There are also several other articles relevant to environmental protection, including Article 45[41] which provides that '[e]very individual shall have the duty to: … e) defend and promote health; f) protect and conserve the environment …'.[42] Article 81[43] provides that '[a]ll citizens shall have the right to popular action in accordance with the law, either personally or through associations for defending the interests in question';[44] with the right consisting of, inter alia, '[t]he right to advocate the prevention, termination or judicial prosecution of offences against the public health, consumer rights, environmental conservation and cultural heritage'.[45]

Finally, Article 117[46] provides that '[t]he State shall promote efforts to guarantee the ecological balance and the conservation and preservation of the environment, with a view to improving the quality of life of its

---

[37] Article 8: Protection from Deprivation of Property; and Article 9: Protection of Privacy of Home and Other Property.

[38] Articles 8(1)(a) and 9(2)(a).

[39] Constitution of the Republic of Mozambique, 2004. See <http://www.refworld.org/docid/4a1e597b2.html> accessed 31 August 2016.

[40] Article 90: Right to a Balanced Environment.

[41] Article 45: Duties Towards the Community.

[42] Article 45(e) and (f).

[43] Article 81: Right of Popular Action.

[44] Article 81(1).

[45] Article 81(2)(b).

[46] Article 117: Environment and Quality of Life.

citizens'.[47] Further, Article 117 provides that, '[w]ith a view to guaranteeing the right to the environment within the framework of sustainable development, the State shall adopt policies aimed at':[48]

(a) preventing and controlling pollution and erosion;
(b) integrating environmental objectives with sectoral policies;
(c) promoting the integration of environmental values into educational policies and programmes;
(d) guaranteeing the rational utilisation of natural resources and the safeguarding of their capacity to regenerate, ecological stability and the rights of future generations;
(e) promoting territorial ordinance with a view to ensuring the correct location of activities, and balanced socio-economic development.[49]

2.10 Namibia

The Constitution of the Republic of Namibia, 1990,[50] contains no specific 'environmental' mention of water, despite water being extremely scarce throughout the country. Article 100[51] is contained in a chapter headed 'Principles of State Policy'[52] and provides that:

> Land, water and natural resources below and above the surface of the land and in the continental shelf and within the territorial waters and the exclusive economic zone of Namibia shall belong to the State if they are not otherwise lawfully owned.

Clearly, this is extremely weak as a right[53] and barely affords environmental protection. The effect of the article is merely to provide that, by policy, the state will have default ownership of land, water and other natural resources where these are not privately owned.

---

47 Article 117(1).
48 Article 117(2).
49 Article 117(2)(a)–(e).
50 Namibian Constitution, 1990, as amended by the First Amendment Act, 1998, and the Second Amendment Act, 2010. See <http://www.orusovo.com/namcon/> accessed 31 August 2016.
51 Article 100: Sovereign Ownership of Natural Resources.
52 Chapter 11.
53 At best, a right to hold the state to implement and abide by policies that conform with the principles provided.

2.11 The Seychelles

The Constitution of the Republic of Seychelles, 1993,[54] does not contain a specific water-related right, but there is an environmental right:

> The State recognises the right of every person to live in and enjoy a clean, healthy and ecologically balanced environment and with a view to ensuring the effective realization of this right the State undertakes –
>
> (a) to take measures to promote the protection, preservation and improvement of the environment;
> (b) to ensure a sustainable socio-economic development of Seychelles by a judicious use and management of the resources of Seychelles;
> (c) to promote public awareness of the need to protect, preserve and improve the environment.[55]

In addition, the Constitution records that '[t]he State recognises the right of every citizen to protection of health and to the enjoyment of [the] attainable standard of physical and mental health'.[56]

Associated rights may be found in the limitation[57] of the right to property[58] 'for vesting in the Republic of the ownership of underground water or unextracted oil or minerals of any kind or description'.[59] Further, it is provided under the heading 'Fundamental Duties'[60] that '[i]t shall be the duty of every citizen of Seychelles ... to protect, preserve and improve the environment'.[61]

2.12 South Africa

South Africa's Constitution[62] has a section which contains a specific right of access to water.[63] It has been argued, in fact, that 'South Africa is one of the few jurisdictions in the world that provides for an explicit right to

---

[54] Constitution of the Republic of Seychelles, 1993; 21 June 1993; as amended in 2011. See <http://www.wipo.int/edocs/lexdocs/laws/en/sc/sc001en.pdf> accessed 31 August 2016.

[55] Article 38.

[56] Article 29.

[57] Article 26(2).

[58] Article 26(1).

[59] Article 26(2)(i).

[60] Part II: Fundamental Duties.

[61] Article 40(f).

[62] Constitution of the Republic of South Africa, 1996.

[63] Section 27: Health care, food, water and social security of the Constitution of the Republic of South Africa, 1996.

water' with the Constitution 'guarantee[ing] the right to access to adequate water'.[64] In terms of this section '[e]veryone has the right to have access to ... sufficient food and water'.[65] The state is then required to 'take reasonable legislative and other measures, within its available resources, to achieve the progressive realisation of [this right]'.[66]

There is also a specific right in respect of the environment[67] which provides that '[e]veryone has the right':

(a) to an environment that is not harmful to their health or well-being; and
(b) to have the environment protected, for the benefit of present and future generations, through reasonable legislative and other measures that –
  (i) prevent pollution and ecological degradation;
  (ii) promote conservation; and
  (iii) secure ecologically sustainable development and use of natural resources while promoting justifiable economic and social development.

South Africa provides an example of a country in which the state's interpretation of the constitutional water-related right has been judicially challenged – in the Constitutional Court's judgment in *Mazibuko and Others v City of Johannesburg and Others* (*Mazibuko*).[68] According to the Constitutional Court, the constitutional right of access to sufficient water does not impose upon the state a duty to provide every person, on demand and without more, with sufficient water. The obligation does not, further, confer on any person a right to claim 'sufficient water' from the state immediately; the constitutional right, instead, requires the state to take reasonable legislative and other measures, progressively and within the state's available resources, to realise the achievement of the right of access to sufficient water.[69]

---

64 Linda Stewart and Debra Horsten, 'The Role of Sustainability in the Adjudication of the Right of Access to Adequate Water' (2009) 24 *SAPR/PL*, 486–505.

65 Section 27(1)(b).

66 Section 27(2).

67 Section 24: Environment.

68 [2009] ZACC 28; 2010 (4) SA 1 (CC); 2010 (3) BCLR 239 (CC). Date of judgment: 8 October 2009.

69 *Mazibuko* paras 50 and 57.

According to the Court, the state's obligations can be enforced judicially by requiring the state to take those steps necessary in order to meet the constitutional standard of reasonableness. A measure would be considered unreasonable if it made no provision for persons in desperate need; and if the state were to adopt policies which contained unreasonable limitations or exclusions, then a court could order such limitations or exclusions to be removed; further, it was held that a duty was imposed upon the state (by the obligation of progressive realisation) to review its policies continually to ensure that the achievement of the right is realised progressively; and, finally, that an obligation rests on the state to set out, clearly, the targets which it hopes to achieve.[70] Although the right to have access to sufficient water is a fundamental right, the judgment demonstrates that the state may circumscribe this right severely through policy.

2.13 Swaziland

The Constitution of Swaziland, 2005,[71] contains several provisions relevant to water, either express or implied. Chapter XII is headed 'Land, Minerals, Water and Environment' and provides that[72] land, minerals and water are[73] national resources;[74] and that, '[i]n the interests of the present and future generations', the state 'shall protect and make rational use of its land, mineral and water resources as well as its fauna and flora, and shall take appropriate measures to conserve and improve the environment'.[75] There is then a specific section dealing with water, which provides that '[t]here shall be no private right of property in any water found naturally in Swaziland'.[76] The section[77] immediately following provides that:

> (1) Every person shall promote the protection of the environment for the present and future generations.

---

[70] *Mazibuko* paras 67 and 70. On the facts, the Constitutional Court found that the policy in question was not unreasonable.

[71] Constitution of the Kingdom of Swaziland Act, 2005 (Act No. 001 of 2005). See <http://www.wipo.int/edocs/lexdocs/laws/en/sz/sz010en.pdf> accessed 31 August 2016.

[72] Section 210: Declaration of Land, Minerals and Water as National Resource.

[73] Subject to the provisions of this Constitution or any other law.

[74] Section 210(1).

[75] Section 210(2).

[76] Section 215: Water.

[77] Section 216: Environment.

(2) Urbanisation or industrialisation shall be undertaken with due respect for the environment.
(3) The Government shall ensure a holistic and comprehensive approach to environmental preservation and shall put in place an appropriate environmental regulatory framework.

There is then a further section which provides that Parliament may make laws:

(a) providing for the management of land and settlement of land disputes and for the regulation of any right or interest in land whether urban or rural and whether privately owned or vesting in the King;
(b) regulating the rights and interests in minerals and mineral oils;
(c) regarding the use of water naturally found in Swaziland; and
(d) for the protection of the environment including management of natural resources on a sustainable basis.[78]

2.14 Tanzania

The Constitution of the United Republic of Tanzania, 1977,[79] contains no specific water-related right – nor does it contain a right specific to the environment more generally. The nearest the Constitution comes to providing such a right is to be found in the duty imposed[80] on '[e]very person' to 'protect the natural resources of the United Republic, the property of the state authority, all property collectively owned by the people, and also to respect another person's property'.[81] The same article then further provides that '[a]ll persons shall be required by law to safeguard the property of the state authority and all property collectively owned by the people, to combat all forms of waste and squander, and to manage the national economy assiduously with the attitude of people who are masters of the destiny of their nation'.[82]

78 Section 217: Further Provisions.

79 Constitution of the United Republic of Tanzania, 1977, as amended 2005. See <http://www.wipo.int/wipolex/en/text.jsp?file_id=216577> accessed 31 August 2016.

80 Through an amendment in 1984.

81 Article 27(1).

82 Article 27(2).

There is currently a longstanding negotiation process underway for the drafting of a new Constitution.[83]

2.15 Zambia

The Constitution of Zambia, 1991,[84] does not contain a direct water-related right. On 11 August 2016 a referendum was held on the question of whether to replace the existing human rights provisions in the Constitution with a Bill of Rights. The proposed Bill provided for a right to 'clean and safe water' and a right to 'decent sanitation'; a general right to 'a safe, clean and healthy environment'; and the right to have the state take 'reasonable measures' toward the 'progressive realisation' of these rights.[85] The referendum, however, failed – with a majority affirmative vote but with insufficient voters from those eligible – leaving the existing human rights provisions unchanged.[86]

At the moment, therefore, the only provisions directly relevant to water rights are provided under the heading 'Directive Principles of State Policy and the Duties of a Citizen',[87] which state that the 'directive principles of State policy' will include that 'the State shall endeavour to

---

[83] See, for instance, <http://www.constitutionnet.org/country/tanzania-country-constitutional-profile>.

[84] Constitution of Zambia, 1991, as amended by Act nr 18 of 1996. See <http://www.refworld.org/docid/3ae6b5610.html> accessed 31 August 2016.

[85] The Constitution of Zambia Amendment Bill, 2016, 23 May 2016. See <https://www.elections.org.zm/media/the-referendum-question-mobileversion.pdf> accessed 31 August 2016. The proposed rights read, under the heading 'Economic, Social, Cultural and Environmental Rights':

> 39. (1) A person has the right, as prescribed, to –
>
> (a) …
> (d) clean and safe water;
> (e) decent sanitation; …
>
> 44. A person has the right to a safe, clean and healthy environment.
>
> 45. (1) The State shall take reasonable measures for the progressive realisation of economic, social, cultural and environmental rights.

[86] See, for instance, 'Referendum vote flops, fails to meet the threshold' (*Lusaka Times*, 19 August 2016) <https://www.lusakatimes.com/2016/08/19/referendum-vote-flopsfails-meet-threshold/> accessed 31 August 2016.

[87] Heading: Part IX. It is provided (in art 110(1)) that the Directive Principles of State Policy set out in this Part shall guide the Executive, the Legislature and the Judiciary, as the case may be, in the

(a) development of national policies;
(b) implementation of national policies;

provide clean and safe water, adequate medical and health facilities and decent shelter for all persons, and take measures to constantly improve such facilities and amenities'.[88] Further, the directive principles include that 'the State shall strive to provide a clean and healthy environment for all';[89] and that 'the State shall promote sustenance, development and public awareness of the need to manage the land, air and water resources in a balanced and suitable manner for the present and future generation'.[90]

There are also a few articles which could arguably be used to support a demand made in respect of water-related rights; such as Article 17 which provides for a right to 'protection for privacy of home and other property', but which is a right that may be derogated from by a law that 'is reasonably required in the interests of defence, public safety, public order, public morality, public health, town and country planning, the development and utilisation of mineral resources, or in order to secure the development or utilisation of any property for a purpose beneficial to the community'.[91]

2.16 Zimbabwe

The Constitution of Zimbabwe, 2013,[92] contains sections which provide fundamental environmental rights[93] and a fundamental right to food and water.[94] Both of these rights are contained in Chapter 4 – in respect of rights in this chapter the state is required to 'take all practical measures to protect' and 'to promote their full realisation and fulfilment'.[95]

(c) making and enactment of laws; and
(d) application of The Constitution and any other law.

[88] Article 112(d).

[89] Article 112(h).

[90] Article 112(i).

[91] Article 17(2)(a).

[92] Constitution of Zimbabwe Amendment (No. 20) Act, 2013, in force 22 May 2013. See <http://www.wipo.int/wipolex/en/text.jsp?file_id=312646> accessed 31 August 2016.

[93] Section 73.

[94] Section 77.

[95] Section 11. In addition, it is provided in section 3 ('Founding Values and Principles') that there is to be 'the equitable sharing of national resources, including land' (section 3(2)(j)).

The right to food and water provides that '[e]very person has the right to – safe, clean and potable water';[96] and to 'sufficient food'[97] and the state 'must take reasonable legislative and other measures, within the limits of the resources available to it, to achieve the progressive realisation of this right'.[98] The environmental right provides that '[e]very person has the right to an environment that is not harmful to their health or well-being';[99] and to 'have the environment protected for the benefit of present and future generations, through reasonable legislative and other measures that – prevent pollution and ecological degradation; promote conservation; and secure ecologically sustainable development and use of natural resources while promoting economic and social development'.[100] It is notable that this right is identical to the corresponding right in the South African Constitution.[101] The Zimbabwean right is, however, qualified by the requirement that '[t]he State must take reasonable legislative and other measures, within the limits of the resources available to it, to achieve the progressive realisation of the rights set out in this section'.[102]

2.17 Kenya

Kenya is not an SADC member state, but has been included here for its relevance in the region and as an influential near-neighbour.[103] The Constitution of Kenya, 2010[104] does not provide a specific water-related clause. However, the Constitution does contain a clear environmental right; providing that '[e]very person has the right to a clean and healthy

---

[96] Section 77(a).
[97] Section 77(b).
[98] Section 77.
[99] Section 73(1)(a).
[100] Section 73(1)(b)(i)–(iii).
[101] Section 24 of the Constitution of the Republic of South Africa, 1996.
[102] Section 73(2) of the Constitution of Zimbabwe. The South African clause does not carry this rider, but there is a general limitations clause (section 36) which would have the same effect.
[103] Kenya also has a relatively well-developed jurisprudence in the field of environmental rights. Further, Kenya cooperates extensively with its East African neighbours, including Tanzania, in environmental matters. See, for instance, the East African Network for Environmental Compliance and Enforcement (EANECE) <http://www.eanece.org/eanece/> accessed 31 August 2016.
[104] The Constitution of Kenya, 2010, enacted 27 August 2010. See <http://www.kenyalaw.org/Downloads/The%20Constitution%20of%20Kenya.pdf> accessed 31 August 2016.

environment, which includes the right'[105] to 'have the environment protected for the benefit of present and future generations through legislative and other measures, particularly those contemplated in Article 69';[106] and 'to have obligations relating to the environment fulfilled under Article 70'.[107]

Article 69, to which Article 42 refers, provides duties for the state.[108] The state is required to:

(a) ensure sustainable exploitation, utilisation, management and conservation of the environment and natural resources, and ensure the equitable sharing of the accruing benefits;
(b) work to achieve and maintain a tree cover of at least ten per cent of the land area of Kenya;
(c) protect and enhance intellectual property in, and indigenous knowledge of, biodiversity and the genetic resources of the communities;
(d) encourage public participation in the management, protection and conservation of the environment;
(e) protect genetic resources and biological diversity;
(f) establish systems of environmental impact assessment, environmental audit and monitoring of the environment;
(g) eliminate processes and activities that are likely to endanger the environment; and
(h) utilise the environment and natural resources for the benefit of the people of Kenya.[109]

There is then a duty placed on every person 'to cooperate with State organs and other persons to protect and conserve the environment and ensure ecologically sustainable development and use of natural resources'.[110]

The environmental right in the Kenyan Constitution is buttressed by an article which sets out the right to litigate of any person;[111] sets out the orders which a court may make;[112] and broadens *locus standi* by

105 Article 42.
106 Article 42(a).
107 Article 42(b).
108 Part 2 – Environment and Natural Resources.
109 Article 69(1)(a)–(h).
110 Article 69(2).
111 Where such person alleges that an article 42 right has been 'or is likely to be denied, violated, infringed or threatened' (Article 70(1)).
112 Article 70(2).

providing that an applicant litigant 'does not have to demonstrate that any person has incurred loss or suffered injury'.[113] A further article then provides that a 'transaction' which, after a certain date,[114] 'involves the grant of a right or concession by or on behalf of any person, including the national government, to another person for the exploitation of any natural resource of Kenya' is subject to ratification by Parliament.[115] The Kenyan Parliament is required to enact legislation to give effect to these rights.[116]

3 CONSIDERATION

3.1 Introduction

All of the Southern African states have constitutions, and all undoubtedly have laws which regulate access to and usage of water. However, only some of the states considered here have included water-related rights within their constitutions. Those that have done so have achieved it in a variety of different ways.

This chapter is concerned only with comparing the different inclusions of water-related rights within constitutions. Some states might indeed have gone further in their dedicated, sector-specific legislation than in their constitutions. The South African Constitution, for instance, only guarantees a right to access to sufficient water,[117] while the National Water Act[118] provides a surprisingly radical 'expansion' on the constitutional right, effectively creating priority use rights by providing that sufficient water must be reserved for a 'basic human needs reserve' and an 'ecological reserve'.[119] Nonetheless, a comparison of constitutionally protected water-related rights remains valuable. Constitutional protection has a symbolic value, in that it is a reflection of the values which a country seeks to protect in its most powerful legal instrument. More

---

[113] Article 70(3).

[114] Article 71(1)(b).

[115] Article 71(1)(a).

[116] Articles 71(2) and 72.

[117] See 2.12 above.

[118] Act 36 of 1998.

[119] Part 3: The Reserve, sections 16–18. The basic human needs reserve 'provides for the essential needs of individuals served by the water resource in question and includes water for drinking, for food preparation and for personal hygiene'. The ecological reserve 'relates to the water required to protect the aquatic ecosystems of the water resource'. Part 3: Introduction.

practically, constitutional protection of water-related rights creates opportunities to incorporate water-related concerns into legislation and to create duties of conduct to be imposed on administrative and executive office-bearers with water-related duties; as well as constitutional control of laws or executive actions, including litigation by individuals.

Although limited in scope and depth, this chapter opens opportunities for further research into, and self-assessment within, each national context, in the light of comparative findings which relate to states from the same region, and which share similar socio-economic, political and cultural traits.

3.2 Explicit Recognition of a Water-related Right as a Fundamental Right

It is possible to include a water-related right as a fundamental right, be the nature of the right ecocentric as in the Ecuadorian Constitution[120] or anthropocentric, as is South Africa's.[121] However, there are very few instances of this amongst the states under discussion here.

The Democratic Republic of Congo's Constitution of 2005 contains a specific right of access to drinking water, as well as a general environmental right.[122] South Africa's Constitution of 1996 contains a specific water-related right ('access to sufficient water') and also a dedicated environmental right.[123] Zimbabwe's Constitution of 2013 provides a fundamental right to 'safe, clean and potable water', as well as a dedicated environmental right very similar to that of South Africa.[124]

It is interesting to note that the only SADC states which explicitly protect a human right to access to water or a right to water, also provide for environment protection rights. This makes for an interesting dynamic especially when, in the future, the courts may be required to balance rights connected to environment protection and rights to water or access to water should a clash between them arise. The obligations of the states will be more nuanced in these constitutional systems, and when clarifying the legal content of each of these rights the courts will need to be mindful of their interaction (which consists of interdependence as well as potential conflict). For the purposes of this chapter it suffices to say that the approach taken by these three states is well-rounded in that it

---

[120] See 2.1 above.
[121] See 2.12 above.
[122] See 2.4 above.
[123] See 2.12 above.
[124] See 2.16 above.

recognises the importance of water for human consumption and general human development while at the same time acknowledging the importance of protecting the environment of which water is an essential element. This approach opens opportunities for individuals to claim that their water-related rights have been violated, and to seek appropriate remedies for addressing the violation.

3.3 Indirect Protection of Water-related Rights

Indirect protection takes place by way of constitutions protecting more general environmental rights, in the absence of explicit and independent water-related rights protected therein. The constitutions of Angola, Kenya, Mozambique, Seychelles and Swaziland fall within this category.

Angola's Constitution, 2010, contains no express water-related right, but there is an environmental right which refers to 'maintain[ing] the ecological balance' and to 'respect for the rights of future generations'.[125] Kenya's Constitution does not contain a water-related right; nor is there even a specific mention of water in its (comparatively lengthy) listing of environmental duties resting on the state, but it does have a strong general environmental right – providing that '[e]very person has the right to a clean and healthy environment', and then indicating further that this right includes the right to have the environment protected 'for the benefit of present and future generations'.[126] The Constitution of Lesotho, 1993, contains a general environmental right[127] intended to 'assure to all citizens a sound and safe environment adequate to their health and well-being'.[128] Mozambique's Constitution of 2004 contains no specific water-related right but there is a general environmental right which provides that '[a]ll citizens have the right to live in a balanced environment'.[129] The Seychelles' Constitution of 1993 contains an environmental right which includes the right 'to live in and enjoy a clean, healthy and

---

[125] See 2.2 above.

[126] See 2.17 above.

[127] We have here considered this to be a right, although it must be conceded that it (s 36) is found not in Chapter II (headed 'Protection of Fundamental Human Rights and Freedoms') but in Chapter III (headed 'Principles of State Policy'). There is overlap between the two chapters; and s 36 is formulated in familiar rights-based language, such as the phrases '... for the benefit of present and future generations ...' and '... assure to all citizens a sound and safe environment adequate for their health and well-being ...'.

[128] See 2.5 above.

[129] See 2.9 above.

ecologically balanced environment', but no specific water-related right.[130] Swaziland's Constitution of 2005 contains a section dedicated to water and also one in respect of the environment; however, its provision in respect of water provides only that there shall be no private ownership thereof.[131] Zambia's Constitution of 1996 does not provide a direct right, but imposed on the state are 'directive principles' which include the obligation to 'endeavour to provide clean and safe water'.[132]

The inclusion in this part of constitutions such as those of Lesotho, Malawi or Zambia is based on the present authors' assumption that if there is a duty or some form of undertaking on behalf of the state to protect water or the environment then there is necessarily a right accruing to a rights-bearer. That may not, however, always be so. The Constitutions of both Lesotho and Zambia have chapters on human rights. The non-inclusion of water-related rights in such chapters, but their coverage in policy directive chapters, has probably been deliberate. The state has consciously avoided creating a legal right in relation to water or the environment, as the case may be. However, the assumption made is that rights *are* created, albeit that they may be of a different (and indirect) order.

Arguably, the effectiveness of indirect protection will necessarily be dependent upon how 'generously' the provisions of national, sector-specific statutes, or national courts adjudicating on challenges, interpret terms such as 'clean and healthy environment' or 'sound and safe environment' or 'balanced environment'. What is required for an effective protection is for it to be acknowledged that accessible and qualitatively sound water is essential for securing an environment of the quality envisaged in the cited constitutional provisions. From a water perspective, the general environmental protection clauses seem to aim at protecting the availability of water. However, those constitutions which protect water in this manner lack provisions protecting access to or entitlements to water by the people. It can be argued that the indirect protection of water-related rights under more general clauses related to environmental protection is a premise of the right of access to water –

---

130 See 2.11 above.

131 See 2.13 above. The point about barring private ownership is that when the state holds water then it presumably does so in the public trust, or at least for the benefit of the populace. While indirect, this prohibition on private ownership does separate the Swaziland Constitution markedly from that of Namibia, which makes provision only for ownership of water to reside in the state by default where it is not privately owned.

132 See 2.15 above.

protecting the health of the environment in order that water will later be available for people to use. To what extent this is meant to protect the access to water for human or household consumption (as opposed to its use in agriculture and other industries) is uncertain in the absence of supporting constitutional or statutory provisions, and in the face of competing development needs of states in this region. It is probably only permissible to argue that a fundamental right of 'access to water' or 'to water' derives from a general environmental right if the text of the environmental right gives, even if only by context, at least some indication that water is to be preserved in order to enable human access thereto.[133] As always, close reading of an individual constitution will be needed in order to determine whether there is a fundamental right or not.

3.4 No Inclusion of Water-related Rights

The constitutions of five states fall into this category: Botswana, Madagascar, Mauritius, Namibia and Tanzania. These constitutions also lack environment-related rights. Botswana's Constitution, 1996, contains no specific mention of water; or, indeed, of the environment generally.[134] Madagascar's Constitution, 2010, likewise, contains no specific mention of water or of the environment.[135] In the same position is Mauritius, where the Constitution of 1968 contains no specific provisions on either water or the environment.[136] Namibia's Constitution of 1990 is another example of a constitution with a specific right to neither water nor environment.[137] Tanzania's Constitution of 1977 provides no specific right in respect of water or of environment.[138] The constitutions of Mauritius and Tanzania are 'old order' constitutions, being drafted before environmental concerns became wide-spread. Thus the absence of water-related or environmental rights is perhaps not surprising. The absence of these rights, however, in the other constitutions is more difficult to explain, especially when the constitutions were drafted or amended after

---

133 In most cases it would not be difficult for a court to find from a general environmental right (for instance, Angola's 'maintain[ing] the ecological balance' or Mozambique's 'balanced environment') that there was an implied necessity to protect water and to provide people with water in order to give effect to the right.

134 See 2.3 above.

135 See 2.6 above.

136 See 2.8 above.

137 See 2.10 above.

138 See 2.14 above.

significant events, such as the United Nations Conference on Environment and Development (UNCED), 1992, where the Rio Declaration[139] was adopted. In terms of the Rio Declaration all states present[140] committed to principles such as that human beings are 'entitled to [] healthy and productive li[ves] in harmony with nature',[141] that the 'right to development must be fulfilled so as to equitably meet developmental and environmental needs of present and future generations',[142] and that 'in order to achieve sustainable development, environmental protection shall constitute an integral part of the development process and cannot be considered in isolation from it'.[143] Principles such as these have been repeated in many subsequent instruments.

The Constitution of Botswana was adopted in the same year as the South African Constitution, and it is highly unlikely that its drafters were not aware of the latter's provisions. The same can be said about the Namibian Constitution, which although adopted in 1990, was amended twice, in 1998 and 2010, without any of the amendments seeking to address the absence of water or environmental rights. In Madagascar, the 2010 Constitution was the product of a government which came to power through arguably undemocratic means,[144] and as a consequence it understandably gives more attention to the human and political rights of people, and to the political and administrative organisation of the state, than to complex and 'less direct' rights such as those that are environment- or water-related. However, that such 'less direct' but still important rights are entirely absent is hard to understand. Botswana, Namibia and Madagascar seem to offer the weakest constitutional protections of the constitutions discussed here in relation to water or the environment.

---

139 Rio Declaration on Environment and Development, 1992, <http://www.un.org/documents/ga/conf151/aconf15126-1annex1.htm> accessed 31 August 2016.

140 172 governments were represented at the UNCED. Only observers attended from South Africa, and Malawi appears not to have attended at all, but Botswana, Madagascar, Mauritius, Namibia and Tanzania were all represented. See United Nations General Assembly *Report of the United Nations Conference on Environment and Development* (Rio de Janeiro, 3–14 June 1992). See <http://www.un.org/documents/ga/conf151/aconf15126-4.htm> accessed 31 August 2016.

141 Principle 1.

142 Principle 3.

143 Principle 4.

144 Following a coup in January 2009.

While it is possible that the drafters of these five constitutions may not have considered water-related or environmental rights of sufficient importance to enshrine them in their states' leading legal document, it may also be that constitution drafters considered that socio-economic rights (such as water-related rights) or third generation rights (such as the environmental rights) would impose too heavy a burden on the states.[145] This does not mean that water-related rights or rights in connection to the environment are not given any protection in these states. Effective protection can be secured through statutes and it may also be possible for other constitutional provisions to be successfully used to provide indirect protection to the mentioned rights.[146] However, there will be a need for creative arguments to be made and this may not be easy in the absence of specialised environmental law practices or activists in these states. This type of indirect protection also rests on a judiciary informed on environmental issues, an ideal which is still to be achieved. Achieving constitutional protection for water-related rights or environmental rights in the context of these five states is, arguably, a matter of chance.

3.5 Duty-based Approaches

Some constitutions provide that the state, or even the citizens, have duties toward protecting the environment. Arguably, these approaches[147] protect water-related rights by preserving water as part of the wider environment, rather than by conferring rights of access to water or a right to water. Two categories of constitutions can be distinguished here: those creating duties for the state and those creating duties for individuals.

In the category of constitutions creating duties for the state, one can include the constitutions of Lesotho, Malawi and Zambia. The clearest formulation of the duty may be found in the Constitution of Malawi, 1994, which imposes duties on the state, *inter alia*, to 'provide a healthy living and working environment' and to 'conserve and enhance the biological diversity of Malawi'.[148] The constitutions of Lesotho and Zambia reflect environment or water-related duties in their guiding principles for the states. One of the principles of state policy in the

---

[145] This remains a matter for further research.

[146] Some examples include just administrative action provisions, access to information, freedom of expression, public participation, separation of powers, etc.

[147] Which may be found within the various categories discussed in sections 3.2, 3.3 and 3.4 above.

[148] See 2.7 above.

Constitution of Lesotho, 1993, provides that the state 'shall endeavour to assure to all citizens a sound and safe environment adequate to their health and well-being'.[149] Zambia's Constitution of 1996 includes in its 'directive principles' an obligation for the state to 'endeavour to provide clean and safe water'.[150]

The second category, which sees constitutions creating individual duties, includes the constitutions of Angola, DRC, Mozambique and Seychelles. For example, Angola's Constitution indicates that '[e]veryone has the right to live in a healthy and unpolluted environment' and also that everyone 'has the duty to defend and preserve it'.[151] The Democratic Republic of Congo's Constitution provides that '[a]ll persons have the right to a healthy environment and (one) propitious for their integral development', but also that '[t]hey have the duty to defend it'.[152] Mozambique's Constitution provides that citizens have the 'right to live in a balanced environment', but also have the 'duty to defend it' – and also the duties to 'defend and promote health' and to 'protect and conserve the environment'.[153] The Constitution of the Seychelles provides that '[i]t shall be the duty of every citizen … to protect, preserve and improve the environment'.[154]

Although a step forward by comparison to the 'silent' constitutions mentioned in section 3.4 above, the duty-based constitutions remain weak. The duties created for the state in the policy directives of the Lesotho and Zambian constitutions are not enforceable by courts, leaving compliance with such directives largely at the discretion of the state. How effective the Malawi constitutional duties have been is not known at this stage. Notwithstanding, a duty-based approach does not create opportunities for constitutional litigation by individuals as no individual right is protected by the constitution. Thus, individuals who are primarily interested in securing water-related rights, or even environmental rights, are deprived of important legal protection tools and remedies. In relation to the individual duties, there is lack of clarity and not much known jurisprudence to shed light on the meaning and significance of such individual duties. An obvious question that arises is to whom such a duty is owed. Will it be owed to the state; to other people; to the environment itself; or to all, or a combination, of these? Is such duty a legal duty or

149 See 2.5 above.
150 See 2.15 above.
151 See 2.2 above.
152 See 2.4 above.
153 See 2.9 above.
154 See 2.11 above.

simply a moral duty, with no legal weight? Are these duties intended to diminish the responsibility of states in protecting the environment by 'burdening' individuals with duties in this regard? Can these duties be used to justify limitations of other human rights, such as property rights or even access to water? What would be the impact of such individual duties on individual claims to access to water?

Answering these questions would be beyond the scope of this chapter, but the relevance of some of these issues will be touched on below.

4 CONCLUSIONS

As suggested in the introduction above, Southern Africa (the SADC region) is one of the most arid regions in the world, with many of its states being water-stressed or in places even water-scarce.[155] All central and most southern African states are considered water-scarce – those that are not water-scarce are either 'approaching water-scarcity' or have not been estimated.[156] In such circumstances, it would not be surprising in the least to find water-related provisions featuring in the constitutions of states in this area – be such inclusion through provisions entrenching rights of access of persons to water; provisions calling for balancing of rights between human access and environmental protection;[157] or even provisions stipulating an environmental protection *per se*.

What *is* surprising, however, is that there are states whose constitutions contain no[158] water-related provisions at all. This does not appear to relate necessarily to how recent the constitution is. Of the five states whose constitutions have no water-related provisions (Botswana, Madagascar, Mauritius, Namibia and Tanzania), two are 'elderly' (Mauritius,

---

[155] According to the UN Department of Economic and Social Affairs (UNDESA), a water-stressed area is one in which annual water supply drops below 1700m³ per person; a water-scarce area is one in which the supply drops below 1000m³ per person; and 'absolute scarcity' means a supply of less than 500m³ (UNDESA, *International Decade for Action 'WATER FOR LIFE' 2005–2015* (24 November 2014) <http://www.un.org/waterforlifedecade/scarcity.shtml> accessed 1 September 2016.

[156] Ibid, referring to map drawn from World Water Assessment Programme (WWAP), 'World Water Development Report 4' (March 2012).

[157] Including access to quantities beyond those required for mere sustenance.

[158] Or virtually none, as in the case of Namibia (see 2.10 above).

1968; Tanzania, 1977); two are fairly recent[159] (Namibia, 1990; Botswana, 1996) and one is very recent (Madagascar, 2010).

Some constitutions acknowledge water as a resource which is conducive to development, or empowerment of people, rather than as something to be protected for its own value. Arguably, Lesotho (1993), the Seychelles (1993) and Mozambique (2004) are states whose constitutions fall into this category. That most states adopt this course is not surprising. It might be expected that, given the history of Southern Africa's states' emergences from colonial usurpation and, often, subsequent political instability, the majority would have constitutions which are largely reflective of human and political rights. Environmental protection therefore becomes incidental – required in order to enable development.

Of the 16 states considered, only three have constitutions which contain dedicated water-related rights; eight have constitutions with general environmental rights framed sufficiently broadly for water-related rights to flow necessarily therefrom; and five have no water- (or environment-) related provisions at all.

Only the Democratic Republic of Congo (2005), South Africa (1996) and Zimbabwe (2013) have *both* direct environmental *and* water-related provisions in their Constitutions. It could be queried whether it is therefore only in those states that the constitution-making body has applied its mind to the balance, the relationship, between access to water and environmental protection. Making this suggestion would be to go too far. One of these (the DR of Congo) and three other states (Angola, 2010; Mozambique, 2004; and the Seychelles, 1993) have gone further than providing duties to protect the environment which rest on the state alone, and have created duties for their citizens also – the Seychelles in particular creating such a duty at an early date. In each jurisdiction, the strength of protection afforded to either peoples' rights of access to water, or the right to have water protected as a component of the environment in its own right, or to both, will need to be determined through study of a complex matrix of legal aspects.

In cases where there is no constitutional inclusion of water-related provisions, it would potentially be useful to consider whether subordinate statutes provide sufficient protection of water-related rights on their own, or whether constitutional inclusion would enhance such protection. Where, as in most cases, there is indirect inclusion of water-related rights

---

[159] In between these two constitutions being adopted, South Africa included a direct water-related right in its Interim Constitution of 1993 (Constitution of the Republic of South Africa Act 200 of 1993).

in constitutions, it would be useful to consider the relationship between such indirect provisions and subordinate statutes. Whether a duty to enact further, water-specific legislation is created will depend on the individual circumstances of each state. In the few instances where there are direct water-related provisions included, it would be especially useful to study whether this in fact enhances the strength of protection afforded – the natural inclination would be to assume that it does, but this might not necessarily be the case. In all cases there will be peculiarities. States will have different understandings of the nature of constitutional rights; and of the relationships between such rights and subordinate legislation. In addition, there will always be different emphases or priorities placed on rights by different governments, and at different times, which will affect these issues.

All of the SADC states and Kenya have been affected by the transition from the colonial era, with political and economic instability characterising the region. This may explain, at least partly, why constitutions in the region tend to focus more on administrative, civil, judicial and political rights than on environmental rights. Further environmental rights probably represent a more recent spurt in jurisprudence. Political and economic stability in Southern African has not yet been achieved, and the quest for these is likely to be rendered more difficult to achieve by the environmental problems which are increasing in complexity in all states – including the huge stresses which lack of water can put on states.

Against this background, increasing the sophistication of our understanding of the jurisprudences of human rights and environmental law, and the interactions between them, is important. This chapter has been written as an initial contribution toward such increased understanding in the area of constitutional inclusion of water-related rights.

3. Private and public ownership of water areas – structures and implications of the Finnish model

Pekka Vihervuori

1 STARTING POINTS IN THE LEGAL SYSTEMS

It may be said that in most legal systems land areas or certain categories thereof are either subject to private ownership or they are not. However, in the case of various surface water areas in state sovereignty,[1] the very diverse legal systems often combine elements of property rights, public law and public access in several ways. Broadly, it may be assumed that the public law ingredients are predominant regarding territorial waters in the seas on the one hand, and probably least visible regarding small brooks and natural ponds on the other hand. But this is, of course, a generalization. Moreover, there are generally significantly more inherent stakeholders than just one vis-à-vis a certain water area or part thereof. This results from the natural characteristics of the water as an element and often also from the mere fact that a property law system based on cadastral survey, typical for land areas, has been extended to water areas. Moreover, the rights of public access and general rights of use are characteristic, as is the possibility for various actors and interested parties to resort to measures of a more or less expropriatory nature, these often being more far-reaching than in the case of land areas.

[1] Water area here means surface water areas, i.e. groundwater areas are not covered by the term.

2 GEOMORPHOLOGICAL AND HYDROBIOLOGICAL BACKGROUNDS OF PROPERTY LAW AND WATER LAW

The Finnish natural watercourse systems typically consist of series and branches of numerous, diversely alternating lakes of various shapes and sizes, rivers, brooks and other formations, each having their individual natural characteristics. These characteristics also define their protection and use. The formations are often complex in shape and dotted with a myriad islands. This often causes the sub-systems and single water bodies to lack any clear discernible individual extension. Hence, the natural watercourse systems in Finland are far from the typical continental river basins which have strongly influenced the EU Water Framework Directive.[2] Consequently, these basins are not easy to master through any management system or a property law system. This has also made the application of the water formation typology provided in the WFD somewhat difficult. In the WFD (Article 2) 'body of surface water' has been defined as 'a discrete and significant element of surface water such as a lake, a reservoir, a stream, river or canal, part of a stream, river or canal, a transitional water or a stretch of coastal water'.

In reality, some discretion and use of imagination has been inevitable here. The same holds true regarding the notion of 'sub-basin' which in the WFD stands for 'the area of land from which all surface run-off flows through a series of streams, rivers and, possibly, lakes to a particular point in a water course (normally a lake or a river confluence)'. 'River basin district' again is more an administrative than a geomorphological notion. According to the WFD it is 'the area of land and sea, made up of one or more neighbouring river basins together with their associated groundwater and coastal waters, which is identified under Article 3(1) as the main unit for management of river basins'. Because the watercourse systems flowing into the sea are very diverse with regard to size and form, the merging of neighbouring basins has been inevitable on the one hand. On the other, some of the natural total basins are geographically so extensive and diverse that they are not very suitable as functioning management and planning units.

---

[2] Directive 2000/60/EC of the European Parliament and of the Council establishing a framework for the Community action in the field of water policy [2000] OJ L327/1 (hereafter WFD).

Notoriously, the integrated management approach provided in the WFD is largely based on management planning, i.e. River Basin Management Plans and the ancillary Programmes of Measures. The actual judicial role of the Plans and Programmes in substantive decision-making, especially on projects affecting a certain level of water quality, has nevertheless remained rather vague or uncertain, in spite of the multitude and detailed nature of WFD provisions regulating the planning system. The problem and the possible contradiction have been observed and discussed by many authors.[3] The normative status of the Plans has been regarded as a major question.[4] In a well-known landmark case, namely The Weser Case,[5] the European Court of Justice (ECJ) settled one of the main problems: the dispute over the legal status of the WFD objectives. The ECJ concluded, in short, that the objectives had to be regarded as legally binding.[6]

In Finland, the most important decisions affecting the use, protection and status of waters are those made under the permit mechanisms of the Water Act (587/2011, as amended later on) and the Environmental Protection Act (527/2014, as amended later on). In these Acts, the legal effect of the national modifications of the River Basin Management Plans has been explicitly provided only for some situations with limited scopes (see below).

Generally, the preconditions of various projects and measures related to water or water areas are decided upon in individual permit procedures

---

3 See, among others, William Howarth, 'Aspirations and Realities under the Water Framework Directive: Proceduralisation, Participation and Practicalities' (2009) 21 *JEL* 391, 410–412, 417; William Howarth, 'Accommodation without Resolution? Emission Controls and Environmental Quality Objectives in the Proposed EC Water Framework Directive' (1999) 1 *Environmental Law Review* 6–26; J.J.H. van Kempen, 'Countering the Obscurity of Obligations in European Environmental Law: An Analysis of Article 4 of the European Water Framework Directive' (2012) 24 *JEL* 499, 520–33; Andrea M. Keessen, Jasper J.H. van Kempen, Marleen van Rijswick, Jan Robbe and Chris W. Backes, 'European River Basin Districts: Are They Swimming in the Same Implementation Pool?' (2010) 22 *JEL* 197–221; and Emilia Korkea-aho, 'Legal Interpretation of EU Framework Directives: A Soft Law Approach' (2015) 40 *EL Rev* 70–88.

4 Jussi Kauppila, 'Pintaveden normatiivinen tila' in YPOV (2011) V 11, 36–40 sees the main reason for this in the difficulty to define the plans clearly as either legal-normative or factual-descriptive. This also makes their status as a source of law obscure.

5 C-461/13 *Bund für Umwelt und Naturschutz Deutschland eV v Bundesrepublik Deutschland* [2015] OJ C294/3.

6 See e.g. the analysis in Tiina Paloniitty, 'The Weser Case: Case C-461/13 BUND V GERMANY' (2016) 28 *JEL* 151–8.

of the Water Act and the Environmental Protection Act, often without any, or any significant, impact or contribution by any prior planning of a more general nature. The main focus is on the individual project plan and the impact report concerning the actual project. The permit procedure with its participatory elements is based on such plans. This applies to both one-time construction projects of extremely varied purposes, kinds and sizes, such as hydroelectric plants, dams, quays, piers, jetties, canals and waterways, timber floating facilities, water abstraction or transfer pipes, sewage outlets, embankments, bridges, road banks, water reservoirs, fish farm constructions, extraction of sand from the seabed, flood-prevention measures, drainage and ditching, filling of water areas with earth, dredging, or clearing of rapids or straits on the one hand, and continuous and possibly permanent activities such as discharge of waste water and other water-polluting emissions, water intake from natural sources,[7] and water-level regulation on the other hand. One-time and continuous sub-regimes may also mix, typically in projects for damming up or impounding water or lowering the surface levels. The various preconditions set forth in a permit decision typically aim to prevent or mitigate detrimental and harmful impacts on other interests, be they either private or public, use or conservation/protection.

The general land-use planning system provided in the Land Use and Planning Act (132/1999, as amended later on) also applies to water areas as such, but its impacts are generally rather vague, especially regarding the core areas of water management and water area management. However, the detailed land-use plans pursuant to the aforementioned act have to be observed, where in force and relevantly applicable. Moreover, according to judicial practice, the more general land-use plans may also, under the same act and among other viewpoints, be taken into account in the permit considerations, particularly in the important weighing of interests. In the past, there have been legislative efforts to create a special and parallel mechanism for general water management, but they have not resulted in any legislation.

Initially, when the WFD was adopted, it was perceived as a re-vitalization of such a general planning approach in Finland. However, for several reasons, and especially due to the many legal uncertainties involved, these expectations did not materialize. The technical and administrative provisions for the planning mechanism are quite extensive and also detailed. This can be seen, especially, in the Act on the

---

[7] Either surface or ground water, but also the impacts of each may be intermingled.

Arranging of Water Management and Marine Waters Management (1299/2004),[8] and the respective Government Decrees (particularly decrees 1301/2004 and 1040/2006).[9] Direct references to the Management Plans in the Water Act and the Environmental Protection Act are, in turn, few. Despite this, and outside the direct scope of application of such provisions, references to the contents of and information contained in the Plans are today quite common in practice, e.g. in the reasoning parts of permit decisions.

On the other hand, nowadays there is also other EU-based planning legislation affecting the same water law and permit regime, such as the planning provided in Directive 2007/60/EC on the assessment and management of flood risks,[10] as implemented in the Flood Risk Management Act (620/2010).

3 PROPERTY LAW DEVELOPMENTS AND WATERS

The proprietary status of water areas has typically evolved as a result of long-time historical development. Among the Nordic countries, Sweden and Finland make up the eastern group, with a common legislative history until 1809. Property law, having its roots in medieval laws, was already rather sophisticated at that time. One of its fundamental features was that inland waters, lakes and rivers, as well as inner territorial waters in the sea, were subject to ownership of the adjacent shore land. More precisely, a water area belonged to that land proprietor whose shore land happened to be located nearest to the water in comparison to any other land proprietor. The right of the water area owners was, due to the natural differences between lands and waters, conceived in a specific manner. Above all, it was a matter of fishery rights, and, regarding waterfalls in rivers, a matter of rights to water power. Also the resources of the bed bottom were included. Water traffic was regarded as free to anyone, where physically possible in existing circumstances. Abstraction of water

---

[8] The Act also covers the implementation of the Marine Strategy Framework Directive, i.e. Directive 2008/56/EC of the European Parliament and of the Council of 17 June 2008 establishing a framework for community action in the field of marine environmental policy [2008] OJ L164/19.

[9] On the legislation, see Pekka Vihervuori, *Environmental Law in Finland* (Wolters Kluwer 2014) 329–36.

[10] Directive 2007/60/EC of the European Parliament and of the Council of 23 October 2007 on the assessment and management of flood risks [2007] OJ L288/27.

was seldom a problem, because there was generally no scarcity thereof. Of course, irregularities and local specialities were often present.

The early influence of public law (in later terminology) had two main expressions, their histories intertwined with each other – and particularly that history where a fundamental medieval document was fabricated by a later (sixteenth-century) Swedish monarch to establish a suitable legal basis, influenced by the continental notion of *ius regale*. This was a matter concerning the valuable rivers. For one thing, in all (proper) rivers (but not in brooks etc.) the middle part was *ex lege* regarded as the so-called *(royal) artery*, or main channel, in order to preserve possibility of open boat traffic (and timber floating), as well as to maintain fish migration. It was consequently not allowed for anyone to close the rivers off by dams of any kind. For another, the fishery rights in the most important rivers hosting migratory salmon and trout – particularly the rivers in the north – were established as a privilege of the Crown. A further early implication of public law (or of the subjective powers of the Crown) was that only the inner belt of the territorial waters in the sea followed the rule on the right of the respective coastland owner.

Also the question of boundaries between different property owners had to be tackled – or not exactly those between the current owners but between the respective property entities, which were very static. Fixed boundaries were, naturally, often indiscernible in large or complex and vaguely shifting water formations. However, even though the existence of property boundaries in water areas was generally acknowledged, judicial proceedings in order to attain more precision and certainty were needed in conflict situations, like those emerging from competing fishery interests and alleged rights. Exact boundaries on official maps are a later phenomenon overall. Today all boundaries are digitalized in the cadastral database of the National Land Survey.

All this, of course, impacted the various forms of utilization and water resource/area management. In addition, any harmful alteration of water bodies was at least in principle forbidden already in the early eighteenth century, as were various polluting activities. However, the scale of the impacting measures in practice was generally rather small until the era of industrialization, due to the lack of technical development. For example, the damming of major rivers became technically possible rather late.

In 1809, the eastern part of the Swedish Kingdom (Finland) became separated from the rest of the Swedish Kingdom as it fell under the rule of the Russian Empire. From there on, Finland was an autonomous Grand Duchy under the Russian Emperor. However, Swedish legislation and the judicial system were preserved. Gradually, the two legal systems adopted some divergent traits, including in property law and water law.

This development was swifter in Sweden, while in Finland the Czar, in the capacity of Grand Duke, did not permit summoning the Finnish legislature until 1863. That was the era of intensive industrialization. In 1809–1863 only home government decrees based on old laws were regarded as possible.

The Watercourse Decree of 1868 was the first codification of water (management) law and a parliamentary law level codification was enacted in 1902. The regulation on proprietary entities in water areas was then, as previously, an integral part of cadastral land (registered conveyance) law, and hence mainly outside the notion of proper water law. The land law regime was very static as such, but due to a certain vagueness concerning old regulation applied in new circumstances, the principles of ownership of water areas aroused major interest at the same time, from both politicians and legal scholars.[11] In spite of these legislative developments, many contested property questions remained unsolved until twentieth-century judicial practice.

In comparison to Sweden, property systems were still very similar in general, because the later Swedish developments had been gradually taken into account in law drafting. Yet, one difference was remarkable: It was at the outset unclear in Finland what the notion of shore land actually meant in the old rule that attached a certain water area to the nearest adjacent shore land. There were two main alternatives for a proprietary entity to own the water area, the very coast land property as such on the one hand, and the whole registered village to which that land property, as a part, belonged on the other. Thus, consequently, these alternatives resulted in different outcomes. According to the latter, the water areas within the outer boundaries of each registered village were, pursuant to a general rule on all village-level common utilities and according to their respective numeric shares, jointly common to all land properties of the village. Whereas, according to the first-mentioned alternative, the water areas would belong to one single land property only. This option had become predominant in Sweden, but in Finland the

[11] For example, in rapids being potential sources of water-power, new technology had, in the late 1800s, opened new needs and new possibilities for hydroelectric projects. The old focus of legislation had been on sharing major rapids for several old-fashioned water-mills, but this soon became history, and the new goal was to combine the falls in several rapids by damming and clearing. In Finnish major rivers, many such projects were also fatal to migratory fish. The old artery rule was often turned aside, through exception laws or otherwise. The remaining natural rapids are now largely protected by a specially targeted law, but they still remain in private ownership.

village-level model eventually prevailed. Hence, after an uncertain period in judicial practice, it was not possible to buy the water power of a water area (rapid) by buying only some narrow shore strips. The relevant property exchange was hence transferred to the level of village shares.

On village-level in general, i.e. between different villages, the same old rule had generally appeared less problematic. To clarify the rights to fishery in a certain area, some early litigation had taken place. Soon after the adoption of the water property legislation of 1902, practically all village-level boundaries in water areas became officially defined, and after gaining legal force once, the boundaries remain permanent. Naturally, due to lack of boundary markers in larger open water areas, the boundaries became, and they still are, only dot lines on the map (or today actually bits in the cadastral part of the electronic Land Information System of the National Land Survey).

Even today the majority of Finnish inland waters and territorial waters are subject to village-level joint (shared) ownership, unless owned by the state. The number of individual shareholders may vary between a couple and several thousands, according to the numeric shares of each shareholding landowner, based on cadastral law. Also the areal extent varies strongly. In principle, each shareholder has the right to separate a certain water area equivalent to his share from the common area, but the legislator has been reluctant here. According to the present Cadastral Survey Act[12] (554/1995, as amended later on), Article 137: sharing shall not cause any other shareholder harm, and it is moreover required that the sharing has to be necessary for a specific use or that another particular reason exists. There is *ex lege* a statutory shareholders' association for each jointly owned area. Today many of them, but not all, have been permanently organized. The previous statutory fishery associations of villages have been merged with the shareholders' associations. Thus, they execute a dual role. Although village-level joint ownership is the prevalent type of property ownership in water areas, there are also water areas belonging to a certain land property, as well as water areas that make up a registered property alone.

In the course of time, property transactions regarding parcelling out of land may have included clauses on the division and transfer of shares in common water areas, and where specifically not agreed upon, they have been default clauses in property law, with different content within different time periods. The shares can also be transferred and registered

---

12 The literal translation 'Real Estate Formation Act' is used in some contexts.

separately.[13] As a result, all property owners in the village are not necessarily shareholders in the village's water areas. Some may also have water areas of their own.

There are of course also other exceptions. One general exception pursuant to an old rule is that a small lake or pond without water connection to another water body via a brook etc. belongs to the surrounding land property only and hence not to the village waters. This rule was, however, only effective in certain historic situations concerning land consolidation, and it is not applicable any more as such. Such an area may become common again, now in relation to a smaller number of properties, where agreed upon in connection with property transactions and when duly registered.

4 STATE OWNERSHIP

State ownership of water areas also derives from old property law. The rights of the villages, which according to the old basic idea follow the right to the adjoining land, are not without areal limits. In such major inland lakes where the open water area in the middle exceeds eight kilometres in length and breadth, the area of private ownership (normally that of the villages) only covers the part of the lake that reaches to the distance of 500 metres from the depth curve of two metres in average water level conditions. According to the respective, legally valid cadastral surveys, there are today nine such lakes. In addition, and even more remarkably, the same regulation applies to Finnish territorial waters in the Baltic Sea. Major islands may be surrounded by their own private water area. All the waters in the sea and the nine lakes beyond the outer boundary of the private areas belong to the state, as public water areas. Previously such areas were often described as *res nullius* or *res extra commercium*, but in 1966 the legislator explicitly declared state ownership. On the other hand, the material powers of the state regarding public water areas are more limited than those of private owners, especially due to public rights to fishing and hunting.

Within the large state forest areas, which are mainly situated in Northern Finland, all surface waters, and there are many, normally belong to the surrounding land area (or the adjoining shore areas), according to the old fundamental rule. These waters, as well as all public

---

[13] This may be motivated by the desire to have (at least some) fishing right, or decimal share of the water power value (without any guarantee of having it realized).

water areas in the sea and the nine lakes mentioned above, are possessed by the Finnish Forest and Parks Service (Metsähallitus), which is a complicated mix of a state enterprise and a public authority. The general tasks of Metsähallitus include the sustainable and profitable management, use and protection of the natural resources and other property possessed by it. In total more than 12 million hectares of land and water areas are governed by it.

5 THE CORE POWER OF A WATER AREA OWNER

Regarding land properties, it is quite natural that no explicit general rule on the powers of an owner has been enacted, neither in Finland nor in other countries. Yet, the (surface) water areas are an exception. In Finland such rules – although the literal object here is the right to water as fluid[14] substance – are contained by Chapter 2 of the Water Act (the Chapter titled 'Public rights, obligations and restrictions'). Due to the overall focus on fluid substance, groundwater and abstracted water are also covered. According to Section 1 of Chapter 2 ('Ownership and property administration of water'), 'the water in a water tank, well and other water abstraction facility is owned by the owner of the tank, well or water abstraction facility. The water in a spring and artificial pond is owned by the owner of the ground.' And then follows the important clause regarding proper (surface) water bodies and groundwater as well: 'Within the limitations provided in this Act, any other water with an open surface and groundwater is administered by the party to whom the water or land area in question belongs, unless otherwise provided by the right of another party.' The expression *vallita* in Finnish and *råda över* in the Swedish-language version, which refers to something less than e.g. *possess*, has been translated as *administrate.*[15] The important thing to note is that the verb *own*, used in the two preceding sentences regarding water as substance, has been intentionally avoided here, although the owner of this very *area* is at stake.

It is moreover provided in the same Section that in a river or brook that is shared half-and-half by two real estates or two jointly owned areas of real estates, the owner of each of the halves has the right to an equal share of the water flowing in the river or brook. This is related to the old property law principle defining the two water-area halves, but in a

---

[14] However, the same applies also to frozen water (ice).

[15] This is similar to the unofficial translation of the Water Act on the Finnish official legislation data website <Finlex.fi>.

different way, in order to avoid the need to define the course and division of actual water streams between the two geographical halves. On the other hand, the rule as such does not create any right to obtain a permit for water-power use or water abstraction, in any quantities. Alongside rivers with different property entities on both sides, the basic right to water power is hence divided in half by water discharge, not by area.

Pursuant to Section 2 of Chapter 2 of the Water Act ('Certain rights and obligations'), 'the rights and obligations provided for in this Act concerning a water area or land area belong to the owner of the area, unless otherwise provided below or separately'. Additionally, provisions on the right of a shareholder of a jointly owned water area to (individually) use such area are laid down in the Act on Jointly Owned Areas (758/1989, as amended later on). The last-mentioned possibility in practice only makes small scale individual measures possible – and only in relation to the rest of the shareholders, and especially not in relation to the provisions of the Water Act.

6 PUBLIC AND GENERAL RIGHTS

However, being only a part of the total management regulation in the Water Act, the whole of Section 2 of Chapter 2 (above) is only the starting-point, which does not guarantee the proprietor any concrete expectations. First, there are rights of public access and use. Regarding passage in a water body (Section 3 of Chapter 2), everyone has, unless otherwise provided by law, the right, without inflicting unnecessary damage, harm or disturbance, to: (1) move in a water body and on its ice-covered surface; (2) anchor in the water body on a temporary basis; (3) float timber in the water body; (4) swim in the water body; and (5) temporarily move traps and other movable objects in a main channel or public channel that hinder passage or timber floating, as well as such movable objects outside the channel that unreasonably hinder passage or prevent timber floating.

The horizontal extent of a water body towards land is defined according to the average water level (based on long-time observations, where there are any, or assessed otherwise). Yet, as a practical solution, it has been explicitly provided that the public rights of access and use, common fishing rights included, also always apply to land areas beyond the average water level when such a land area (often a narrow belt) is actually covered by water.

Everyone also has the right (Section 4 of Chapter 2) to abstract water or take ice from a water body for a personal need on a non-permanent basis. The more or less narrow land belt between the average water level line, if lower, and the actual water level line, if higher, is included here. Everyone also has the right to occasionally take a small quantity of water or ice from a streamlet belonging to another subject or from a spring (fountain) that is not in regular use by its owner or by another person with the permission of the owner. It is required, however, that the abstraction of water or the taking of ice may not cause harm or any more than a minor disturbance to the owner of the area or to other right holders.[16]

Moreover, certain rights of general character extend to horizontal relations between property units. Regarding abstraction of surface water, it is provided (Chapter 4, Section 3) that the owner or possessor of a land property is, without any consent or authorization by public authority, *ex lege* entitled to use alien property so that he or she may abstract surface water from the water area of another subject, but only for ordinary household use on the property. Typically a shoreland property and the adjoining water area are here at stake. It is also required that the water intake does not result in consequences which create the obligation to obtain a water management permit. In addition, the abstraction shall not cause harm to those abstracting water from the same water area on the basis of a permit issued by a permit authority or ownership or occupancy of the water area.

Also a shoreland owner or possessor to whom the adjoining water area does not belong and who does not possess shares to the water area, is nonetheless entitled to place certain minor constructions in such offshore water area belonging to someone else (Section 5 of Chapter 2 of the Water Act). He or she has the right to place an anchor post or mooring buoy for private use, or to build a jetty, boathouse or another comparable construction on his or her shore that extends to an alien water area. Water areas taken into special use are excluded. It is, however, necessary that the building or use of the construction does not require a permit under the Water Act and that the whole undertaking can be done without inflicting damage or causing substantial harm to the owner of the water area.

Moreover (Section 6), anyone who suffers from sludge, shallow water or a similar nuisance regarding the use of a water body may, without the consent of the owner of the water area, carry out a measure necessary for

---

[16] Irrespective of this, all water springs in natural or natural-like state are protected for conservation reasons against measures that might endanger the natural status, by the same Water Act.

removing the nuisance in order to improve the state and possibilities of use of the water body. In such a case it is further required that a permit under the Water Act is not needed and no substantial harm to the owner or polluting impacts in the water area are caused. The same applies to the placing of dredged material in the water area owned by someone else. However, the position of the adjoining shoreland is different, and the protection of the owner stronger: The placing of dredged material on alien land area is subject to the owner's consent. However, a permit authority may grant the right to place dredged substance on land, instead, if this does not cause any notable harm to the use of the area and if no permit under the Environmental Protection Act, due to pollution risk, is needed.

7 GENERAL RESTRICTIONS

A general rule on general obligations in the use of water resources and water areas is provided in Section 7 of Chapter 2 of the Water Act. Pursuant to it, a water (or water resources) management project – that being an extremely extensive expression that covers practically the whole variety of different activities, measures and constructions under the Water Act – shall be implemented, and water resources and water areas otherwise used, in such a way that the project does not cause any avoidable infringement of a public or private interest, as far as the purpose of the project or use can be attained without an unreasonable increase in the costs relative to total costs and without any detrimental consequences to be caused. Both public and private interests are relevant here, and hence not only the private interests of the active party but also the private interests of the outsiders to be impacted.

This general rule is only the starting point; the important permit system of the Water Act is the most crucial factor affecting the sphere of freedom of owners and others. As in Section 7 of Chapter 2, the permit regulation as a whole can be considered as a mix of public and private law; the application and interpretation of law is, however, mainly vested in the administrative bodies and administrative courts. Criminal proceedings are very rare.

First (Section 2 of Chapter 3), a prior permit by the (state) permit authority is always needed if a project may cause certain physical changes and impacts on other interests. The provided impact test stage by stage is the following: there is (I) a likely change in the state, depth, water level or flow, shore, or aquatic environment of a water body or the quality or quantity of groundwater; (II) this change (1) results in a risk of flooding or general shortage of water; or (2) results in detrimental changes in the natural environment and the way it functions or deterioration in the ecological status of a water body or groundwater body; or (3) significantly reduces the beauty of nature, causes deterioration in the amenities of the environment or in cultural values or the suitability of the water body for recreational use; or (4) poses a risk to human health; or (5) substantially reduces the yield of an important or other groundwater body suitable for use for water supply purposes, or otherwise impairs its usability or causes other damage or harm to the water abstraction or the use of water as drinking water; or (6) causes damage or harm to fishing or fish stocks; or (7) causes damage or harm to waterborne traffic or timber floating; or (8) jeopardizes the conditions for a brook channel to remain in a natural state; or (9) violates the public interest in another manner similar to the above; or (10) results in loss of benefit for the water area of another person, or fishing, water supply, land, real estate or other property. However, a permit is not needed due to (10) if the loss of benefit in (10) is caused to certain private interest(s) only and the interested subject(s) has (have) consented to the project.

There are several more reasons which demand a prior permit, however, and all of them cannot be reviewed here. These are mainly situations where a permit is required irrespective of the alleged consequences and impacts. This category (Section 3 of Chapter 3 of the Water Act) contains, among others, any closure or narrowing of an artery or public channel or timber floating channel and placement of a device or another obstruction that hinders the use of the channel; construction of a bridge or a transport device over a public or main channel and a tunnel or water, sewer, power or other line under such a channel, transforming a land area permanently into a water area by raising the water level in a water body; and any construction of a hydropower plant. Also for dredging, placing of dredged materials and removal of usable bottom material, a water management permit is largely required irrespective of harmful consequences. A permit is also required for altering an existing project for which a permit has already been granted, if the change violates relevant public or private interests.

Although groundwater as such is not discussed here, it is worth mentioning (also regarding same-time indirect impacts on surface waters) that a prior permit is always (that is, irrespective of the other applicable criteria) required also for abstraction of groundwater for water supply and for other abstraction of groundwater when the quantity exceeds 250 m^3 per day, as well as for any other measure (typically extraction of gravel) where at least 250 m^3 per day of water is, as a side-effect, continuously removed from a groundwater formation. Typically, Finnish groundwater areas are extensive and often exceed horizontally several land property boundaries, as do the formation and horizontal movements of groundwater. Thus, the measures and impacts on groundwater at a certain point can very easily have effects somewhere else. The right of the area owner to administrate groundwater[17] is consequently very narrow.

8 PRECONDITIONS FOR WATER MANAGEMENT PERMITS

As a starting point, a water management permit pursuant to the Water Act *shall* be granted if the project does not significantly violate public or private interests. This takes place in practice only regarding 'innocent', typically small-scale projects. Otherwise, the weighing of interests rule (Section 4 of Chapter 3) is applied. Here a pluralistic view on (various, perhaps even conflicting) public interests and private interests has been adopted. To achieve a positive outcome for the permission, it is required that the benefit gained from the project to public or private interests is considerable in comparison with the losses incurred for public or private interests.[18]

As a private benefit gained from the project is at stake, the increase in the utility value of property resulting from the improved productivity or usability of a land or water area or other property and any immediate other benefits gained from the project are taken into account.

Relevant private-interest losses again encompass the following: (1) right of use or right to purchase another's property granted to the applicant as a part of the permit; (2) costs incurred from damage and right of use that the applicant has separately agreed on in order to implement the project, and the costs of acquisition of areas voluntarily handed over to the applicant; and (3) other losses incurred by a subject not participating in the project or by a (compelled) passive partner in a joint ditch drainage project.

When it comes to weighing the interests regarding the benefits and losses caused to public interest, a general assessment has to be made. A monetary value may be used in the assessment only if it is possible to define the amount of the benefit or loss in monetary terms.

---

17 See Section 5 above.

18 See in general e.g. Niko Soininen, *Vesioikeudellinen perusteluvelvollisuus* (2016); with an English summary 'Transparencies in legality: A legal analysis of the reason-giving requirement in water management permitting in Finland'.

In addition, and without a distinction between private and public interests, the matters included in a Management Plan under the Act on the Arranging of Water Management and Marine Waters Management with respect to factors related to the state and use of waters in the area impacted by the project shall be taken into account in the assessment of benefits and losses.

In the permit deliberation, a locally valid, detailed plan pursuant to the Land Use and Building Act shall be taken into account (if there is such a plan). In addition, the provisions of the same act on the legal effects of regional plans and municipal master plans shall be taken into account. When deciding on a permit, it shall be ensured that the permit does not complicate the preparation of a land use plan in any significant way.

Nevertheless, and irrespective of the result of interest weighing, a permit shall not be granted at all if the project jeopardizes public health or safety, causes considerable detrimental changes in the natural state of the environment or the aquatic environment and its functions, or causes considerable deterioration in the local living or economic conditions. There is only one case where this rule (more accurately its predecessor in the previous Water Act) has been applied with a result negative to the project.

In water abstraction, the weighing of interests is complemented by provisions in Section 5 of Chapter 4, in order to coordinate, where necessary, the different needs for water intake and use. If there is not enough water to meet the needs of all users, the following order of precedence applies: (1) abstraction of water for use near the abstraction site for ordinary household use of the properties there; (2) abstraction of water for the water supply of the local community; (3) abstraction of water for the use of local industry or otherwise for use in the locality and abstraction of water serving the water supply of a community outside the locality; and (4) abstraction of water to be conducted or transported for use elsewhere for a purpose other than supplying water to a community.

9 RIGHTS OF USE AS A PROPERTY LAW TOOL IN PERMITTING CONTEXT

In practice, it is often the case that the applicant does not have prior ownership or possession of the impacted area or of all or even part of the exact area that is to be subject to construction or otherwise directly used. Even a total outsider from the property point of view may be granted a water management permit in several circumstances. However, regarding the total outcome of a permit procedure that results positively for the applicant, the applicant, or now more properly the permit receiver, shall, and will, hold rights of use to all the areas required for the project. 'Use' in this context stands for direct use and certain crucial immediate impacts, not including the areas of various detrimental consequences only. In practice, 'use' as a rule means placing of constructions in an alien water or land area, filling a water area or part thereof with soil substance and thus making it a new land area, or elevating the average water level of a water body thus altering all shoreland areas of the water body into water areas up to the new calculated average water level.

If the applicant does not own the area or possess a permanent right of use, a permit may only be granted if the applicant is granted the right of use to the area as a part of the very permit decision. Sometimes it may suffice for granting of permit that the applicant presents a reliable account of how the right of use will be arranged otherwise (Section 4 of Chapter 3).

In short, if the applicant does not have all the necessary permanent rights of use beforehand, such rights may, provided that the preconditions for the permit are otherwise met, be granted (Sections 12 and 13 of Chapter 2) only if the applicant owns or permanently possesses more than one half of the required area. Regardless of this requirement of prior possession, a similar right of use may be granted for certain minor measures in someone else's area: (1) to build a minor construction in a water body in the area of another subject or jointly owned area provided that the water body is not substantially changed; (2) to cause water to rise temporarily into an alien area; (3) to carry out cleaning work in an alien area; (4) to straighten or widen the bed to a minor extent; or (5) to place removed bottom substance on alien land or in an alien water area that is not taken into special use. Also rights to remove or alter alien structures may be included in a permit decision.

Totally irrespective of prior ownership or possession, and regarding a water management project of any kind or scope, the necessary rights to use alien property, compulsory purchase of property included, may be granted to the applicant by the permit authority, if the project is regarded to be required by a public need (Section 13 of Chapter 2). Here a state or municipality authority as a permit applicant is most often involved.

Regarding water abstraction, different rules on rights of use are applicable. If the level of general abstraction rights is exceeded, the permit authority may grant the applicant a right to the abstraction of surface water in the area of another subject (Section 3 of Chapter 4). A permit decision may here and otherwise include the right to place a water main for the purpose of abstracting water in an alien area (Section 8 of Chapter 4). The placing of a water main in an alien area shall be done in a way that the nuisance caused by it is minimized. The area in which the water main is located shall not be used by the owner or by others in a way that might damage the water main or cause unreasonable difficulties to its maintenance.

In all cases, full compensation for losses caused by the granted right of use is provided for, and also decided upon by the permit authority.

10 THE PERMIT MECHANISM AS A FRAMEWORK FOR PARTICIPATION AND COOPERATION

The Finnish permit mechanism as a framework is also used for several side-regulations with different purposes, among others to master the effects of scattered property and ownership structures. One of them regards utilization of jointly owned hydropower. Only the basics may be presented here: A subject who owns at least one fifth of hydropower in a part of a water body that can be used in an appropriate manner in one and the same plant may initiate the construction of a power plant necessary for taking water-power into use. Such an initiative may also be taken jointly by those hydropower shareholders who own at least one fifth of the hydropower to be taken into use. If an agreement can be reached on taking the power into use, the initiative-maker (who is normally the permit applicant) shall make an offer with the intent to participate, to all those who own at least one hundredth part of such hydropower. The initiative-maker and those entitled who wish to participate in the project may be jointly granted a water management permit[19] to construct a power plant and, in connection with this, the necessary permanent right to use against compensation regarding the shares of the passive shareholders, provided of course that the ordinary preconditions for a permit otherwise exist.

In Finnish water law,[20] a typical permit for a water management project is far from being a mere consent or authorization. In fact, a permit is often an instrument for balancing of interests, rights and expectations, and for mitigation. The main tool for this is the imposing of different tailored provisions ('conditions') in the permit in order to protect various public interests and alien private interests. The same idea, as a starting point, has been expressed already in Section 7 of Chapter 2 of the Water

[19] A permit is always needed for a hydroelectric plant (Section 3 of Chapter 3 of the Water Act).

[20] This applies to Finnish environmental law in general, pollution prevention law included.

Act,[21] but the more explicit requirements in Chapter 3 are crucially important in practice.

Each permit decision shall issue the necessary regulations on (1) avoiding any nuisance resulting from the project and its implementation; (2) landscaping and other elimination of traces of measures; and (3) measures and devices necessary for preserving the status of water bodies and groundwater. Where the project affects the water level or water flow in a water body, the permit shall also include the necessary provisions regulating the maximum and minimum water levels and water flow. If the project at stake simultaneously causes pollution of surface water or poses a risk thereof, the provisions of the Environmental Protection Act that set the requirements for provisions (or regulations) to be imposed in individual permits are additionally applicable.

Moreover, there are specific provisions on the imposing of various monitoring and reporting obligations, obligations to maintain or arrange accessibility and transport connections, as well as fish stock and fisheries management obligations and fisheries management fees. In all cases, both public and private interests are taken into account.

The project-category-specific chapters (Chapters 4–10 of the Water Act) include additional rules on provisions to be set forth in permits. For example, regarding abstraction of water, a permit shall also contain provisions on the location of the necessary structures, monitoring the quantities of water abstracted and measures that may be taken to secure water supply in special situations.

In principle, all expected losses and damages resulting from the permitted project are compensated. The decision on pecuniary compensations is *ex officio* decided upon by the permit authority.

A permit is, as a rule, permanent in principle, but the provisions set forth in a permit may be altered later on, depending on the category and nature of the project. For special reasons, a permit may be granted for a fixed term only. A permit decision may also include an order on a periodic review of the permit provisions regarding the aquatic environment and its use in case this is necessary to avoid significant harm. Upon application by a party, the permit authority may in certain circumstances review the provisions in the permit and issue new ones. In certain legally determined situations, a permit may also be ordered to expire.

All permit decisions can be appealed against in administrative courts by various private and public stakeholders, local and environmental

---

[21] See Section 7 above.

associations included. The courts may also directly amend the contents of a permit where necessary.

11 CONCLUSIONS

In Finnish law, the water management permit is the main instrument for the arranging of relationships between various property owners and other stakeholders. The group of subjects is determined by the actual project, natural conditions and the property structure of the area. Regarding pollution of waters, the same may be said of the environmental permit pursuant to the Environmental Protection Act. The two permits also make up the main tool to implement the Plans and Programmes pursuant to the Act on the Arranging of Water Management and Marine Waters Management, and, indirectly, the WFD.

4. Human right to water: Argentine cases, human rights – are they enforceable?

Maria Onestini

1 HUMAN ENVIRONMENTAL RIGHTS IN ARGENTINA: LAW AND POLICY

1.1 Constitutionally Enshrined Environmental Rights: Constitutional Reform of 1994

Constitutionally-based environmental rights in Argentina are drivers for defining and enforcing actionable entitlements in policy and in law regarding human rights vis-à-vis the environment. The Argentine Constitution, since its 1994 reform, recognizes environmental rights and directly links them to human development via its Article 41.[1] The Article establishes the 'right to a healthy, balanced environment, apt for human development'.[2] This recognition falls under the realm of what are defined

[1] Many provincial and municipal constitutions do also incorporate some sort of rights definition in their environment sections, in particular within reformed sub-national constitutions that were updated in the late 1980s and in the 1990s. These are not mentioned specifically given that an analysis goes beyond the scope of this chapter. Nevertheless, the reforms are concordant with the National Constitution's spirit and letter of the law.

[2] The full article of the Argentine Constitution that includes environmental rights is as follows:

> *Artículo 41 – Todos los habitantes gozan del derecho a un ambiente sano, equilibrado, apto para el desarrollo humano y para que las actividades productivas satisfagan las necesidades presentes sin comprometer las de las generaciones futuras; y tienen el deber de preservarlo. El daño ambiental generará prioritariamente la obligación de recomponer, según lo establezca la ley. Las autoridades proveerán a la protección de este derecho, a la*

as 'New Rights and Guarantees' given that these are newly conceptualized rights incorporated in this reform.[3]

Albeit some class actions suits were successful in promoting the recognition of environmental rights before the last constitutional reform, in 1994 there was explicit recognition. For instance, in 1983, and before the formulation of this right was introduced at the constitutional level, an administrative court established that 'the right of any citizen to preserve his or her habitat amounts to a subjective right' and that such right would entitle any person to initiate an action for environmental protection. A year before the constitutional reform, in a legal suit concerning environmental harm to a lake's fisheries and wildlife courts proclaimed that 'the right to live in a healthy and balanced environment is a fundamental attribute of people. Any aggression towards the environment ends up becoming a threat to life itself and to the psychological and physical integrity of the person – which is based on ecological balance'.[4] Therefore, the constitutional reform enshrining environmental rights builds upon other entitlements recognized by law and policy.

Argentina joined the regional trajectory in Latin America, on returning to democratic rule, in linking constitutional reforms to the update of rule of law instruments. Along with these restructurings and modifications, most Latin American countries have included additional environmentally-related sections in their reformed constitutions. In the aftermath of the United Nations Conference on Environment and Development of 1992, some constitutional and legislative modifications at the regional Latin American level dealt with environmental rights. Some other reforms (broader, such as in the Argentine Case) included wording and visions linking environment with development issues by stating the right to a healthy, balanced environment, suitable for human development, as in the

---

utilización racional de los recursos naturales, a la preservación del patrimonio natural y cultural y de la diversidad biológica, y a la información y educación ambientales. Corresponde a la Nación dictar las normas que contengan los presupuestos mínimos de protección, y a las provincias, las necesarias para complementarlas, sin que aquellas alteren las jurisdicciones locales. Se prohíbe el ingreso al territorio nacional de residuos actual o potencialmente peligrosos, y de los radiactivos.

[3] Capitulo II, Nuevos Derechos y Garantías. Translation: Chapter II, New Rights and Guarantees.

[4] The cases where these proclamations were made were: Margarita v. Copetro (1993), ruling of 1993, Civil and Commercial Chamber of La Plata, citing an earlier decision in V.H. Morales *et al.* v. Mendoza Province (1986), Amparo Action, Civil Court of First Instance (2 October 1986).

Argentine constitution: 'All citizens enjoy the right to a healthy, balanced environment, suitable for human development and for productive activities that meet present needs without compromising those of future generations ...'[5]

This reform also included significant modifications regarding human rights and international law since updating of the Constitution includes updating human rights concepts in the country. It should be pointed out that the 1994 constitutional reform was the first revision of the National Constitution since Argentina had experienced a period of massive human rights violations during military rule. Having this as a historical background, international norms (primarily human rights instruments) were embedded into the National Constitution and defined them to be above national laws. Hence, the new reformed Constitution indicates, in its Article 75 (section 'Declarations, Conventions, and Complementary Covenants of Rights and Guarantees') that 'Treaties and concordats have a higher hierarchy than laws'.[6] Furthermore, the same section of the Constitution lists seven international human rights instruments that were given constitutional hierarchy at the time of reform. Among the international instruments having constitutional status since 1994 are: the Universal Declaration of Human Rights,[7] the American Convention on Human Rights,[8] the International Covenant on Economic, Social and Cultural Rights,[9] and the Convention on the Rights of the Child[10] (*inter alia*). The reform also presents the possibility to include other international human rights norms with constitutional status in the future.

The assertion that all the inhabitants of the country have the 'right to a healthy, balanced environment, apt for human development' has provided a fertile ground to sustain policy development and enforceability of rights

---

5 Article 41 of the Argentine Constitution, added as part of its 1994 reform.

6 This is indicated in Article 75 of the Constitution as translated in the text (Artículo 75, Inciso 22. Aprobar o desechar tratados concluidos con las demás naciones y con las organizaciones internacionales y los concordatos con la Santa Sede. Los tratados y concordatos tienen jerarquía superior a las leyes).

7 Universal Declaration of Human Rights, General Assembly Resolution 217 A, 10 December 1948.

8 The American Convention on Human Rights, adopted on 22 November 1969, in force 18 July 1978.

9 The International Covenant on Economic, Social and Cultural Rights, General Assembly Resolution 2200 A (XXI) of 16 December 1966, in force 3 January 1976.

10 The Convention on the Rights of the Child, General Assembly Resolution 44/25 of 20 November 1989, in force 2 September 1990.

regarding the human right to the environment in Argentina. Notwithstanding, and as will be indicated later in this chapter, the understanding of what the definition of these rights are and what they entail has been vague or broad at best.

Nevertheless, one of the most concrete areas of work where the right to the environment construct is being summoned is within the right to water realm. Several cases have been successfully brought to courts (from municipal courts to the country's Supreme Court) where the right to water is not only actionable as such but also acts as a keystone in building a wider right to the environment concept as a whole.

1.2 Actionable Rights in National, Regional and Municipal Norms and Jurisprudence

The country has a tradition of maintaining that social and economic human rights, not only political and civil rights, are actionable and therefore enforceable in many areas. However, the right to the environment has proven to be an elusive construct and has been formed in the last two decades. This section will briefly examine how this new rights-based paradigm is being developed by norms and jurisprudence and how the concept is being operationalized; the latter in particular through actionable cases, jurisprudence, and paradigms related to the right to water.[11]

Although, as stated earlier, some court cases before the constitutional reform of 1994 attempted to draw out issues related to the right to the environment, this construct was explicitly recognized after the constitutional reform, with varied interpretations. These interpretations go from the broader extended stances and points of view to more practicable or concise recognitions. Regarding the former, for instance, immediately after the reform, in 1994, courts contemplated the right to defend the environment as a 'natural human right', which permits putting grievances

---

[11] As indicated earlier in this chapter, many of the provincial and municipal constitutions do also incorporate some sort of right to the environment section within the reformed constitutions and norms that were generally introduced in the late 1980s and 1990s at the sub-national level in Argentina. These are not mentioned specifically given that an analysis goes beyond the scope of this chapter. Nevertheless, the reforms are concordant with the National Constitution's spirit and letter of the law. The cases highlighted here nevertheless do specifically abide by national and supra national legal instruments and therefore can be conceptually and normatively linked to the national concept of right to the environment.

before government agencies regardless of the previous existence of a specific administrative procedure (right to environmental justice). Namely, the right to the environment entails the right to challenge government agencies in this respect by any resident. Furthermore, in 1999, a court decision indicated an entirely more expansive interpretation. It was established in that ruling that the right to live in a suitable environment and to enjoy an adequate standard of living warrants a wide-ranging interpretation, comprising also the widest number of environmental offences. In the decision, the right to enjoy a view was impaired by the construction of infrastructure.[12]

The aforementioned broad interpretations and the subsequent more tangible outcomes that will be displayed here indicate how the right to environment is being defined through several different interpretations (e.g. broad right to the environment per se as indicated above) and different components (i.e. specific ones such as the right to water, the right to environmental information, the right to health) and how these different components establish and construct a broad embodiment of what 'environmental rights' mean in terms of public substantive obligations.

Arguably the most extensive environmental right component sustained by law, policy and recognition in Argentina is *the right to water*. It can be interpreted that the right to water, albeit as a partial construct and to a fractional degree, dates back previously to the 1994 reform given that there is an implicit recognition of it in the National Constitution's Article 14 Bis of the 1955 reform. There it is indicated that inhabitants have the right to access decent housing (i.e. '*el acceso a una vivienda digna*'). This axiom within the constitution has been interpreted as an indication that 'decent housing' implies a right to housing that complies with international criteria, which include adequate services, which, in turn, include water. However, the right to water has been strengthened and expanded as an interpretation in law, policy and court rulings following the greater understanding of environmentally-related rights, going beyond the dominion of housing and habitat. Here three cases will be analysed where court rulings have sustained the right to water within the right to environment at national, provincial and municipal scale: these are the Comunidad Paynemil Case; the Valentina Norte Colony Case (both in Neuquén Province, Patagonia Region); and a case in the capital city of Buenos Aires.

---

[12] 'Don Benjamín, S.A. v. Ente Nacional de Regulación de la Electricidad' (24 February 1999).

The Comunidad Paynemil and Valentina Norte Colony (both in Neuquén Province, Patagonia Region) cases are interlinked. Cases were presented by the Minors Ombudsman of the Neuquén province on behalf of children and youth from indigenous communities in order to guarantee their health due to lead and mercury contamination of fresh water. Conflicts in the community were related to pollution in an area with large scale oil extraction and other activities that contaminated water sources. The case presents requirements to the State as it is understood that the State is obliged to provide drinking water needed for survival, the adoption of measures and diagnosis and treatment of affected children, and appropriate measures to prevent soil and water contamination, given the right to water principle. Several rulings and appeals (extending up to the Provincial Supreme Court level) indicated corrective actions and demanded that the Provincial Executive authority establish corrective measures. The corrective measures indicated related to the provision of sufficient safe drinking water as well as the search for longer term corrective measures in relation to pollution sources.

The Valentina Norte Colony Case (in the same region as the previous case) also concerns an action filed by the Minors Ombudsman where this office instructs the Provincial Government to provide children and their families safe drinking water and the means for low-income families to store water under the same condition in which it was delivered. This was professed as an emergency measure while a more permanent one was sought in relation to the contamination of their water sources. At different stages of the trial and appeals, the courts (from local to provincial) specifically recognized the right to the environment and to health (related in this case to contaminated water, i.e. a health issue as it relates to environmental variables) as the base for their legal ruling. The Provincial Supreme Court based its arguments on Article 41 of the National Argentine Constitution which refers to the 'right to a healthy, balanced environment' as well as on the Convention on the Rights of the Child of 1989. This Convention mentions environmental issues regarding children's right to health, for instance, in its Article 24, which indicates that parties shall take appropriate measures to combat disease and malnutrition 'through the provision of adequate nutritious foods and clean drinking water, taking into consideration the dangers and risks of environmental pollution'.

In sum, the two rather similar cases use the argument of environmental rights and international human rights agreements to enunciate that minimum requirements of water supply through remedial temporary action should be provided. They also require that steps should be taken to prevent water contamination and to determine actual health impacts related to the existence of heavy metals in drinking water. Lastly, they require that medical treatment should be provided when water contamination is proven to inflict health problems on humans.

It is worth noting that the Convention on the Rights of the Child is incorporated directly into the Argentine domestic legal system via the assimilation of international human rights agreements as part of the National Constitution since the 1994 reform. This is one of the seven human rights covenants specifically mentioned in the Constitution as a treaty with a higher hierarchy than national laws. This Convention (which clearly includes social, economic, and cultural rights) is one of the first international instruments to integrate environmental issues with a human rights-based domain. With regard to these cases, it is noteworthy that an international human rights instrument, which includes an environmental right aspect, is used (with the correct normative architecture) to enforce the right to the environment at the national level. Critics of international environmental and human rights instruments often underline that the instruments are either not enforceable or have few practical consequences. Nevertheless, in these cases as well as in many others, it is demonstrated that they are enforceable and do have practical consequences through implementation at the national level.

A third instance highlighted here is the case brought against the City of Buenos Aires which also invoked the right to water and to the environment. Although the City of Buenos Aires is very well provided with urban infrastructure in a developing country context, including water, areas remain with little or no adequate provision of safe water. The case was brought against the municipal government for failing to provide water to a poor neighbourhood within the city limits. The court declared that the Government of the City of Buenos Aires had to provide water in a continuous manner to hundreds of families in the poorest neighbourhoods, explicitly indicating that this pertains to the right to water. The court also specifically linked the underlying interdependence between two sorts of rights that make up a broader and more encompassing concept than the right to water by itself. That is, in this case, the right to water and the right to health (the latter as associated with safe non-polluted water) have been associated in a broader notion conceived as 'the right to the environment'.

These highlighted cases point out that there is, at the national level in Argentina, a cluster of instances where courts (at the national and provincial as well as at the municipal levels) recognize the constitutionally enshrined principle of a right to the environment. Constructed from a series of more perceptible rights (e.g. the right to water, the right to health linked to environmental variables when morbidity and mortality are related to pollution – for instance) or from more general concepts (such as the right to environment itself), rulings and jurisprudence exist and this anchors the human right to the environment in rather substantial terms.

1.3 Actionable Rights within Argentina's Supreme Court's Rulings

The hierarchy and standing that the right to water is given in policy as well as law in Argentina is denoted by recent Supreme Court rulings related to this matter. An Argentine Supreme Court ruling recognized water as a human right (and actionable therefore within national law and policy) in a verdict of December 2014.[13] This ruling not only recognizes water as a right but it also explicitly and implicitly acknowledges and supports sub-national processes dealing with the right to water within the country, such as the actionable rights in national, regional and municipal jurisprudence mentioned above.

The December 2014 ruling is also noteworthy since it goes beyond the mere declaration of access to water and also deals very specifically with water safety and water quality issues related to this matter. The ruling resulted from a case presented in lower courts and elevated to the nation's Supreme Court. The case dealt with arsenic contamination of drinking water. The Community of 9 de Julio City (in the Province of Buenos Aires) presented a case before the provincial courts due to high unsafe levels of arsenic present in piped household water provided by the provincial water works company. The case was eventually heard by the Supreme Court of Argentina, which ruled that the human right to water 'is essential to live in dignity and is a prerequisite for the realization of other human rights'. Here the highest court in the country acknowledged that the right to water is not only a distinct human right but also a keystone to other related human rights (right to the environment, right to health, etc.). The Supreme Court also pointed out in the case that 'undoubtedly the human right of access to drinking water is at hand; health and lives of a number of people who are threatened by the acts of the Aguas Bonaerenses S.A. company which provides water to neighbours with proportions of arsenic exceed those allowed by the Argentine Food Code'.

The ruling also acknowledges international norms and decisions dealing with water access, providing once more an interlinked arena between

---

[13] 'Recurso de hecho deducido por Aguas Bonaerenses S .A. en la causa Kersich, Juan Gabriel y otros c/ Aguas Bonaerenses S.A. y otros si amparo', December 2014.

international and national frameworks for water law and the right to water. Several international accords are mentioned specifically in this ruling (such as the aforementioned Convention on the Rights of the Child) as well as broader dogma such as the United Nations Millennium Development Goals and the Human Right Council Resolution A/HRC/RES/27/7, the latter specifically stating that

> the human right to safe drinking water and sanitation entitles everyone, without discrimination, to have access to sufficient, safe, acceptable, physically accessible and affordable water for personal and domestic use and to have physical and affordable access to sanitation, in all spheres of life, that is safe, hygienic, secure, socially and culturally acceptable and that provides privacy and ensures dignity.

The mention of and explicit interlinkage to international norms, accords and dogma is also a key issue in the right to water within the realm of international human rights in Argentina. And, as seen in the Supreme Court ruling supporting the possibility of actionable and enforceable human rights constructs, it falls within the domain of enforceable national law and policy.

1.4 Rights-based Policy

Policy has also followed the conceptualization of the right to the environment, and hence also the right to water. Two instances will be briefly analysed: the right to water as a policy and the right to environmental information (which includes environmental information related to water law and policy).

Regarding the right to water as a concept, the Argentine State has recognized this right specifically in policy. In the national Decree 303 of the year 2006, whereby the national government takes charge of the concession of water works in the capital city of Buenos Aires and 17 surrounding cities in the greater metropolitan area, this concept is specified. It is indicated that the State intends, without prejudice of establishing water as an economic good, to have the resource valued and managed as a social and cultural good 'which in legal terms is translated as a human right'.

As implied in Article 41 of the National Constitution, the declaration of the 'right to a healthy, balanced environment, apt for human development' must be articulated via norms and policies in order to make this assertion effective. The same article indicates that it is the obligation of the Nation to dictate standards containing admissible minimum protection standards. Therefore, a series of norms and policy instruments have been drafted, debated and approved following the general principle of right to the environment subsequent to the inclusion of this concept in the National Constitution.

One of the cases is the right to an environmental information norm. This national law (Regimen De Libre Acceso La Información Pública Ambiental, Ley 25831) regulates the right to request and obtain public environmentally-related information that is held by government at the national, provincial and municipal level, as well as from public service companies (private, public or of mixed ownership). The norm and subsequent policies clearly frame access to environmental information within a human rights framework.[14] The right to environmental information is extended to the state of the environment or one of its components, natural or cultural, including their mutual interactions, as well as to the activities and works which affect them or might affect them significantly. The right to environmental information also extends to policies, plans, programs and actions relating to the management of the environment. Additionally, public authorities are compelled to disseminate environmental information in their possession.

For a country that did not have until recently (2016) a law that would grant a general access to public information, and where the right to information at an overall level was rather limited, the right to environmental information (and the ensuing specific policies that originate from the right to environmental information law) was quite pioneering and forward looking. Framing access to environmental information within a rights framework as a result of the recognition of the right to the environment is another specific case where human rights concepts and environment are constructed in the realm of Argentine policy.

---

[14] The human rights framework within which the right to the environment as well as the right to water is framed within the country follows the international concept for this sort of framework: The human rights framework protects civil, political, economic, social and cultural rights and it includes basic principles that are always part of human rights standards and implementation. These principles are universality, and indivisibility: these are defined as follows. Universality: Human rights must be afforded to everyone, without exception following the premise of the framework that people are entitled to these rights simply by virtue of being human. Indivisibility: Human rights are indivisible and interdependent, which means that in order to guarantee civil and political rights, a government must also ensure economic, social and cultural rights (and vice versa).

2 CONCLUSIONS

Slightly over two decades have passed since an inclusive constitutional reform took place in Argentina. Some of the most substantial changes include the recognition of so-called 'new rights' such as the right to the environment and the incorporation of international human right instruments in the Constitution itself and within national interpretation of classified norms, thereby giving them a hierarchy above national law. This, in turn, strengthens the binding force of the human rights instruments. This juxtaposition has proven to be a significant step for the human rights/environment connection in the policy and normative realm.

In the case of Argentina, the incorporation of the right to the environment has had important results. These include, *inter alia*, more robust environmental norms enclosed within a rights framework and court decisions defending the right to the environment from violations.

Nevertheless, 'the right to the environment' has been at times an elusive construct to compose. Understanding that this concept is collectively formed by society and its institutions has resulted in several approaches. First, a mosaic-like approach where the right to the environment is assembled from parts and operationalized from the more specific to the more general concepts. For instance, this takes place when the right to water or the right to health is invoked. Furthermore, in a more general approach in a few cases the right to environment is summoned as a self-contained construct, without specificity. Yet in the latter cases the affirmations are not as clear.

However, it must be pointed out that the concept expressed in the National Constitution after its most recent reform goes beyond strictly the right to the environment per se and incorporates sustainable development notions. The Constitution, as indicated in this chapter, states that: 'All citizens enjoy the right to a healthy, balanced environment, suitable for human development and for productive activities that meet present needs without compromising those of future generations ...' Nevertheless, what is suitable for human development and what are environmental intergenerational rights has been even more elusive to define and is even, in some cases, avoided by courts as well as by policy. Therefore, this is a pending issue regarding the operationalization of this Constitutional article.

The second broad constitutional reform whereby international human right instruments became an intricate part of national policy and norms is also a fundamental aspect of the incorporation of the right to the environment within Argentine policy, norms and jurisprudence. Recognition of a rights framework and human rights instruments and obligations in policy tools at national, provincial and local levels has been of help in establishing the right to the environment and in exercising this right in concrete forms using international instruments. It is often exercised by and on behalf of local communities, furthering accountability and furthering the use of international instruments in a very concrete enforceable manner and, again, very specifically when dealing with water rights. These tools, when used, have contributed to the exercise of environmental rights within the economic, social and cultural rights sphere. Therefore, there is a substantial case to be made for strengthening the international environmental rights instruments agenda (including the right to water) within the human rights platform given that they are deemed applicable with the appropriate institutional machinery at the national and sub-national levels.

From a survey of the court cases, as well as for policy and norms presented here, three important lessons can be highlighted:

1. The courts are promoting the right to a healthy environment (and within this realm the right to water) within the hierarchy of human rights by recognizing it as a fundamental issue.
2. The courts and policy-making institutions are defining the content and nature of the right to a healthy environment through landmark decisions and robust policy instruments.
3. The cases indicate which methods are most effective to protect these rights.

Therefore, in conclusion, although an elusive construct, the right to the environment within a human rights framework has, in several instances, not only been claimed but has also been successfully used to redress environmental violations, such as the lack of access to clean safe water. As with all instruments of the kind, the enunciation and its internalization strategy has been successful when it is understood in clear terms what 'the right to the environment' or the 'right to water' entails. Furthermore, this strategy also functions when national instruments are strengthened and supported by international rights-based instruments.

Plainly, Argentina has and continues to have a series of environmental issues, many of them dealing with water and water policy. Regarding access to water for human consumption, for instance, over twenty per cent of the population does not have in-house access to drinking water through pipelines. Furthermore, water quality and piped water aptness for human consumption is questionable in many instances (as the cases indicated here reveal) given that piped water and water accessed in households is in many instances not apt for human consumption. Nevertheless, the use of the right to water as a construct to improve access is a powerful tool to be used in many instances in the road to safe and adequate water access for human consumption. Human rights tools, when used, can promote the exercise of these rights. Although water-related problems and access to safe drinking water still remain a crucial problem in Argentina, the highlighted cases show that a rights framework can successfully be applied to social, economic and cultural rights, including all rights related to the access to water, to a healthy environment and to development.

5. Struggle for water rights between the natives and the European settlers: a case study of Njoro Kubwa Springs (1947–56)

Ezekiel Nyangeri Nyanchaga

1 INTRODUCTION

When looked at critically, almost all of Earth's water is saline and is found in the oceans and only 2.5 per cent of Earth's water is freshwater. Gleick observes that Earth's water resides in a very thin slice all around the Earth's surface and almost all of it is locked up in ice while another 20.9 per cent is found in lakes. Only a little more than 1.2 per cent of all freshwater is surface water. Rivers make up 0.49 per cent of surface freshwater. Although rivers account for only a small amount of freshwater, this is where humans get a large portion of their water from.[1]

As Sikes, the Chairman of the Select Committee on the first ever Kenya Water Bill in 1928 observed: '... there is no single matter which gives rise to so much unfriendliness between the farmers as uncontrolled water dispute ...[2]

Allocation of water resources needs to recognize that water is a scarce resource, critical to life, and that enhancing universal accessibility is therefore essential. Water resources are under a serious pressure from degradation of catchments, rising demands from other sectors and climate change. To live, human beings must have access to water, whether

[1] Gleick, P.H. (ed.), *Water in Crisis: A Guide to the World's Fresh Water Resources*. Pacific Institute for Studies in Development, Environment and Security. Stockholm Environment Institute (Oxford University Press 1993).

[2] The Secretary. Report of the Select Committee Appointed to Consider and Report upon the Provision of the Water Bill. Kenya National Archives. W1/3/3, 22 August 1928.

formally or otherwise and regardless of quality. Conflict will therefore emerge wherever allocation of water rights is or seems to be unfair.

2 THE HISTORY OF WATER RIGHTS LAW IN KENYA

At the turn of the nineteenth century, traditional African societies in Kenya exercised universal water rights through customary law, which rationally responded to uncertainty in the physical environment. Customary law sustained and ensured effective and efficient water use through flexible water rights. There are no records of punishment or non-cooperation and therefore we assume this was effective. However, with the immigration of the Imperial British East Africa Company, the responsibility for water supply changed, becoming the responsibility of a single institution, the Hydraulic Branch.

Meinzen-Dick and Nkonya note that this centralization disorganized the socially accepted water controls and informal institutional structures, rendering them insignificant and ultimately leading to inequitable distribution of water resource.[3] What brought about this warped distribution was unwillingness or inability to enforce the legislation guiding water resource allocation and, as a result, the wealthy and the European settlers had uninterrupted access while Africans struggled for the same. In this chapter, I focus mainly on the struggle for water rights between the Indigenous Kenyans and the European settlers at Njoro Kubwa Springs in the Taveta region of Kenya.

The first water legislation was contained in Section 3 of the Crown Lands Ordinance of 1902 and only covered the issuance of water permits. The Crown Lands Ordinance of 1902 was repealed and re-enacted as the Crown Lands Ordinance of 1915 which denied the person buying, leasing or occupying Crown Land express right to the spring, river, lake or stream on such land except in the case of abstraction for domestic use. It also prevented the damming of spring, river, lake or stream on Crown Land acquired in the above-mentioned ways. The section also provided

---

[3] Meinzen-Dick, R. and Nkonya, L. 'Understanding Legal Pluralism in Water Rights: Lessons from Africa and Asia', Paper Presented at International Workshop on 'African Water Laws: Plural Legislative Frameworks for Rural Water Management in Africa' (Johannesburg, South Africa 2005), 26–8 <http://198.173.106.43/docs/watsan/050126_understanding_legal_pluralism_rmdln.pdf> accessed 6 January 2017.

for the formulation of rules on the issuance of water permits if someone wanted to construct a dam.[4]

The Crown Lands Water Permit Rules of 1919 were enacted, giving the Director of the Public Works Department the power to consent or refuse to permit the abstraction of water from spring, river, lake or stream. The rules also prescribed the method of acquisition of a water permit.[5] This legislation was however inadequate and efforts were made to move towards comprehensive water legislation.

The first was a solo effort by the then Director of the Public Works Department, Sir McGregor Ross, in 1916, which led to the Draft Water Ordinance of 1916. It was not drafted into a bill due to World War I, which took precedence at the time. Disputes regarding water matters were on the increase and those charged with the responsibility for guarding water resources in the country had a very difficult task.[6]

The draft water legislation did not make progress to becoming a bill until 7 July 1928[7] when it was put into the official gazette as awaiting final assent. His Majesty's assent was denied because the rights of the Africans living on Native Reserve Lands were not sufficiently protected. The Secretary of State demanded that the Ordinance be revised. All activities affecting waters in Native areas were subject to the approval of the Native Lands Trust Board, the authority in charge of protecting the rights of 'natives' as laid down in the Native Lands Trust Ordinance.[8] In addition, the draft water legislation was supposed to clarify that the Native Lands Ordinance was superior to the Water Ordinance. This introduced a dual legal regime.

The issue was referred to the Kenya Lands Commission who, in 1934, clarified that there was no distinction with regards to Native or Crown lands with regards to the powers and procedures of the Water Ordinance. All areas were under the jurisdiction of the Water Board, and the Native

---

4 Sikes, H.L. Modern Water Legislation. Public Works Department (Nairobi 1922).

5 Ibid.

6 Nyanchaga, E.N. *History of Water Supply in Kenya 1985–2005: Lessons and Futures* (Tampere University Press 2016).

7 Sikes, H.L. *Sewerage Scheme-Kisumu Township*. Kenya National Archives, BY/29/29, Ref: San.686/17. 10 March 1928.

8 Secretary of State for the Colonies to the Governor of Kenya, Dispatch dated 1930-7-8. Kenya National Archives, Ref: AG/43/79.

Lands Trust Board, which acted legally as a trustee for the Africans, was to be regarded like any other riparian landowner.[9]

The first water legislation was enacted on 1 July 1935.[10] The Water Ordinance was revised in 1951, and the Minister for Agriculture was given the overall mandate over water development policy while it was again revised in 1972, when it was renamed Cap 372 of the Laws of Kenya.[11] This Act was in force for a long period, until 2002[12] when a new act was introduced to accommodate the increasingly complex administration of the water sector. This newest legislation is now entitled the Water Act 2016[13] and takes into account the change of the constitution in 2010.

3 ACQUISITION OF WATER RIGHTS

In the early years of colonialism, the British administrators felt the need for legislation to guide water allocation to avert the emerging conflicts. Most conflicts were due to high demand and inadequate supply, leading to illegal allocations and abstraction by farmers. The period was characterized by consistent violation of native rights, alienation of natives' land for white settlement and the forced reallocation of natives. Thus, denying them their rights to natural resources had an impact on the pre-existing systems which led to loss of indigenous rights to natural resources. For instance, in 1904–1905, the British forcibly moved certain sections of the Maasai out of their favourite grazing grounds in the central Rift Valley into two reserves to make way for white settlement.[14]

---

9 Director of Public Works, letter to the Colonial Secretary, dated 1934-3-7. Kenya National Archives, Ref: AG/43/79.

10 Colony and Protectorate of Kenya. *The Official Gazette*, 25 June 1935. Kenya National Archives, Ref: AG/43/79.

11 Ministry of Water Development, Kenya and Swedish International Development Authority (1992). *Delineation Study, the Water Sector in Kenya*, July 1992.

12 Republic of Kenya (2002). *The Water Act 2002*. Kenya Gazette Supplement No. 107 (Acts No. 9). Government Printers, Nairobi, 24 October 2002.

13 Republic of Kenya (2016). *The Water Act 2016*. Kenya Gazette Supplement No. 107 (Acts No. 9). Government Printers, Nairobi, 24 September 2016.

14 Hughes, L. *Moving the Maasai, A Colonial Misadventure* (Palgrave Macmillan) (in association with St Antony's College Oxford 2006).

4 PROCEDURES FOR APPLICATION FOR A LICENCE

In 1926, the Director for Public Works established a procedure to be followed when one applied for water rights, with measures to ensure that each applicant knew the conditions, giving as much security of tenure as possible and at the same time safeguarding the government and other users. The steps leading to issuance of a water permit were as follows.[15]

(a) An application, giving details of the proposal to the Executive Engineer on a form (which also stated the conditions on which permits were issued) accompanied by a sketch map illustrating the proposals.
(b) When the application was in order, the Executive Engineer sent an 'Objection Form' to each riparian holder, who was likely to be affected if the application were granted.
(c) The Executive Engineer received the views of the riparian holders as expressed in the Objection Forms.
(d) If there was no objection or if the objection was resolved, the executive engineer would submit the file to the chairman of the District Committee with his view stated orally or in writing. If an objection existed, the engineer would visit to investigate or seek more clarifications. The file was later sent to the chairman of the District Committee where applicant and objectors were given an opportunity to attend and state their cases.[16]
(e) A hearing by the District Committee, which was expected to issue a permit, to refuse it or to issue it in a modified form.
(f) Dispatch of the file with the Executive Engineer's comment to the Director of Public Works.
(g) If the Director of Public Works agreed with the recommendations of the Executive Engineer and the District Committee, the permit was issued in those terms.

5 DEVELOPMENT OF POLICY FRAMEWORK

In August 1928, The Legislative Council appointed a Select Committee to consider and report on the provision of the Water Bill of 1928 with an

---

[15] Sikes, H.K. *Report by Mr. A.D. Lewis on Irrigation etc. in Kenya Colony*, 1925. Kenya National Archives, Ref: AG/43/87, 18 July 1926.
[16] Ibid.

intention to control water rights through establishment of a Water Board on behalf of the government.[17] The water bill was intended to ensure that the rights already enshrined in land titles were safeguarded. However, the majority of water rights had been granted by lease or agreement and the board preferred them to be included in the Water Ordinance to re-establish in more definite terms the quantity, point of diversion, method of diversion and use of water.[18]

The water bill was also intended to establish a Water Appeal Board, which would hear appeals in cases where the Water Board cancelled a licence. It was proposed that the Water Board should be constituted and advised by professionals who had the technical knowhow for making final decisions. Establishment of a Water Appeal Board was deemed necessary to consider exceptions and extreme cases where a licence was cancelled by the Water Board and a review of the decision by an independent authority was necessary.[19]

Native rights and interests were adequately safeguarded by: the presence of the Chief Native Commissioner and three other Heads of Department on the Water Board; the statutory need for publication of application for water rights; and the right to appeal to the Governor in Council and if necessary to the Secretary of State. The committee felt that cases could arise when a native who might be affected, or even the District Commissioner of the district in which that person resided, would not be aware of the published notice. Consequently, the committee recommended an inclusion of a new clause 27 in the bill relating to diversion, obstruction or use of a body of water within, abutting on, or flowing under native land. A notice was supposed to be sent to the District Commissioner who was required to inform any native within his district who might be affected by the terms of the application. This gave the natives an opportunity to lodge an objection and represent their case to the authorities before a licence was granted.[20]

After the first legislation, conflicts related to water rights persisted and diverse opinions and priorities in various departments, with duplication of duties among different institutions entrusted with water development,

---

[17] The Secretary, *Report of the Select Committee Appointed to Consider and Report upon the Provision of the Water Bill.* Kenya National Archives, Ref: W1/3/3, 22 August 1928.

[18] Ibid.

[19] Ibid.

[20] Ibid.

undermined the process.[21] Mombasa in 1934,[22] Kajiado in 1953[23] and Nyeri in 1960[24] experienced conflicts at departmental level that greatly jeopardized water and sewerage improvement. By 1955, there was no single department responsible for overall water development issues; the responsibilities were shared among the Public Works Department, Ministry of Health and the Ministry of Agriculture.[25] Inter-tribal battles between the Kenyan tribes themselves over watering and grazing rights were also a major source of conflict, especially among the pastoral tribes.[26]

Reforms of the water resources and services sectors began in the late 1990s. The reforms were based on international best practice principles, which were being promoted at the time. The other reforms focused on the development and management of water in an integrated approach and separated water resource management from provision of water services. A Water Services Trust Fund was created as a funding mechanism aimed at extending water services to the poorest and least served communities in the country. The Water Resources Management Authority was established by the Water Act of 2002 with the aim of developing guidelines and procedures for allocation of water resources, development of water resources management and development strategies at national levels, issuing and monitoring of water permits, regulation and managing water quality, protection of catchments, gathering and publishing information and determining charges and levying water resource fees. This is done through the establishment of water user associations that allow the stakeholders to play more active roles in water resources management.[27]

---

21 Town Clerk. *Water for Public latrines*. Kenya National Archives, BY/29/1, Ref: 33/4, 13 November 1934.

22 Ibid.

23 District Commissioner. *Sanitary Services in the Townships*. Kenya National Archives, BY/21/107, 6 August 1953.

24 Davies, E.R. Nyeri Urban District Council Housing. Kenya National Archives, BY/21/73, Ref: H.4/20/3, 23 May 1960.

25 Colony and Protectorate of Kenya. Sessional Paper No.98 of 1957. *The Manzoni Report on the Public Works Department*. Government Printer, Nairobi 1957. Kenya National Archives, MSS/66/111.

26 Negley, F. *Last Chance in Africa* (Victor Gollancz, London 1949).

27 Ministry of Water and Irrigation (2009). *Kenya National Resources Management Project: Consultancy Services for Irrigation and Drainage Sector Institutional Reform*. Final Study Report Volume – Main Report, Euroconsult Mott MacDonald & Samez Consultants.

6 NJORO KUBWA SPRINGS CONFLICT OVER WATER RIGHTS (1947–56)

6.1 Background

Until the 1940s, there was no written national plan on how to mobilize the country's water resources. Water development was in most cases ad hoc and usually undertaken at township levels. Each township had its own water development plan.[28] The first formal water development plan known as the 'Dixey scheme' was put forward in 1943 and it covered the water scarce areas of the Northern Frontier District, namely Moyale, Marsabit, Wajir, Garissa, Lodwar, Isiolo, Samburu and Mandera. The report covered various aspects including the existing water supplies, necessity for improvement, topography, geology and, most importantly, the recommendations for improvement.[29]

Njoro Kubwa is in Taita Taveta District which is one of the seven districts that form the Kenyan coast province. It occupies the south west part of the coast province. The neighbours include Tana River, Kitui and Makueni District to the north, Kwale and Kilifi District to the east, Kajiado District to the North West and the Republic of Tanzania to the south and south west.[30]

Groundwater emerges from Mt Kilimanjaro forming the three-fold Njoro springs: Njoro Tatu, Njoro Ndogo and Njoro Kubwa. Therefrom waters flow directly into the River Lumi that flows right through the middle of this dry region. The River Lumi feeds Lake Jipe, is shared by both Tanzania and Kenya, and covers an area of approximately 28 km^2 with a length of 12 km and depth of three metres. There are other temporary streams from the Pare Mountains that also act as sources to Lake Jipe.

---

[28] Nyanchaga, E.N and Ombongi, K.S. 'History of Water Supply and Sanitation in Kenya, 1895–2002' in Petri, S., Juuti, Tapio S. Katko and Heikki S. Vourinen (eds), *Environmental History of Water*, ch. 21 (IWA Publishing, London 2007).

[29] Dixey, F. *Hydrological report on the Northern Frontier District, Samburu and Turkana and Report on the hydrology of the Uaso Nyiro*. Kenya National Archives 1950.

[30] Tertiary Consulting Engineers Limited. *Environmental Impact Assessment Project Report for Rehabilitation Works of Njoro Kubwa Irrigation Canal in Taita Taveta District*. National Water Conservation and Pipeline Corporation (NWCPC), Nairobi Kenya 2007.

There were several native land units on the left bank of the River Lumi, where the natives lived and farmed. The inhabitants of this area had practiced rough casual watering of crops, herding and fishing since time immemorial.

The struggle for water rights in Njoro Kubwa Springs was a protracted nine-year battle over water resource allocation between the natives who were farmers and herders and a European settler who was investing in large scale sisal farming and required water to irrigate his 1930 acres of land called Jipe Sisal Estate.

Col. Grogan's Left Bank Canal (Canal No. 1) embodied the whole breadth and width of the dispute involving the water right holders along the River Lumi. Being the main irrigation canal which was to carry the highest amount of water from the river to his farm, and by-passing the Native Reserve Land, it attracted the greatest attention and interest. This canal, passing east of the River Lumi was the heart and soul of the dispute between the natives and the European settler. Tracing its history, its evolution and the ultimate licensing will give us a glimpse of the workings of the Water Bill and the interaction between the representatives of the water right holders and the administration.[31] The proposed Canals 3 and 2, were to extract water from swamps in the low lying areas and were to traverse highly alkaline areas of the Taveta Reserve, and the proposed extraction was so small that they would have had virtually no effect on the water level of the nearby swamp.[32]

Col. Grogan's practice in the past, when abstracting water from swamps, had been by direct tapping without headwork regulation. Jipe Sisal Estate was an expansive 1930 acres next to the natives' land which did not adjoin the River Lumi. The owner, Colonel Grogan, was a European settler. By virtue of the acquisition of the land near the river, he became a riparian right holder to the waters of the river. His huge farming interests and the consequent need to use a substantial amount of the waters of the River Lumi put him at odds with native African residents, the other right holders of the same river.[33]

African areas suitable for farming were subsequently developed and water made available through the Kimorigo, Kiteve and Kaleza Schemes in 1945. Separate from the main stream of development was the 'Goodwillies' area lying between the River Lumi, the railway and the Voi

---

31 Malone, B.A. Notes on the Lake Jipe and Lumi Delta. Kenya National Archives, CA/17/116, 18 May 1955.

32 Edwin Bellis. Application for Water Rights, Jipe Estate Ltd. Kenya National Archives, Ca/17/116, 2 November 1948.

33 Ibid.

Road, where the River Lumi fed three canals before draining into Lake Jipe at the southern tip.

6.2 Application, Reactions and Rejections

When Jipe Sisal Estate applied for a water right vide General Notice No. 1563 of 1949, the main item was Canal No. 1 which had been proposed to carry around 40 cusecs of water a day for general irrigation. The figure of 40 cusecs was later reduced to 30 cusecs.[34] The amount of water in the River Lumi would deteriorate if water rights were to be granted to Jipe Sisal Estate for extraction of the 40 cusecs it had requested via Canal No. 1 and this would adversely affect other water users near the abstraction point and downstream. The natives who were natural right holders to the waters of the Lumi pointed out the dangers through the provincial administration.

The District Commissioner for Voi, was not satisfied with the use to which the water applied for by the Jipe Sisal Estate would be put and was concerned that there was a likelihood of high wastage and misuse.[35] He observed that abstraction of such a huge quantity of water by Jipe Sisal Estate would preclude development of any native irrigation scheme on the Left Bank of the River Lumi. The government had proposed such a programme and the completion of the programme would have been compromised by the approval of the application. He observed that the total normal flow of the River Lumi was insufficient for the purpose for which the application was made.[36]

The Provincial Commissioner sought to obtain some expert advice on the canals before granting a wayleave in Taveta Division. Therefore, he asked the agricultural officer to study the effects of the application on the natives' rights. The report indicated that Canal No. 1 would cause an impact in the swamp at the lower parts which would adversely affect the original layout of a government irrigation scheme.[37] In addition, the report observed that the requested amount of water (40 cusecs) could not be raised at that point. He pointed out that the most optimistic gauging had shown that less than 30 cusecs was available at the proposed headworks site and that no substantial increase in the normal flow of the

---

[34] District Commissioner, Voi. Ref: No. PW. 8/17/3/03. Kenya National Archives, Ca/17/116, 1948.

[35] Ibid.

[36] Ibid.

[37] Bellis (n 32).

River Lumi was expected. The agricultural officer cautioned against movement of headworks of Canal No. 1 to the vicinity of Njoro Kubwa, since it would affect the area's proposed irrigation.[38]

The Wataveta elders observed that the abstraction of water by the Jipe Sisal Estate would affect them in different ways. These were: the canal for Jipe Sisal Estate would drain water from various streams which was used for their farms, thus compromising their food security, and since water from River Lumi would be reduced, the elders would be not be able to irrigate their farms. In addition, they needed sufficient water for their livestock and the application by the Jipe Sisal Estate for the water rights was bound to disrupt human and livestock movement in the area. All the canals were to pass through grazing pastures, roads and passages.[39] The canal would generate unanticipated difficulties in using both sides of their land and cause the separation of neighbours.

The elders noted that the presence of a canal would prevent them from using both sides of their land with ease since the canal would run through the land and separate families. Equally, it would necessitate ferrying of loads and the crossing of cattle from one side of the canal to the other.[40] The excavation of the canals could lead to release of crocodiles which was dangerous for the human and social life of the native community. The excavation of Canal No. 1 would affect the ecology of the forest belt by drying the water course between Canal No. 1 headwaters and Njoro Tatu accretion. However, it was accepted that the presence of innumerable springs in the area ameliorated this claim.

Despite these objections and restrictions, Col. Grogan started digging the canal to extract the water he had applied for.[41] The District Commissioner investigated the matter and reported that there was no digging in progress on the Native Land but some clearing was being done. This had been necessitated by the fact that more detailed survey work was needed for the proposed underground pipeline than was originally deemed necessary in the proposal.

The Director of Public Works pointed out that since Jipe Sisal Estate was also a riparian to Lake Jipe and the River Lumi at the outfall, it was therefore entitled to some consideration.[42]

---

[38] Ibid.

[39] The Taveta Elders. *The Water Ordinance, Lumi River, Taveta District.* Kenya National Archives, Ca/17/116, 23 May 1949.

[40] Ibid.

[41] Ibid.

[42] Ag. Director of Public Works. *Jipe Sisal Estate – Application for Water Right – Lumi River.* Kenya National Archives, Ca/17/116, 14 October 1948.

It was apparent that the water rights of the natives would be affected adversely in one way or another, if the application was granted in full. Since the Jipe Sisal Estate was also a riparian right holder, and was therefore entitled to access to water, the Water Board faced the challenge of granting the licence, while, at the same time, maintaining the rights of the natives.

The excavation of Canal No. 2 would have led to a more severe segmentation of the available grazing land, and even if adequate crossings were provided, severe trampling would occur along the tracks which had to be developed between crossing points and the village.[43]

6.3 The Solution/Compromise

These challenges had to be faced and a solution had to be found. The major issue was that Col. Grogan, being a riparian holder, had to be allowed to build the three canals but with several restrictions. The administration had to strike a balance between satisfying the needs of the estate and those of the natives, and to bring least inconvenience to both parties while at the same time satisfying the needs of both the estate and the natives. The following conditions were agreed upon:

1. There would have to be a change of point of extraction. The Agricultural Researcher observed that Jipe Sisal Estate's application in respect of Canal No. 1 was to be opposed and, in its place, an application for 62 cusecs from the upstream of Njoro Kubwa had to be made. This was because the Jipe Estate Canals would not affect native interests upstream of the Njoro Kubwa group of springs.[44]
2. The Provincial Commissioner indicated that there would be free access to water in the canals by the local inhabitants for both domestic purposes and stock. In addition, the company would have to construct and maintain three bridges over each canal for the passage of stock, and those bridges were to be sited at points agreed by the District Officer, in consultation with the natives.
3. Additionally, the company would have to stake the canals in such a way as to ensure the prevention of crocodiles passing up or down their courses.[45]

---

[43] Bellis (n 32).

[44] Ibid.

[45] Provincial Commissioner. *Application for Water Rights – Jipe Estate.* Kenya National Archives, Ca/17/116, 2 March 1949.

4. The Water Board condition was that only half of the low flow of the river should be extracted at the point of intake.[46]

There was a great reluctance by the locals to concede the right to cut this canal. The locals needed a clearly set out account of the means by which their cattle would cross the canals and how they would be compensated. They suspected that Col. Grogan had ulterior motives – i.e. to have control over the land on which the canal passed.[47] However, the Native Lands Trust Board took a long time to endorse the water sanction/right of the Jipe Sisal Estate, taking into consideration protests from the Africans. The Board initially sanctioned Canal Nos. 2 and 3 and requested some time to study the application. The Provincial Commissioner, Mombasa sought the professional opinion of Mr. Edwin Bellis, a researcher from the Ruiru based Coffee Research Station. The proposed Canals 2 and 3, wrote Mr. Bellis, were to extract water from swamps in the low lying areas and traverse highly alkaline areas of the Taveta Reserve, and the proposed extraction was so small that it would have virtually no effect on the water level of the nearby swamp.[48]

In May 1949, the Board eventually approved the application provisionally[49] and in June 1949 the Secretary of the Water Board wrote to Col. Grogan cautioning him that the approval was only conditional upon meeting the set conditions.[50]

The importance of coordination and collaboration among agencies that are entrusted in running water use rights cannot be overemphasized; conflict among them is detrimental to the accessibility of the water resource. The Njoro Kubwa case study illustrates a scenario whereby roles conflicted and were duplicated, leading to water use permit delay. Despite the objections from the Provincial Commissioner and the agricultural researcher, the Director of Public Works approved the canal and pointed out that the Chief Native Commissioner had recommended that the application of Jipe Sisal Estate for water rights from the River Lumi

---

46 Chief Native Commissioner. *Application by Jipe Sisal Estates for Water Rights on the Lumi River.* Kenya National Archives, Ca/17/116, 20 August 1949.

47 District Commissioner. *Application for the Extraction of Water on the Left Bank of the River Lumi by Jipe Estate.* Kenya National Archives, Ca/17/116, 25 May 1949.

48 Bellis (n 32).

49 Secretary, *Native Lands Trust Board. Application for Water Right – Jipe Sisal Estates Limited.* Kenya National Archives, Ca/17/115, 25 May 1949.

50 Secretary, Water Board. *Application for Water Right – Jipe Sisal Estate.* Kenya National Archives, Ca/17/116, 20 June 1949.

go ahead, provided that 10 cusecs were reserved for the Goodwillies and 10 cusecs for future development on the Left Bank of the native reserve.[51]

The Chief Native Commissioner expressed his concern to the Provincial Commissioner that the approval was taking too long. This was in spite of the Trust Board having approved the Jipe Estates' application for water right as long as half of low flow water was left in the river at the point of intake.[52]

Impatient and in need of a permit, Col. Grogan, while addressing the District Officer, intimated that the Native Lands Trust Board had no legal standing in the matter and that it had completely exceeded any powers it may have had in determining the amount of water allowed to him, and that his only remaining concern was with the Water Board, which alone had powers to apportion water.[53]

The Provincial Administration and Col. Grogan engaged in extended deliberations about the wayleave and the conditions attached to it. Some of the conditions included in the wayleave demanded that Col. Grogan construct a permanent bridge over the Mboghoni Road to provide means of vehicular access and two footbridges constructed to enable human traffic. Natives had to have unimpeded access to water for domestic and stock use – with a fair compensation for the landowners.[54]

6.4 Delayed Decision Making

Four years after the application, Jipe Estate had still not been granted a wayleave for Canal No. 1 although the government had technically acceded to the request. Several issues were still outstanding including:[55]

(i) The area Col. Grogan demanded for the canal (100 feet for the canal and bank) was too big and there was nothing in the

---

[51] Ag. Director (n 42).

[52] Chief Native Officer. *Application by Jipe Sisal Estates for Water Rights on the Lumi River*. Kenya National Archives, Ca/17/116, 26 August 1949.

[53] District Officer, Taveta. *Application by Jipe Estate for Water Rights on the Lumi River*. Kenya National Archives, Ca/17/116, 21 September 1949.

[54] District Officer, Taveta. *Wayleave – Agreement – Col. Grogan's Left Bank Canal*. Kenya National Archives, Ca/17/116, 6 November 1950.

[55] District Officer. *Confirmation of Agreed Course of Action Concerning Outstanding Affairs at Taveta*. Kenya National Archives, Ca/17/115, 29 February 1952.

correspondence with the Chief Secretary which suggested that any such provision should be made in a wayleave.

(ii) The degree of responsibility to be assumed by Africans damaging the canals with their stock required definition.

(iii) The Africans were to be granted the right to extract a maximum of 5 cusecs from the canal for watering stock and irrigation.

(iv) The licensee was to construct one permanent vehicle bridge and one cattle and goat bridge at a place to be decided by the District Officer.

(v) The compensation should be £10 per acre.

These conditions were to be satisfied through the incorporation of special conditions in the wayleave, including the right to extract water for domestic and stock use by the natives. Col. Grogan was to build a fence on one or both sides of the canal to prevent African cattle from trespassing onto the banks of the canal and causing damage.[56]

After a protracted and prolonged process, Col. Grogan's Jipe Sisal Estate was granted a wayleave by the Provincial Administration on September 1952. To maintain and defend the rights of other water right holders, especially the native Wataveta, and to safeguard government control over this critical resource, the following points were included:[57]

(a) The Provincial Commissioner of Coast Province granted Jipe Sisal Estate a wayleave licence for making canals, aqueducts, weirs, dams, and other water works namely headgates, concrete tunnels, and transition outlets and spillways required for the supply of water from River Lumi together with the rights always to enter upon the lands for inspecting, maintaining, or repairing the water works.[58]

(b) The licence could be cancelled by the grantor in case of breach of the conditions.

(c) The position and the construction of the water works were to be in accordance with the plans and specifications duly described.

(d) The licence was subject to restrictions contained in the Water Act.

(e) The Africans should not be held responsible for damage done to the canal by stock.

(f) The licensee was to take measures to deal with soil erosion.

---

[56] Provincial Commissioner. *Wayleave License under Section 37(4) of the Native Lands Trust Ordinance, 1938*. Kenya National Archives, Ca/17/116, 5 September 1953.

[57] Ibid.

[58] Ibid.

(g) The licensee was to maintain the bank in good condition to prevent the flooding of the Native Land Unit.
(h) The licensee was to construct a one-foot bridge at a place to be identified by the District Officer.
(i) The licensee was to allow the residents unrestricted access to water from the canal for both domestic and stock use.
(j) The licensee was to build and maintain two bridges; one for vehicles and another one for human beings.[59]

6.5 Conclusions

The anticipated loss of locals' water rights to some extent occurred. Road access to Mata village was hampered. This happened in 1952, after the Jipe Sisal Estate had constructed the canal but not yet made the necessary arrangements to facilitate movement from Mata village to the Native Reserve. The manager of the Jipe Sisal Estate refused to allow vehicles to use the only alternative route, which ran through the estate, to the Native Reserve. The District Officer's intervention failed to produce any conclusion and he lamented that the government should either forcibly acquire right of access through the estate or force the Jipe Estate to build bridges to allow access.[60] The Water Board stated that it would not issue the wayleave licence until Col. Grogan built a road access to the village.[61]

Battle ensued over how the implementation of these recommendations was to be applied; on one side were the natives and the provincial administration, with Col. Grogan on the other. The issue of compensating the natives over damages that would be caused by the construction of the canal came up and further delayed the approval of the wayleave. Col. Grogan made a strong protest over the delay in being awarded a licence to continue the construction of the canals. The Acting Coast Provincial Commissioner expressed concern about the delayed issuance of the wayleave for Canal No. 1 of the River Lumi in Taveta.[62]

---

59 Ibid.

60 District Officer, Taveta. *Road of access to Mata through Jipe Estate.* Kenya National Archives, Ca/17/116, 4 September 1952.

61 The Secretary. *Notes of a meeting at Girigan house on the 22 December 1952, to discuss the Wayleave on Messrs. Taveta Sisal Estates.* Taveta. Kenya National Archives, Ca/17/115, 22 December 1952.

62 Provincial Commissioner, Coast. *Wayleave – No.1 Left Bank Canal, Lumi River, Taveta.* Kenya National Archives, Ca/17/116, 4 April 1952.

After the native issues had been addressed, in September 1955, the Acting Coast Provincial Commissioner granted the wayleave. The text of the wayleave was as follows 'Wayleave License under Section 37(4) of the Native Land Trust Ordinance, 1938.'[63]

In 1956, Col. Grogan brought on board Major Conrad Walsh, the Sisal King of Tanganyika, whose involvement as special adviser to the investors propelled sisal farming to new heights, with plantations soon spreading over large swathes of the former Taveta bushlands. The sisal farms employed large numbers of Taitas, Tavetas and upcountry people from the Kavirondo Gulf, now Nyanza Province.

As the struggle for independence gathered momentum, Col. Grogan and his fellow settlers were worried about the future of the colony and withdrew from farming, sold the sisal estates and started other businesses. Jipe Sisal Estate was sold on the eve of independence and again transferred to a different owner.

Currently the land is owned by natives who carry out small irrigation schemes, using the major canal whose excavation they had previously objected to. The residents are mainly farmers. Much of the land is within the forests or hills, only 35 per cent of the land is arable. The main crops are sisal, bananas and maize.[64]

---

[63] Acting Provincial Commissioner, Coast Province. *Wayleave license Under Section 37(4) of the Native Land Trust Ordinance 1938*. Kenya National Archives, Ca/17/116, 5 September 1955.

[64] Tertiary Consulting Engineers Limited. *Environmental Impact Assessment Project Report for Rehabilitation Works of Njoro Kubwa Irrigation Canal in Taita Taveta District*. National Water Conservation and Pipeline Corporation (NWCPC), Nairobi Kenya, 2007.

PART II

Models of water allocation

6. Towards a sustainable, balanced and equitable allocation of water use rights

Liping Dai, Marleen van Rijswick and Bram Schmidt*

1 INTRODUCTION

In times of increasing water scarcity due to climate change, urban development, growing populations and economic development, the allocation of water use rights becomes an issue that needs further discussion. Transparent allocation of water rights is crucial to a sustainable, balanced and equitable use of water as well as to guarantee the right to water and to contribute to the achievement of the United Nations' Sustainable Development Goals.[1] Water scarcity will not only be a problem for arid areas, but also for those countries that are not yet familiar with water scarcity but may become so in the near future.[2] Current allocation mechanisms may need re-thinking.[3] The re-allocation of existing water use rights is a sensitive topic, as it may lead to shifts in water property rights or water use rights that have been granted in the past and thus to a shift in profits, risks and burdens. Water allocation has therefore strong normative aspects.[4] The mechanisms and instruments that a government

---

* With research contributions from Bianca Mogos, Irene Akinyi Otieno, Annisa Noyara Rahmasary, Merlijn van Selm and Rebecca Van Weesep.

1 UN Sustainable Development Goals <http://www.un.org/sustainable development/sustainable-development-goals/> accessed 14 January 2017.

2 Keessen, A. and Ernst, W. 'The adaptiveness of Dutch water law put to the test – dealing with water scarcity in a water-rich country' (2015) 24 *Journal of Water Law*, 239.

3 Hendry, S. *Frameworks for Water Law Reform* (Cambridge University Press 2015).

4 Driessen, P. and van Rijswick, H. 'Normative aspects of climate adaptation policies' (2011) 2 *Climate Law*, 559.

or legislator chooses[5] to allocate water use rights will depend on worldviews, values, climatic and geographical circumstances,[6] the legal context in a country and political perspectives.[7] Shifts in the allocation of water use rights and thus shifts in burdens and benefits often warrant compensation in one way or another, to avoid a lack of legitimacy in newly developed policies.[8] Therefore, the allocation of water use rights calls for procedural and distributive justice.

2 APPROACH

Most legal literature available in English on water allocation stems from common law countries and/or former colonies of Britain.[9] This can be explained by the fact that a legal analysis is mostly based on legislation in the national language of the country subject to the analysis. This leads to an emphasis on discussions of riparian rights,[10] public trust,[11] a

---

[5] Mees, H., Dijk, J., van Soest, D., Driessen, P., van Rijswick, M. and Runhaar, H. 'A method for the deliberate and deliberative selection of policy instrument mixes for climate change adaptation' (2014) 19 *Ecology and Society*, 58.

[6] Termeer, C., Dewulf, A., van Rijswick, H., van Buuren, A., Huitema, D., Meijerink, S., Rayner, T. and Wiering, M. 'The regional governance of climate adaptation: a framework for developing legitimate, effective, and resilient governance arrangements' (2011) 2 *Climate Law*, 159.

[7] Keessen, A., Hamer, J., van Rijswick, H. and Wiering, M. 'The Concept of Resilience from a Normative Perspective: Examples from Dutch Adaptation Strategies' (2013) 18 *Ecology and Society*, 45.

[8] Haisman, B. 'Impacts of water rights reform in Australia' in Bryan R. Bruns, Claudia Ringler and Ruth Meinzen-Dick (eds), *Water Rights Reform: Lessons for Institutional Design* (International Food Policy Research Institute 2005).

[9] Getzler, J. *A History of Water Rights at Common Law* (Oxford University Press 2004).

[10] Evans, B. and Howsam, P. 'A critical analysis of the riparian rights of water abstractors within England and Wales' (2005) 16 *Journal of Water Law*, 90.

[11] Scanlan, M. 'A comparative analysis of the public trust doctrine for managing water in the United States and India' in Alistair Rieu-Clarke, Andrew Allan and Sarah Hendry (eds), *Routledge Handbook of Water Law and Policy* (Routledge 2017), 23–37 (hereafter *Handbook of Water Law and Policy*); Craig, R. 'Adapting to climate change: the potential role of state common law public trust doctrines' (2009) 43 *Vermont Law Review*, 781.

rights-based approach[12] and so on.[13] This chapter takes a different approach, looking instead at the ways in which the allocation of water use rights can contribute to a sustainable, balanced and equitable use of water resources. In turn, this may help the realization of the right to water and the achievement of the Sustainable Development Goals.[14] It builds on earlier work published in the *Journal of Water Law* in which Van Rijswick analyzed the allocation of water use rights in international, European and Netherlands law from a general public law perspective.[15] As this general public law perspective has a strong economic focus, she concluded that a different evaluation criterion is necessary for the assessment of allocation mechanisms for water use rights. The allocation of water use rights should also take social and ecological aspects into account.

In this chapter, we use a method to assess the governance capacity for water management[16] in order to discuss the ways in which allocation mechanisms have been developed in several selected States around the globe. We look at what policy instruments – such as planning, permits and priority systems – are in place and which are able to contribute to a sustainable, balanced and equitable use of water resources in order to implement the right to water and to achieve the Sustainable Development Goals. The focus lies on values, principles and policy discourses,[17] public participation or stakeholder involvement, policy instruments, enforcement

---

[12] Schlager, E. 'Getting the Relationship Right in Water Property Rights in Water Rights Reform: Lessons for Institutional Design' in Bryan R. Bruns, Claudia Ringler and Ruth Meinzen-Dick (eds), *Water Rights Reform: Lessons for Institutional Design* (International Food Policy Research Institute 2005), 27, 37–43.

[13] Tarlock, A. 'Water allocation and management during drought' in *Handbook of Water Law and Policy*, 149–61. Note that Tarlock also gives examples from non-common law countries. See also Dellapenna, J. and Gupta, J. (eds), *The Evolution of the Law and Politics of Water* (Springer 2009).

[14] See the following goals in particular, nr. 2: Zero Hunger, nr. 3: Good Health and Well-being, nr. 5: Gender Equality, nr. 6: Clean Water and Sanitation and nr. 12: Responsible Consumption and Production.

[15] van Rijswick, H. 'Mechanisms for water allocation and water rights in Europe and the Netherlands – lessons from a general public law perspective' (2015) 24 *Journal of Water Law*, 141.

[16] van Rijswick, M., Edelenbos, J., Hellegers, P., Kok, M. and Kuks, S. 'Ten building blocks for sustainable water governance: an integrated method to assess the governance of water' (2014) 39 *Water International*, 725.

[17] See for a comparison of values on the international, the EU and the Dutch national level: Ambrus, M., Gilissen, H. and van Kempen, J. 'Public Values in Water Law: A Case of Substantive Fragmentation?' (2014) 10 *Utrecht Law Review*, 8.

and conflict resolution. Other elements that are part of the assessment method are water system knowledge, authority and means, as well as engineering and monitoring. We discuss the tradeoffs by looking at how countries balance the different interests.

A distinction should be made between the allocation of water between States or regions, and allocation between stakeholders such as agriculture, industry, energy, ecology (environmental flows) and societal groups, including vulnerable and disadvantaged groups. We discuss water allocation between States and regions, looking at the international and European law approach as well as the allocation between regions or provinces within China.[18] We discuss the allocation within a State but between several stakeholders by looking more closely at allocation mechanisms in Indonesia, Kenya, the Netherlands (as an example of an EU country) and China. South Africa will be discussed because of its specific policies to re-allocate water use rights as part of its redress policies after the apartheid regime.[19]

3 WATER ALLOCATION BETWEEN STATES: INTERNATIONAL LAW

The management of transboundary watercourses addresses the allocation of water use rights between States. The UN Watercourses Convention[20] is the overarching legal framework. Furthermore, there are numerous multi- and bilateral treaties and conventions that govern transboundary watercourse management. They are too many to discuss in detail in this chapter. Therefore, we take the European level as an example. For European river basins, the UNECE Water Convention (Treaty of Helsinki)[21] is the most important. From February 2013, the Convention has been open for universal accession.[22]

---

18 van Rijswick (n 15).

19 Information used in these sections has been provided by researchers from the Utrecht Centre for Water, Oceans and Sustainability Law and Masters students from the Utrecht University Masters 'Water Science and Management' as well as the 'Legal Research Masters'.

20 UN Convention on the Law of the Non-navigational Uses of International Watercourses (1997) (entered into force 17 August 2014).

21 Convention on the Protection and Use of Transboundary Watercourses and International Lakes (1992) (entered into force 6 October 1996).

22 See UNECE TWC <http://www.zaragoza.es/ciudad/medioambiente/onu/en/detallePer_Onu?id=862> accessed 30 January 2017.

3.1 Values and Principles

General Comment No. 15 of the International Covenant on Economic, Social and Cultural Rights states that water should be treated as a social and cultural good, and not primarily as an economic good.[23]

At the international level, parties are obliged to ensure that transboundary waters are used in a reasonable and equitable way.[24] The UN Watercourses Convention adds that, in particular, an international watercourse shall be used and developed by Watercourse States with a view to attaining optimal and sustainable utilization thereof and benefits therefrom, taking into account the concerned Watercourse States' interests consistent with adequate protection of the watercourse. Watercourse States shall participate in the use, development and protection of an international watercourse in an equitable and reasonable manner. Such participation includes both the right to utilize the watercourse and the duty to cooperate in its protection and development, as provided in the present Convention. Thus, sustainable utilization and participation, as well as cooperation, are also core values at the international level.

The Convention on the Protection and Use of Transboundary Watercourses and International Lakes, in its Article 2, adds the aim of ecologically sound and rational water management, conservation of water resources and environmental protection as well as the conservation and, where necessary, restoration of ecosystems. It refers explicitly to some environmental principles such as the obligation that measures for the prevention, control and reduction of water pollution should be taken, where possible, at source; the obligation that a transfer of pollution to other parts of the environment should be avoided; the precautionary principle; the polluter-pays principle; the requirement that water resources should be managed so that the present generation's needs are met without compromising the ability of future generations to meet their own needs (the sustainability principle); as well as the obligation to cooperate based on equality and reciprocity.

To achieve these goals, allocation mechanisms are needed and should be developed by riparian States. Article 6 of the UN Watercourses Convention states the following: 'Factors relevant to equitable and reasonable utilization' that may be useful for further development of allocation criteria include

---

[23] General Comment No. 15 on the International Covenant on Economic, Social and Cultural Rights (1966).

[24] UN Watercourses Convention, art.5; UNECE Water Convention, art.2.

(a) geographic, hydrographic, hydrological, climatic, ecological and other factors of a natural character;
(b) the social and economic needs of the concerned Watercourse States;
(c) the population in each Watercourse State dependent on the watercourse;
(d) the effects of the use or uses of the watercourses in one Watercourse State on other Watercourse States;
(e) existing and potential uses of the watercourse;
(f) conservation, protection, development and economics of the use of the water resources of the watercourse and the costs of measures taken to that effect; and
(g) the availability of alternatives, of comparable value, to a particular planned or existing use.

The weight to be given to each factor is to be determined by its importance in comparison with that of other relevant factors. In determining what is a reasonable and equitable use, all relevant factors are to be considered together and a conclusion reached on the basis of the whole. The distribution should be implemented after consultations in a spirit of cooperation between the States involved and further regulated in agreements.

4 WATER ALLOCATION BETWEEN STATES: EUROPEAN UNION LAW

The UN Watercourses Convention and the UNECE Water Convention have been implemented by the European Union under the Water Framework Directive (WFD)[25] and the Floods Directive.[26,27] However, allocation and distribution are not regulated under European law, nor have there been any criteria developed for an equitable and sustainable distribution or use of freshwater among the various countries sharing a river basin.

---

[25] Directive 2000/60/EC of the European Parliament and of the Council of 23 October 2000 establishing a framework for Community action in the field of water policy [2000] OJ L327/1.

[26] Directive 2007/60/EC of the European Parliament and of the Council of 23 October 2007 on the assessment and management of flood risks [2007] OJ L288/27.

[27] van Rijswick, H. and Keessen, A. 'The EU approach for integrated water resource management: transposing the EU Water Framework Directive within a national context – key insights from experience' in *Handbook of Water Law and Policy*, 51–64.

The WFD only mentions a very general requirement – 'the control of abstractions within the general objectives of the WFD' – for the allocation at the national level. It is for the Member States to establish a water allocation process and to grant rights to abstract water, for example by way of licences or permits. However, Member States do not fulfill this obligation yet.[28] Therefore, it is necessary to take a specific example of how the allocation works at the level of an EU Member State. In this chapter, we use the Netherlands as an example.

4.1 Values and Principles

In EU legislation, one cannot find an explicit recognition of the right to water. This right has to be found in several pieces of legislation and case law.[29] Principles in European law can be found in the EU Treaties themselves or in the preamble or text of directives in the field of water management. Some principles apply to EU law generally, such as the principle of conferral. According to this, the Union may act only within the limits of the powers conferred on it and the principles of subsidiarity and proportionality which govern the exercise of Union competence (Art. 5 TEU). The subsidiarity principle also plays an important part in water management. Article 191 TFEU elaborates certain environmental law principles, and provides that Union policy on the environment shall contribute to preserving, protecting and improving the quality of the environment, protecting human health, and prudent and rational utilization of natural resources.[30] Union policy on the environment is based on the precautionary principle[31] and on the principles that preventive action should be taken, that environmental damage should as a priority be rectified at source and that the polluter should pay. These overlap to a great degree with the principles of the UNECE Water Convention. These EU principles are detailed in European legislation, as far as water is

---

[28] Report from the Commission to the European Parliament and the Council COM/2012/0670 on the Implementation of the Water Framework Directive [2012].

[29] van Rijswick, H. 'Searching for the Right to Water in the Legislation and Case Law of the European Union' in Henri Smets (ed.), *The implementation of the right to safe drinking water and sanitation in Europe: Rapport préparé pour le 6ème Forum Mondial de l'Eau* (L'Académie de l'Eau 2012), 87–113.

[30] Macrory, R. (ed.), *Principles of European Environmental Law* (Groningen, Europa Law Publishing 2004); Jans, J. and Vedder, H. *European Environmental Law* (Europa Law Publishing 2008), 35–46.

[31] See Communication from the Commission COM(2000)1final on the precautionary principle [2000].

concerned, largely by means of directives. In the European Union, water has a special status, being 'not a commercial product as any other, but rather a heritage that must be dealt with and defended as such' (preamble to the WFD).

4.2 Stakeholder Involvement

EU law requires that individuals must be involved in the formation of water policy, they must have access to the necessary relevant information and the States must supply this information. Individuals have a right to participate in the decision-making process and to judicial protection after the decision has been taken, to the extent that the legislation confers rights on individuals. This follows from the Aarhus Convention and its implementation in EU law[32] as well as from specific requirements in Article 14 of the WFD.

4.3 Policy Instruments: Regulations and Agreements Including Planning and Permits

The WFD requires that the Member States collaborate, draw up transboundary river basin management plans together, confer with each other and exchange information in order to be able to meet the goal of a sustainable, balanced and equitable use within the entire river basin. However, it does not stipulate the criteria or allocation mechanisms on the basis of which water is to be distributed among the States. Van Kempen, amongst others, states that, although a reasonable and equitable use is one of the main goals for transboundary water management stemming from international water law, European Union water legislation is not clear what reasonable and equitable use or distribution actually means. The concept is not defined in either legislation or jurisprudence. He suggests using criteria set by the UN Watercourses Convention.[33]

Member States can make use of treaties, new or existing, or adopt informal methods of collaboration, in order to agree the transboundary

---

[32] Council and European Parliament Directive 2003/4/EC of 28 January 2003 on public access to environmental information [2003] OJ L041/26; Council and European Parliament Directive 2003/35/EC of 26 May 2003 providing for public participation in respect of the drawing up of certain plans and programmes relating to the environment [2003] OJ L156/17.

[33] van Kempen, J. '*Europees waterbeheer: eerlijk zullen we alles delen*?' (Boom Juridische Uitgevers 2012); van Kempen, J. 'Good River Basin Management in Europe: Distribution and Protection' (2014) 4 *Water Governance*, 31.

allocation, distribution and regulation of water rights. However, making use of international treaties does not release Member States from their individual obligation to meet the objectives of a directive.[34] Ultimately, the relationship between European directives and international treaties is not fully clear.[35]

In the European approach, there is a Communication by the European Commission on water scarcity and droughts. This emphasizes the importance of pricing water use, allocating water efficiently, managing drought risk, integrating water interests in other policy fields, efficient water technology, stimulating saving water, improving knowledge and, only as a last resort, building water infrastructure works. Water allocation and dealing with water scarcity issues should be tackled mainly within the regime of the WFD, but they have not been further developed in EU legislation.

4.4 Enforcement and Conflict Resolution

Agreements for transboundary river basin management between States can be enforced in a number of ways. Disputes concerning the interpretation or application of European law are heard before the European Court of Justice. Otherwise, international arbitration is an obvious option, which is also applicable where the interpretation or application of a treaty to which not only the Member States but also the European Union is a party, such as the Treaty of Helsinki.[36]

The WFD (like the Floods Directive) provides for arbitration and mediation by the European Commission. Member States can request the Commission to act as mediator. Although the Directive obliges the Commission to respond, it cannot make a binding decision because the Directive makes no provision for this. On the other hand, the Commission can decide that the conflict gives cause for initiating treaty infringement proceedings before the Court of Justice. However, given the fact

---

34 Suykens, C. 'EU Water Quantity Management in International River Basin Districts: Crystal Clear?' (2015) 24 *European Energy and Environmental Law Review*, 134; Gilissen, H., van Kempen, J. and van Rijswick, H. 'The need for international and regional transboundary cooperation in European river basin management as a result of new approaches in EC water law' (2010) 11 *ERA Forum*, 129.

35 Hey, E. and van Rijswick, H. 'Transnational water management' in Oswald Jansen and Bettina Schöndorf-Haubold (eds), *The European Composite Administration* (Intersentia 2011), 227.

36 Case C-459/03 *Commission v Ireland* [2006] ECR I-04635.

that there are no criteria for the distribution or restriction of the right to use water, it is not clear on what the Court could then rule. Member States can initiate treaty infringement proceedings against each other, but they seldom do so.

Another possibility lies in allowing national authorities to take legal action in another Member State, for example in a dispute concerning the granting of a permit for water abstraction. However, there is no obligation for national authorities to take cross border legal action under national law.[37] The question thus arises as to whether in any given instance it is the central or the local authorities that have access to the courts. The answer will depend in particular on the national procedural law of the Member State in question.

Under the Aarhus Convention, local authorities may have opportunities to gain access to the courts in all EU Member States and in other countries that are parties to the Convention if they can demonstrate that they have a sufficient interest or that a right has been infringed. However, unlike environmental organizations, local authorities do not have a privileged position. Thus, access to justice before the administrative courts in another country is not necessarily certain.[38]

5 WATER ALLOCATION AT THE REGIONAL RIVER BASIN LEVEL: THE BRANTAS BASIN, INDONESIA[39]

Water policy in Indonesia considers several factors such as population growth, contribution of each sector (agriculture, drinking, and environment), and the potential of conservation of watercourses.[40] The Brantas basin is one of the most studied areas within the subject of water allocation in Indonesia. Furthermore, the basin is regularly used as the

---

[37] Case C-178/97 *Banks* [2000] ECR I-2005.

[38] Keessen, A., van Kempen, J. and van Rijswick, H. 'Transboundary river basin management in Europe, Legal Instruments to comply with European water management obligations in case of transboundary water pollution and floods' (2008) 4 *Utrecht Law Review*, 35.

[39] This paragraph is based on the research of Annisa Noyara Rahmasary and Merlijn van Selm, 'Water Allocation Governance in the Brantas Basin Indonesia: Assessment Using The Ten Building Blocks' (Utrecht University 2016).

[40] Sosiawan and Subagyono, 'Strategi PEembagian Air Secara Proporsional Untuk Keberlanjutan Pemanfaatan Air' (2009) 2 *Pengembangan Inovasi Pertanian*, 299.

implementation target of integrated water resources management (IWRM), which pursues comprehensive basin development in a decentralized manner with the support of local knowledge and public participation.[41] The basin is located in the province of East Java, Indonesia. Water diversion from the river into canals supports agricultural activity in the vicinity, the Porong River acts as flood diversion during the rainy season and the Surabaya River serves as primary source for raw water for the City of Surabaya.[42] The principal water uses in Brantas are irrigation (70–80 per cent of utilized water abstraction), domestic water supply, industrial processes, hydroelectric generation, and brackish fish or shrimp ponds.[43] Economically, the basin is perceived as having a potentially great benefit if it is properly managed. Therefore, the area itself has become one of national interest with many development and management strategies applied, including for water allocation governance.

5.1 Values and Principles

The Indonesian constitution of 1945 states that the earth, water, and any wealth therein are governed by the State and utilized as far as possible for the welfare of the people. This places rights and responsibilities of water resource management with the State. The State shares this authority with local governments (in accord with decentralization policy) and other (private) parties. As a result, the implementation of a water allocation system in one region may differ from that in others. Over the years, the paradigm of water as a (free) social good has shifted to water as economic good with a social function. The principles of public utility, harmony, and conservation are widely accepted in Indonesia. Others include the principle of universality (in contrast to exclusivity) of property rights.[44] Values based on Islam, which is the majority religion in Indonesia, explain that water is a gift from God that needs to be conserved collectively and shared based on a basic need priority.[45]

---

[41] Bhat, A., Ramu, K. and Kemper, K. 'Institutional and policy analysis of river basin management: the Brantas River Basin, East Java, Indonesia' (2005) World Bank Policy Research Working Paper 3611.

[42] Sarwan, S., Subijanto, T.W. and Rodgers, C. *Development of water rights in Indonesia. Lessons for Institutional Design* (2005), 237; Bhat *et al.* (n 41).

[43] Bhat *et al.* (n 41).

[44] Sarwan *et al.* (n 42).

[45] See also for an Islamic approach Misiedjan, D., van Rijswick, H. and Kwoei, A. 'A human right to water while the well runs dry: analysing the legal

Meanwhile, decentralization and democratization allow people and NGOs to monitor implementation processes of any government act.[46]

In general, prevailing values, principles, and policy discourses in Indonesia are affected by the history of governance, in which the New Order regime provoked distrust toward the government during 1966–1998.

5.2 Stakeholder Involvement

At the national level, several ministries and agencies are responsible for the water allocation system. They participate in the basin level through their department representatives at the province and regency level. In the Brantas basin, Balai Pengelolaan Sumber Daya Air, Water Resource Management Agency (BPSDA) is authorized to manage the river basin, including having responsibility for existing data related to water resources, while Perusahaan Daerah Air Minum, Drinking Water Regional Company (PDAM) organizes the supply of treated drinking water. State-owned water management company PJT-I is the main stakeholder, given the biggest responsibility for operating and maintaining infrastructure and also other management activities related to existing water resources. It is a public corporation that is appointed to manage five large basins in Central and East Java Province.[47]

Most of the Brantas Basin management programs are initiated and funded by the Japan International Cooperation Agency (JICA), which cooperates with government officials to track the involvement of farmer communities in the basin.[48] Other involved stakeholders are The World Bank and the Asian Development Bank (ADB), which financially support these projects in the development of water allocation mechanisms. In the Brantas area, farmers have become significant stakeholders, as their water use is the largest within the allocation. There are a number of water

---

and regulatory framework of Yemen water law' (2015) 24 *The Journal of Water Law*, 199; Huntjens, P. 'The Political Economy of Water Management in Yemen: Conflict Analysis and Recommendations' (2014) The Hague Institute for Global Justice.

[46] Bakker, L., Nooteboom, G. and Rutten, R. 'Localities of value: ambiguous access to land and water in Southeast Asia' (2010) 38 *Asian Journal of Social Science*, 167.

[47] Sarwan *et al.* (n 42).

[48] Jamadar, N. 'The Community and Beneficiaries' Participation in Comprehensive Water Resource Management' (2015) 49 金沢星稜大学論集 41.

user associations (WUAs) where farmers can register to be involved in irrigation management.[49]

The involvement of these stakeholders is generally well-executed, mainly due to the fact that the cooperation of PJT-I and JICA provides platforms to communicate between stakeholders. However, not all of the farmer groups in the various smaller areas in the Brantas basin are committed continuously to getting involved in the existing water allocation system. Jamadar (2015) highlighted that the farmer groups from the areas of Blobo, Lodoyo, Warujeyang, Widas and Turi Tunggorono had decreasing engagement in the regular meetings of beneficiaries. In addition, there is no fishery group available in existing WUAs. Nevertheless, Rahmasary and Van Selm consider overall participation by the various stakeholders as adequate in many aspects.

5.3 Policy Instruments: Regulations and Agreements Including Planning and Permits

In general, the water allocation in the Brantas basin is based on the predetermined *Master Plan*, where water allocation priority goes to agriculture, domestic use, industry, and brackish ponds. Water for irrigation is considered an important social objective due to that sector's importance. Although we see a shift in the Master Plan's emphasis, the water allocation is made based on historical practices where the agriculture sector was given the highest priority.[50] Agriculture becomes even more significant in the Brantas basin because 32 per cent of the rice production in East Java comes from this area.[51] PJT-I annually reviews the prediction of water use calculation of each of the main beneficiaries. There is a practical problem in Java because middle-class farmers, who own agricultural land, are not able to fulfil their required irrigation measures.[52] In the contemporary management system, many farmers share water that is often stored and delivered from remote locations via public infrastructure, and use it for a wide variety of purposes. The risk that poor farmers would lose access to water vital to their livelihoods has been a major concern in water law debates in Indonesia.

---

49 Bhat *et al.* (n 41).

50 The World Factbook CIA (2016) <https://www.cia.gov/library/publications/the-world-factbook/geos/id.html> accessed 30 January 2017.

51 Bhat *et al.* (n 41).

52 Sosiawan and Subagyono (n 40).

Besides the farmers, other beneficiaries that use water are obliged to pay for each water service. For a long time, this was accepted as a social norm supported by regulations; Law No. 11/1974 and Law No. 7/2004. Currently, PJT-I is striving to build a 'beneficiary-pays' system that considers the social settings of the target group, such as poverty, low income, uncertainty, etc. This is confronting the mindset of farmers, who retain the old belief that the State should supply water to them freely.[53]

At least four Master Plan phases that are recorded as a part of the implementation of the Brantas River Development Project show a shift from flood control, in which reservoirs were constructed with the secondary function to provide irrigation and hydropower (1961), irrigation (1973), and support of industry and municipalities (1985), towards an emphasis on conservation and management (1998).[54] In line with allocation for multiple purposes, the idea of water use rights has emerged as a formal procedure to grant the right to use water from the public ownership to a user that the government licences. Sarwan et al. proposed that transferable water rights could lead to more efficient use of water, yet could also trigger water abuse.[55] The water sharing principle (i.e. water rights for everyone) is governed by Law No. 7/2004, but this law lacks protection for lower class societal groups, and promotes privatization. A framework of community and beneficiary participation in the Brantas Basin has been implemented to monitor the performance of comprehensive water management in the area.[56]

In addition to the Constitution of 1945, the newest legislation in the field of water management and allocation was Law No. 7/2004 which governs two distinct water rights: *Hak Guna Pakai Air* (usage for basic needs and non-commercial uses) and *Hak Guna Usaha Air* (commercial exploitation). However, this law has been revoked and Law No. 11/1974 on Water Resources is in place again. Corporations, associations, and individuals need government approval (a permit). It is the primary responsibility of central and local governments to ensure that water and water resources are used beneficially. The water law also requires that water allocation is subject to established priorities.

PJT-I has to pay a concession fee to the national government. It generates revenue from water sales to industrial users, from the energy generated in hydropower plants and through sales to water suppliers.[57] It

53 Jamadar (n 48).
54 Bhat *et al.* (n 41); Sarwan *et al.* (n 42).
55 Sarwan *et al.* (n 42).
56 Jamadar (n 48).
57 Bhat *et al.* (n 41).

does not receive any revenue from farmers. The farmers pay a relatively small Irrigation Service Fee (ISF) to the local water utility to contribute to the costs of distributing irrigation water. This ISF is related to the land area and is approximately $1.5 per hectare for rice fields.[58] The larger users, industrial or big agricultural firms, pay a volumetric rate and pay indirectly for the costs incurred by PJT-I in supplying small-scale farms with water. However, the system has a double edge: Small-scale farmers pay no volumetric fee for their water, but they are the lowest in priority when water is scarce.[59]

The cost-recovery system is based on the solidarity principle, in which the community and richer users pay the costs for financially weak small-scale users. The fact that small-scale farmers, usually the most vulnerable when compared with larger companies, are the first to experience cuts in water supply is certainly not equitable and is arguably undesirable.

5.4 Enforcement and Conflict Resolution

The amount of public participation is increasing in Indonesia, increasing support for the enforcement of rules. The small-scale farmers that use the most water are not adequately represented. However, farmers take steps to organize themselves in water user associations that take a seat in the regional councils.[60] PJT-I and JICA have also initiated a participatory approach to monitor farmer involvement. The legislation governing water is generally consistent with important values and principles in Indonesia, by defining water as a public good and taking into account the needs of vulnerable groups like farmers that receive water for free. In short, the participatory approach in the Brantas Basin facilitates enforcement. Through the system of master plans and the existence of PJT-I, the whole policy process from goal-setting to the actual achievement of goals is covered. Legislation is still imposed upon the basin by national government, but implementation of the policy is secured through detailed master plans and decentralization. The fact that there is management at the basin level also contributes to coherent and complete policy-making.

A complicating factor in enforcement, however, is the law itself. The newest law in the basin, Law 7/2004, was recently revoked and the

---

58 Rodgers, C. and Hellegers, P. 'Water Pricing and valuation in Indonesia: case study of the Brantas River Basin' (2005) International Food Policy Research Institute.

59 Ibid.

60 Bhat *et al.* (n 41).

current legislation is outdated and no longer sufficient. As for the laws that are clear, not all are enforced due to the importance given to economic growth and corruption.[61] Explicit policy to prevent conflicts is not in place in the basin. The integrated water resources management and the participatory approach, which tries to incorporate every stakeholder in the process, is an important first step to avoid conflicts. Changing the water allocation drastically to achieve a financial optimum would stir significant social turmoil, since a large number of people would no longer be able to generate sufficient income. This needs to be considered when putting an economic label on water. If impacts, such as those on the livelihood of small-scale farmers, and socio-political situations were incorporated in water prices, optimization of water allocation from an economic point of view could be an option.

6 WATER ALLOCATION AT THE NATIONAL LEVEL: KENYA[62]

Kenya is classified as a water scarce country with 670 m3 of available renewable freshwater per capita, less than the 1000 m3 per capita recommended by the United Nations.[63] Water resources endowment are unevenly distributed both in space and time. The bulk of the renewable water resources is derived from rainfall. Also, the country experiences repeated droughts and floods that occur in a cycle of three to four years. Because of the ever increasing demand and competition for available water resources, water use conflicts in the country have escalated in recent decades. The principle reason for regulating water use is to secure a fair portion of the accessible water resources for different contending needs in a sustainable way.[64] The history of Kenya's water governance dates back to 1927, when the first water-related law was the Water Ordinance 1927. After independence came the Water Act 1974 and currently the Water Act of 2002 is in force.

---

61 Ibid.

62 Paragraph based on the research of Bianca Mogos, Irene Akinyi Otieno and Rebecca van Weesep, 'Water allocation in Kenya' (2016) Utrecht University.

63 UN World Water Development Report, 'Case study: Kenya, Kenya National Water Development Report' (2006). Retrieved via <http://unesdoc.unesco.org/images/0014/001488/148866E.pdf> accessed 30 January 2017.

64 Akech, M. 'Governing Water and Sanitation in Kenya: Public Law, Private Sector Participation and the Elusive Quest for a Suitable Institutional Framework' (2009) Conf. International Environmental Law Research Center.

6.1 Values and Principles

The human right to water is relevant to water access in water governance. Kenya's water governance has evolved incrementally with the focus based on the human right to water and an equitable share of water resources. The focus of the water sector reforms (the Water Act of 2002) lies with the concept of equitable water allocation, which should take into account environmental, economic and basic needs. The right to water is not explicitly provided for in the Water Act 2002 but is explicitly recognized in the Kenyan Constitution. Article 2(6) of the Constitution recognizes the right to water as a constitutional right and in Article 43(1)(d) as a socio-economic right. The right to water was also later formally recognized in other national water sectors policies (National Water Service Strategy, National Water Resource Management Strategy, Water Services Regulatory Board tariff guidelines and the Pro-Poor Implementation Plan). The underlying principles are that the water resources are vested in the State; that water resources are finite and vulnerable; that allocation may pose risks to the environment, economy, and social well-being; and that the precautionary principle should be applied. Other principles include a command of highest priority for basic reserves and public notification to provide opportunity for stakeholder participation and to ensure transparency. Water use conflicts can be a source of social and economic disruption, and equity and efficiency should be the rationale for giving priority to water use with highest public benefit. The main issue is the implementation of the already established policies of the water system. In particular, the National Water Master Plan 2030 takes into account demand and effects of climate change in the water system and also aims to strengthen institutional capabilities. The ministry has progressively enacted measures to address the human right to water and sanitation. More explicitly, the National Water Services Strategy (NWSS) 2007 and the Pro-Poor Implementation Plan (PPIP) 2008 embrace human rights approaches in the provision of water services, calling for citizen participation and empowerment.

6.2 Stakeholder Involvement

The Water Act of 2002 recognizes, and has enhanced, the involvement of stakeholders in the provision and management of water resources (section 16, subsection 3). These include contact with government ministries and public institutions.[65] Other stakeholders are tourism operators, fishermen, pastoralists, farmers, research institutions, NGOs, development partners/ international organizations, community managed water projects and organizations, and citizens.

Citizen participation has been shown in various ways through which the various stakeholders try to involve the citizens in their decision-making.[66] Though citizen participation is recognized in the Water Act through community groups organized through the Water Resources User Associations (WRUAs) and at individual levels, the recognition does not mean that they are always given an opportunity to participate in decision-making. Generally, stakeholder involvement is recognized but there is no sufficient balance of interests, values and concerns, especially in the decision-making process. For marginalized groups and stakeholders at the lowest institutional levels, seeking to redress injustices or inequalities in the water sector is an inherent feature of struggle because, in most cases, they only receive notifications rather than being involved.

6.3 Policy Instruments: Regulations and Agreements Including Planning and Permits

The primary aim of the 1974 National Water Master Plan (NWMP) was to ensure availability of potable water within a reasonable distance of all households by 2000. The legal frameworks are the Water Act 1974 followed by the Water Act 2002 and Chapter 372 of the Laws of Kenya. The main focus of the Water Act 2002 was to decentralize the water system and separate water policy formulation from regulation and service provision. It also introduced commercialization of water resources as part of the decentralization process and stakeholder participation in the management of water resources.[67]

The NWMP identifies several frequent problems, such as weak enforcement of law and an insufficient amount and inadequate allocation of financial and human resources at national and regional levels.[68] The current decentralization of water management is likely to improve overall

---

65 Draft National Irrigation Policy (2015) Ministry of Agriculture, Livestock and Fisheries.

66 Moraa, H., Otieno, A. and Salim, A. 'Water governance in Kenya: Ensuring Accessibility, Service delivery and Citizen Participation' (2012) iHubResearch Nairobi.

67 Moraa *et al.* (n 66).

68 National Water Master Plan (2012); Progress Report 4 on the development of the National Water Master Plan 2030, Ministry of Water and Irrigation

water allocation and management, but is currently not well-organized and integrated enough to achieve this purpose. The Water Act 2002 section 8(1)(a) mandates the responsible authority to develop guidelines, principles and procedures of water allocation. The guidelines and procedures focus on the main considerations that affect the water allocation, while the principles guide the formulation of policy and management. As identified by WRMA 2009, the basis for water allocation consists of the underlying priorities: planning of water allocation to support long-term planning of water resource management, management of allocation and prioritization, re-allocation and proportional allocation, and efficiency, compliance and demand management.

The water resources in Kenya are insufficient to meet demand, so there is the need to allocate resources in a manner that serves both public and ecological interests. This need calls for prioritization of the allocation system to circumvent substantial risks to the environment, economy, and social well-being. Although it is mentioned in many policies (i.e. National Water Policy of 1999, the Water Act 2002 and Water Resources Management Rules (2007)), water resources allocation in the country is still not clear.[69] Two approaches can be said to form the basis of the allocation: 'public benefits' and so-called 'explicit approaches'. Where priority is given to the water use with higher 'public benefits', the social, economic, environmental and commercial returns per quantity of water are therefore acceptable criteria in prioritizing water for allocation.[70] The 'public benefits' are not directly defined in the Water Act. In section 19 of the Water Act, 'public purpose' is defined as 'the supply of water, or electrical energy derived from the energy of the moving water, to the public or any section of the public'. Still, in sections 20 and 21, the state scheme is given a special status, precedence and priority above all other schemes of water use. This 'public benefit' priority concept when defined as in section 19 of the Water Act can be said to be in conflict with the principle of equity that calls for fairness in the allocation process. The

---

and the Water Resources Management Authority in cooperation with the Japanese International Cooperation Agency. Retrieved via <http://www.wrma.or.ke/images/presspdf/ProgressReport_4_%20NWMP%202030.pdf> accessed 30 January 2017.

[69] Draft National Irrigation Policy [2015] Ministry of Agriculture, Livestock and Fisheries.

[70] Water Sector Reform in Kenya and the Human Right to Water [2007] Ministry of Water and Irrigation. Retrieved via <http://www.worldwaterweek.org/documents/WWW_PDF/Resources/2009_17mon/Kenya_Water_Human_Rights_brochure-final.pdf> accessed 30 January 2017.

explicit approach as a method of allocation of water resources and the right to water is mainly by use of permits. The permit provision bases are provided in the water rights and works from sections 25–43, and also in the Water Bill 2014.

Land ownership is related to water rights.[71] As the permit provision is also premised on a land ownership system, it marginalizes some stakeholders, especially the rural poor and the pastoralists, in the acquisition and exercise of the right to water use. Inadequate water supply has clear gender dimensions too, especially in the rural areas where women are tasked with the responsibility for collecting water. In this sense, the right of water can be recognized as a crucial component of gender equity and female empowerment.

Values and principles can conflict. Therefore, re-evaluation and clear formatting of policies should take existing policy discourses into account, because these discourses might bring about conflicts in the allocation system. There is a good base of rules and guidelines for water allocation in Kenya: the Water Act (2002), the Water Resources Management Rules (2006), and the Final Water Allocation Guidelines (2009). However, the regulations lack legal certainty, which causes problems in enforcement. It is, for example, not clear on which criteria permit requests for the use of a water resource are assessed. According to the WRMA (2009), the allocation of water is based on four demanding priorities (ecological need, basic human needs, international obligations and international basin transfers; all other water uses are authorized according to the criteria of equitable allocation). Water outside the reserve is allocated using a permit system which is regulated by the WRMA. Thus, water use rights are arranged by the government. Correspondingly, a system of permits for water issues has been issued by the WRMA. The permits system is designed in four categories, from A to D, where A is the lowest water demand and D the highest. It is based on demand and availability of resources, taking into account the reserve of water that is required for basic human needs and the environment. However, water should not be abstracted before the permit is obtained, even though this is common practice, which makes enforcement of this condition very difficult. Another downside of this system is that it reinforces the predominance of landowners when it comes to the use of water resources, as one must be a landowner to apply for permits. Therefore, through the Water Act, collectivities in poor rural areas are marginalized, and disadvantaged in

---

71 Moraa *et al.* (n 66).

using the available water resources efficiently, while this also influences the effectiveness of the implementation of the new water law.[72]

When water resources cannot meet demand, the approach of the WRMA in water allocation is characterized by prioritization, re-allocation and proportional allocation. For prioritization, the reserve (i.e. basic human needs and ecological need) always has priority. According to section 32(1) of the Water Act (2002), of all other types of water demands, domestic water has a higher priority. Furthermore, the Water Allocation Guidelines (2009) state: 'With respect to all the other types of demand, the Water Act 2002 is silent with respect to priority, although various considerations must be made'. These considerations consist, for example, of efficiency, public benefit and potential impacts on other water users and resources.

With respect to reallocation and proportional allocation, the Water Act (2002) 'essentially ... provide[s] the Authority [WRMA] with the powers to review and amend or cancel a permit or permits where there is reasonable cause so as to promote sustainability, equity, public beneficial interests, efficiency and to protect the water resource quality' as summarized by the WRMA (2009). This shows that the underlying values and principles (e.g. equity and sustainability) are reflected in regulations and agreements. However, the fact that there are no precise regulations on how water resources are further allocated and the authorization of permits shows that the rules are not fully transparent. It is not clear on what criteria permit applications are assessed and how water quantities are determined, which testifies to a lack of legal certainty. Therefore, problems arise in the permit system regarding enforcement.

Even though the legislation concerning the main infrastructure is set by the Water Act and the NWCPC is legally responsible for it, the distribution system and the connections to households do not seem to be regulated in any way, and communities have to rely on themselves.[73]

6.4 Enforcement and Conflict Resolution

The Water Act (2002) is the legal basis in Kenya regarding management, conservation, use and control of water resources. Stakeholder associations, called Water Resources Users Associations (WRUAs), have been

---

[72] Mumma, A. 'Kenya's New Water Law: an Analysis of the Implications of Kenya's Water Act, 2002, for the Rural Poor', in Barbara C.P. Koppen, Mark Giordano, John Butterworth (eds), *Community-based Water Law and Water Resource Management Reform in Developing Countries* (CABI 2007).

[73] Mumma (n 72).

created to enforce the rules. WRUAs use a set of guidelines that ensure that objective and suitable decisions are taken regarding water allocation issues. They appeared in 2005 and their numbers are increasing by the year. Their legal status is obtained through registration at the office of the Attorney General.[74] The Water Service Regulatory Board has the obligation to monitor the compliance of service providers, using the norms regarding human rights to water and sanitation. This monitoring will be done according to standards and developed indicators. These indicators should try to provide an overview of the efforts made by the Government. However, if the rights are not met, authorities can be held responsible.[75] Due to the above-mentioned inconsistencies, the enforcement system is inefficient, a fact addressed by the government itself in 2006, in the National Water Development Report. The laws are inefficient, illegalities and conflicts between citizens are still ongoing, water has not been sustainably used, and water allocation practices are weak. Therefore, stronger enforcement is necessary to ensure compliance and reduce the number of water abstractions without authorization.

Because the user pays principle is applied in Kenya, it is challenging to implement policies because people view water as a social good, not an economic one. Also, conflicts might arise concerning the ownership of community water systems, due to the poor implementation of the law.[76] To deal with these issues, according to the Water Act (2002), the WRUAs' role is also to help resolve conflicts that might arise.[77] WRUAs, together with other associations, can act as a forum for dispute resolutions and cooperative management of water resources. Where associations do not exist, they need to be formed.[78] It can be said that the conflict-solving means are rather vague, and rely mostly on the community, and they are a result of cultural views and poor legal management. Even though the stakeholder associations are involved, which is a good base for conflict solving, a stronger legal basis needs to be created, adhering to the provisions of the Water Act (2002) and the National Water Resources Management Strategy (NWRMS).

---

[74] Water Resources Management Authority, 'WRMA Performance Report 2' (2012).

[75] Mumma (n 72).

[76] Mumma (n 72).

[77] WRMA Performance Report (n 74).

[78] Mumma (n 72).

7 WATER ALLOCATION AT THE NATIONAL LEVEL WITHIN THE EUROPEAN UNION: THE NETHERLANDS[79]

The Netherlands is a water-rich delta in North-West Europe. In this chapter, it serves as an example of a Member State of the European Union, which means that the allocation mechanism has to comply with EU legal requirements. Drinking water is sufficiently available to all, affordable and of high quality. However, the chemical and ecological statuses of waters are suboptimal. The Netherlands does have sufficient freshwater, but not always at the right time or in the right place. In dry periods, certain areas of the country can suffer shortage, which adversely affects safety in relation to flooding, drinking water and energy supplies, nature and economic functions. The increasing salinity of freshwater is an even greater problem. The new problem of water scarcity confronts the Netherlands, and the country has to learn how to deal with this. The challenge will be to further develop mechanisms for a fair allocation, based on a sustainable use and solidarity. Due to population density there are many water users, such as individuals, agriculture, drinking water companies, energy supply, industry, tourism, shipping, and fishing.

7.1 Values and Principles

In the Netherlands, water is a public good. There is no enforceable classic fundamental or human right to water.[80] The authorities' responsibility for water management is enshrined as a social fundamental right in Article 21 of the Constitution, which provides that it shall be the concern of the authorities to keep the country habitable and to protect and improve the environment. Water management relates – as is apparent from the above description of the public responsibility – to both aspects of this fundamental right. The authorities' responsibility is worked out in more detail in legislation, principally in the *Waterwet* (Water Act). Solidarity is an important value,[81] but goes hand in hand with a well-developed system of cost recovery of water services based on the

---

[79] See van Rijswick (n 15).
[80] van Rijswick (n 29), 377.
[81] Keessen, A., Vink, M., Wiering, M., Boezeman, D., Ernst, W., Mees, H., Van Broekhoven, S. and Van Eerd, M. 'Solidarity in water management' (2016) 21 *Ecology and Society*, 35.

polluter pays principle.[82] There are hardly any principles found in legislation, but EU legal principles are relevant for Dutch water law too. Since water scarcity is a rather new phenomenon, people still think that they have an unlimited right to (almost) free water and rely on the government to supply clean, fresh water as well as drinking water. However, discourses regarding water allocation are changing, with a greater role for shared responsibilities of public and private parties.[83]

7.2 Stakeholder Involvement

Planning and decision-making has a long tradition of stakeholder involvement. The Dutch 'polder-model' is famous and aimed at consensus building.[84] Stakeholders are allowed to participate in the preparation of the plans, a right that emanates directly from European water directives and the directives that implement the Aarhus Convention but were already implemented in Dutch water law and in general administrative law. The same goes for the application for permits. The regional water authorities have a long tradition of involving all stakeholders in their planning and decision-making procedures, not only in a formal way but even more in an informal yet rather effective way.[85] In general, citizens are not much interested in participation, which can be explained by the fact that water management and governance in the Netherlands are of a very high standard. This causes a new problem, which the OECD mentioned as a serious awareness gap for water issues.[86]

---

[82] van Rijswick, H. and Havekes, H. *European and Dutch Water Law* (Europa Law Publishing 2012); OECD, 'Water Governance in the Netherlands, Fit for the Future?' (2014).

[83] See <http://www.deltacommissaris.nl/deltaprogramma/inhoud/gebieden-en-generieke-themas/zoetwater> accessed 30 January 2017.

[84] OECD (n 82).

[85] Keessen, A., van Daalen, L. and van Buuren, A. 'Participatie in watermanagement – goed voorbeeld doet volgen?' (2014) 89 *Nederlands Juristenblad* 2117; Reinhard, S. and Folmer, H. (eds), *Water Policy in the Netherlands, Integrated Management in a Densely Populated Delta, Issues in water resource policy* (RFF Press 2009).

[86] OECD (n 82).

7.3 Policy Instruments: Regulations and Agreements Including Planning and Permits

In the Netherlands, the distribution and licensing of abstraction rights is based on planning, a permitting system and a priorities list that applies only in times of water scarcity.

A National Water Plan is drawn up by the Ministers of Infrastructure and the Environment (I&E) and of Economic Affairs (EA).[87] Section 4.7 of the Water Decree stipulates that the National Water Plan must state which national waters are suitable for the 'abstraction of drinking water'. Provincial Executives draw up regional plans that also are part of the implementation of EU river basin management plans (RBMPs) and programmes of measures showing which regional waters can be used as drinking water.[88] The Water Act allows additional functions to be allocated in the management plans of central government and the water authorities.[89] The designation of functions (shipping, drinking water, bathing water, water for industrial uses and energy plants, agricultural use, etc.) is of particular importance in the allocation and distribution of abstraction permits. This is because the function determines not only which objectives apply to specific waters and how these objectives will, through the adoption of measures, be achieved, but also who are the appropriate claimants to the use of the water. The plans therefore lay down how the water will be distributed among the various activities, although this is implied rather than clearly stated. It has to be deduced from the combination of objectives and function allocation how the water will be distributed. It is the government that distributes water from rivers and lakes across the country by means of channels and other water works.

For certain abstractions, a permit is required. Under section 6.4 of the Water Act, a permit is required for the abstraction of large quantities of groundwater, the competent authority being the Provincial Executive. For smaller amounts of groundwater, the competent authority is the regional water authority. Abstractions from a body of surface water require a permit from the Minister of I&E or the water authority (s 6.5 Water Act). Only in times of (potential) shortage does the list of priorities (below) overrule permit holders' rights. Reviewing a permit application for conformity with the statutory grounds for refusal is seen as a permitted distribution criterion. In the Netherlands, water abstraction permits are

---

[87] Water Act, section 4.1 and Water Decree, section 4.1.

[88] Water Decree, section 4.12(1)(b).

[89] Water Act, section 4.6(2)(b).

granted on a 'first come first served' basis and, notably, have no time limit. Conditions and restrictions can be attached to the permits and contravening the conditions is prohibited.[90] The granting of permits is further subject to the normal requirements of administrative law based on the General Administrative Law Act (GALA) and principles of good governance.[91] There is also a regime of general rules applicable to a large number of users where a permit is not required provided that the users report their water use to the competent authorities. Once a general regulation is in force, everyone has equal rights and has access to water to an equal degree.

The Water Framework takes the approach that cost recovery for water services is an important instrument to contribute to a sustainable, balanced and equitable water use. However, the Dutch financial arrangement does not recover all costs for all water use. Allocation is certainly not equal. Every Dutch citizen pays for drinking water in proportion to their consumption. However, for water used for other purposes, there is no direct relationship between the amount of water used and the tax levied by the water authorities. Use of waters managed at the State level is free of charge.[92] The use of fresh water in agriculture for irrigation or watering cattle and the use of a body of water for shipping is free of charge. The judgement of the European Court of Justice in a case against Germany seems to approve this approach, as the Court ruled that Member States have wide discretion in how they recover the cost for water services (Article 9 WFD).[93]

The Water Act contains a list of priorities for the distribution of fresh water in the event of a drought. Section 2.9(1) stipulates that the social and ecological priorities that will determine the distribution of the available surface water in the event or threat of a water shortage are to be stated by administrative order as stipulated in the Water Decree. Under subsection 2, further rules can be brought in by administrative or

---

[90] Water Act, section 6.20. A restriction can also be one on the term of validity.

[91] Water Act, section 6.16 states that in principle section 3.4 of the General Administrative Law Act and section 3.12 of the Environmental Management Act apply.

[92] See Lindhout, P.E. '*Cost recovery as a policy instrument to achieve sustainable and equitable water use in Europe and the Netherlands*' (Uitgeverij 2VM 2015); Lindhout, P.E. 'Application of the Cost Recovery Principle on Water Services in the Netherlands' (2013) 10 *Journal for European Environmental & Planning Law*, 309.

[93] Case C-525/12 *Commission v Germany* [2014] ECLI:EU:C:2014:2202.

provincial order regarding the priority list and may also be applied to the distribution of groundwater. Section 2.1 of the Water Decree details the list of priorities; those listed in the last two categories can be further specified at regional or provincial level.

1. Guaranteeing flood protection and averting irreversible damage;
2. Public utilities, with drinking water having the highest priority, followed by the power supply, each as far as delivery reliability is concerned;[94]
3. Small-scale high-grade use, prioritized as follows:
   a. temporary sprinkler irrigation of capital-intensive crops,
   b. processing industrial process water, and
   c. the quality of water in urban areas;
4. Other needs, with the following order of priority: shipping, agriculture, natural environment (other), industry, water recreation, inland fishing, drinking water (for the other needs) and energy (for the other needs); and
5. 'Other interests'.

Despite the fact that in the Netherlands permits are, in principle, granted on a permanent basis, various strategies exist for amending a distribution scheme or for the re-allocation of permits. However, it is questionable whether these strategies for achieving redistribution are aimed at a sustainable, balanced and equitable water use. Re-allocation can be made in the interests of similar companies, but can also be made for different types of activities, e.g. a redistribution between agriculture and drinking water supply, or between industry and agriculture on the one hand and nature on the other.

New policies, properly communicated through inclusion in the water plans, can change the allocation process by, for example, restricting the granting of permits for groundwater abstractions for low-grade use, such as irrigation in the agricultural sector. Many Dutch provinces have opted for this policy after it became clear that the dwindling amount of clean groundwater needed better protection. Groundwater abstractions are now only granted for high-grade use such as for drinking water. When new policies have been communicated to all stakeholders well in advance, no compensation has to be paid.

[94] At European level, highest priority is given to the drinking water supply, but given the vulnerability of the Netherlands to flooding it is not strange that, because of the guarantee of flood protection in the Netherlands, the safety interest heads the list of priorities.

A more radical step is the amendment, supplementation or revocation of existing permits under section 6.22 of the Water Act. There are only a limited number of reasons under Dutch water law on the basis of which a permit can or must be amended, supplemented or revoked. A permit can be revoked if it has not been used for three years. The withdrawal of existing permits is usually accompanied by financial compensation, based on the Water Act. These decisions are usually the result of new scientific understanding, new case law or new policy. Changes in the status of water bodies can also be a reason for amending a permit. Examples include when periods of drought lead to low water levels and – under the regime of general rules – when the appropriate abstraction levels from a water body may be exceeded because no individual assessment of the users' conformity with norms and objectives – the criterion for distribution – has taken place.

7.4 Enforcement and Conflict Resolution

Stakeholders in the Netherlands have few opportunities to apply for a permit referring to a need for re-allocation of permits, or to bring matters of allocation that are part of the planning system before the administrative courts. They are, of course, able to engage at the planning stage, but this is not the same as legal protection before the courts, which under Dutch law is only available where individual decisions affect private parties. We argue that it is desirable to enable challenging in the courts the allocation mechanisms and the norms and objectives set by the legislature. However, no legal protection is available under Dutch administrative law against generally binding regulations, nor is there a system in place for challenging water plans. For permit-holders and other parties with a direct interest (but not third parties), the only possibility of challenging a decision on allocation or distribution in the administrative courts is when a permit is granted, tightened up or revoked or customized regulations are imposed. Stakeholders can request from the water authority a redistribution of permits. However, the Water Act provides only a limited number of grounds on which a permit can be revoked or tightened up. A re-allocation of permits is not one of them, even if proposed new permits do conform to the obligation to be compatible with the objectives of the Dutch Water Act. Allocation and rights that do not meet the requirements of a sustainable, balanced and equitable water use, or are not in conformity with general legal principles governing the allocation of other public rights, do not appear to be a reason for amending or revoking permits or granting new ones if new permits are incompatible with these objectives. In the Netherlands, if no legal protection is available via the administrative courts, a matter can always be brought before the civil courts. The civil courts, however, have as yet little experience in reviewing water plans, general rules and amendments or revocations of water permits. Criteria for the review of these administrative instruments by the civil courts need to be developed. A principle such as a balanced and equitable allocation of water use rights does not yet exist as a legal rule.

8 WATER ALLOCATION AT THE NATIONAL, INTERPROVINCIAL AND LOCAL LEVEL: CHINA

In terms of the total volume of water resources, China is water-rich. However, several conditions prevent China from enjoying its substantial water resources. These include the massive population of over 1.3 billion, the high variability of seasonal rainfall and the uneven geographical distribution of water resources.[95] The United Nations has listed China as one of 13 countries that is experiencing serious water scarcity,[96] with only 25 per cent of the world's average water resources per capita. This scarcity and the enormous threat of serious water pollution are increasing tensions between various stakeholders and administrative regions.

8.1 Values and Principles

An important Chinese philosophy states that 'the unity of heaven and men' is the basis of the relations between man and nature. The human being is a living body of nature and it is part of nature. Human beings should not wish to conquer, dominate or control nature.[97] Although there is very little academic discussion on water values and there is no legal definition or formal recognition of the human right to water in China,

---

[95] Dai, L. 'Identifying and understanding the main challenges for sustainable water resource management in China' (2015) 24 *Journal of Water Law*, 249.

[96] Dow Water and Process Solutions 'A different look at water: Part I: Water for cities and the water-energy nexus' (2011) retrieved via <http://msdssearch.dow.com/PublishedLiteratureDOWCOM/dh_08cd/0901b803808cdb25.pdf?filepath=liquidseps/pdfs/noreg/609-03016.pdf&fromPage=GetDoc> accessed 30 January 2017.

[97] Dai (n 95) referring to Y. Tong 'Healing the Earth: Confucianism and its environmental protection perspective' (2010) retrieved via <http://www.chinarujiao.net/w_info.asp? PID=6476> accessed 30 January 2017.

most of the public agree that water resources are precious.[98] Chinese governments have consistently devoted themselves to improve access to water. Water provided by water suppliers should be sufficient to satisfy personal and domestic uses; of a quality good enough to prevent water-related diseases; and sufficiently accessible so that not only can people get enough safe water within safe physical reach, but also so that water supplies are acceptably priced and non-discriminatory. Furthermore, an important value in China is the socialism system, with a strong role for the Communist Party, further guided by the Constitution. Principles in Chinese environmental and water policies offer important guidance for further policies and legislation and refer to a priority to protect, an emphasis on prevention, integrated governance, public participation and liability for damages. China strives to build a 'harmonious society', which aims to change its focus from economic growth to an overall societal balance and harmony. The legal system in China is based primarily on the civil law model, and the primary source of law is the statutes. In addition to these statutes, the Communist Party is another principal actor in water resources management, as in other domains of social development management.[99] The Communist Party, for the first time, focused on the rule of law in its Fourth Plenary Session in 2014, with the aim to promote the modernization of the country's governing system and capabilities. The role of government is changing and the role of society is increasing.

8.2 Stakeholder Involvement

During the past decade, China has developed a skeleton structure for public participation, although it is more at the theoretical level. The laws stipulate that the public has the legal right to participate in certain environmental activities, but the measures to realize such rights are lagging behind.[100] One of the reasons is that Chinese society lacks a historical tradition of public participation. Indeed, there is a lack of experience in organizing and promoting public participation. Furthermore, public participation is often promoted in formal legislation but is

---

[98] Dai (n 95).

[99] Dai, L. 'A new perspective on water governance in China – Captain of the River' (2015) 40 *Water International*, 87.

[100] Spijkers, O., Li, X. and Dai, L. 'Public Participation in China's Water and Environmental Governance' (2016) *International Journal of Water Resources Development* (under review).

not possible within the Communist Party system. Although the Communist Party is developing public participation as well, it is only applied in a general way.[101]

8.3 Policy Instruments: Regulations and Agreements Including Planning and Permits

The Water Law (2002) requires the State to make a *National Water Resources Strategic Plan* (NWRSP) for investigating national water development and its utilization status in order to provide the country with a foundation for macro-scale water resource allocation, exploitation, utilization management and protection.

The process of drafting the NWRSP is interactive between the central and local levels. At the decentralized level, river basins and administrative regions develop plans. River basin plans include a River Basin Comprehensive Plan and River Basin Special Plans. A comprehensive plan is an overall scheme that covers water development, utilization, conservation, protection and pollution prevention. A special plan focuses on one theme, such as irrigation or water supply. The State Council sets out strategies, objectives, plans, guidelines and procedures. All River Basin Commissions and provinces must develop their own plans. For the plan of trans-provincial rivers and lakes, the River Basin Commissions are required to cooperate with the competent departments of the provincial governments.

The NWRSP aims to answer the question of how much water is available for allocation and the *Water Allocation Plan* then determines how, within a basin, water should be allocated between different regions, sectors or users.[102] At the national level, the 2002 Water Law has introduced and strengthened water resource allocation and the total amount of control as new concepts. The Ministry of Water Resources issued a Water Quantity Allocation Measure in 2006, which defines water quantity allocation as 'the allocation of the total usable amount of water resources or the total distributable water quantity to administrative divisions on a level by level basis, so as to determine the shares of

---

[101] See Dai (n 95) and Dai (n 99) for more information on public participation in China.

[102] Dai (n 95) referring to Robert Speed, Yuanyuan Li, Tom Le Quesne, Guy Pegram and Zhou Zhiwei, 'Basin Water Allocation Planning: Principles, procedures and approaches for basin allocation planning' [UNESCO 2013].

consumable water quantity for life and production as well as the shares of water taking and its quantity' (Article 2).[103]

The *Water Allocation Measure* is applied at two levels: the trans-provincial level and within a provincial region. The river basin is taken as a unit when the allocated water is shared by different provincial regions, whereas the administrative region is taken as a unit when the allocated water is located within a province. All water allocation plans are required to combine and harmonize different mixes of management strategies, such as the Five-Year Plan, the Division of Water Function Zones and the Most Stringent Water Management System.[104] To date, there are only a few sub-basin allocation plans and regional allocation plans available in China.

Article 48 of the Water Law (2002) requires a *water taking licence* for any unit or individual that takes water and uses water resources directly from a river, a lake, or underground and the payment of a water resource fee. The licence, combined with the fee, provides the right to take water. These requirements are not needed when only a small amount of water is taken for domestic use or for drinking by poultry and livestock reared outdoors or in pens. The Regulation on the Administration of the License for Water Drawing and the Levy of Water Resource Fees (2006) further specifies these requirements by defining 'water drawing' as the drawing of water resources directly from rivers, lakes or underground with the use of water drawing engineering structures or facilities, which relates to water gates, dams, channels, artificial watercourses, siphons, pumps, wells and hydropower stations.

The Water Administrative Department and Drainage Basin Authorities at the localities of important rivers and lakes are the competent authorities for organizing, implementing, supervising and administering the institution of permits for water drawing within their respective jurisdictional scopes. The water administrative department, the administrative department of public finance and the price administrative department of a people's government at the county level or above are responsible for levying, administering and supervising water resource fees.

A permit certificate for water drawing is not necessary in the following cases:

---

103 Interim Measure for Water Quantity Allocation 2006, Order No. 32 of the Ministry of Water Resources (2008).

104 Dai (n 95).

1. A rural collective economic organization or any of its members uses the water in the pond or reservoir of the said organization;
2. A small amount of water is drawn for domestic life, or as drinking water of livestock or poultry under sporadic or enclosed breeding, etc.;
3. The water is drawn (discharged) in response to temporary emergencies in order to guarantee the engineering or work safety of underground structures such as mines, etc.;
4. The water is drawn in response to temporary emergencies in order to eliminate the harms endangering public safety or public interests; or
5. The water is drawn for use in temporary emergencies in order to fight an agricultural drought or maintain ecology or environment.

The drawing of a small amount of water as prescribed under (2) shall be set forth by the people's government of the competent province, autonomous region, or municipality directly under the Central Government; the water drawing as prescribed under (3) or (4) shall be reported in a timely manner to the water administrative department or the drainage basin authority of the local people's government at the county level or above for archival filing; and the water drawing prescribed under (5) shall be subject to the consent of the water administrative department or the drainage basin authority of the people's government at the county level or above. A permit for water drawing shall first satisfy the needs of the urban and rural inhabitants in their domestic use of water and give concurrent consideration to the agricultural, industrial, ecological and environmental need for water as well as to the needs of navigation. The competent authorities may determine the specific sequence of the various items of water use when the actual situation so requires. Besides the above-mentioned regulations, there are Measures for the Administration of Water Abstraction Licensing (2008), prescribing the application, approval and supervision of water drawing.

In China, there is a shift from allocation of water resources dominated by the government towards trans-provincial allocation of water resources jointly guided by River Basin Commissions and provincial governments. This cooperation and consultation is also taking place for allocation of water resources between provinces and local governments. In the field of reallocation, the market has become more active during the last few years, but is still subject to extensive interference from government.

8.4 Enforcement and Conflict Resolution

Despite the huge amount of laws and local regulations related to environmental and natural protection, their enforcement can be improved. The Communist Party therefore has taken the unprecedented step of emphasizing and promoting an approach based on the rule of law, leading to the revision of environmental legislation in 2014. The new Environmental Protection Law expands the range of public interest by providing certain organizations with the right to sue. In addition to the Communist Party internal reform and the Environmental Protection Law amendments, the judicial system has recently further developed special tribunals for handling cases related to environmental problems. The application of political power by means of the Government Objective Responsibility System will further enhance better governance of water issues. This system could help to distribute responsibilities more efficiently in the short term.[105]

Local competition for economic development and fragmented institutional arrangements is the main causes of water disputes in China. The new round of ongoing reform and rebalancing economic and environmental matters, together with the creation of environmental tribunals, is expected to mitigate these conflicts to a certain extent. Citizens in China are used to resolving their conflicts through administrative channels, and they can apply to a higher-level administrative institution to reconsider the challenged decision. Judicial channels are less popular because local governments often influence them, as local judiciaries are dependent on local governments for funding.

9 SOUTH AFRICA AS AN EXAMPLE OF A POLICY AIMING AT THE RE-ALLOCATION OF WATER USE RIGHTS IN THE INKOMATI-USUTHU CATCHMENT

In this chapter, water allocation in South Africa is not discussed in general but serves as an example of a country that explicitly pays attention to re-allocation of water use rights to achieve more equitable water use. The section is based on literature review and a period of field research in the Inkomati-Usuthu Catchment, enabling insights from practice based on additional interviews.

---

[105] Dai (n 99).

9.1 Values and Principles

A core value in South Africa is 'water for all', complemented by values such as sustainability, efficiency and equity. The National Water Resources Strategy aims for 'Sustainable, equitable and secure water for a better life and environment for all'. In South Africa, redress of past racial discrimination is an important factor in the re-allocation of water use rights. The obligation of redress entails a compulsory licensing scheme, which should benefit Historically Disadvantaged Individuals (HDIs) who are disadvantaged as a result of the apartheid regime. There are many South African legal norms that could result in a more equitable distribution of water rights, providing HDIs with a chance to have access to water. Because it allows for a redistribution of scarce water rights, taking equity into account, South Africa can be an example for other countries.

The key issue of redress, according to Joseph Mabunda, the IUCMA's director of stakeholder participation, is 'access to water for everyone'. The Water Allocation Reform (WAR) is the most important redress instrument. Johan Boshoff, the head of the IUCMA's legal department called the provision on compulsory licensing the 'affirmative action clause' as it enables HDIs to get water rights.

9.2 Policy Instruments: Regulations and Agreements Including Planning and Permits

Although it does not explicitly use the word 'redress', section 9(2) of the South African Constitution is the foundation of South African law of redress on past racial discrimination. Section 9 describes the right of equality before the law for all people. The second paragraph states that, in order to promote equality, legislative measures may be taken to advance the position of disadvantaged persons. In the *Bato Star* judgment, the Constitutional Court of South Africa emphasized that, in order to achieve the desired equality, positive measures need to be taken to eliminate the inequalities stemming from the apartheid regime.[106] This legal imperative also applies to water, reflected in section 25(8) of the Constitution. This section mentions the need for land and water reform in order to redress the results of past racial discrimination. Reallocation of water is closely related to South Africa's Land Reform. The Land Reform

---

[106] *Bato Star Fishing (Pty) Ltd v Minister of Environmental Affairs and Tourism and Others* (CCT 27/03) [2004] ZACC 15; 2004 (4) SA 490 (CC); 2004 (7) BCLR 687 (CC).

programme gives people whose ancestors have been dispossessed of their land under apartheid the opportunity to make a claim to the government to repossess that land. Redress plays a part here in two ways: the provision of water, and the efficient use of the water allocated. Land Reform beneficiaries generally want to use their land for agriculture.

The South African National Water Act (NWA) is the country's main statute on water law, and contains important provisions on redress. Section 2 states that the NWA aims to ensure that South Africa's 'water resources are protected, used, developed, conserved, managed and controlled in ways which take into account amongst other factors … redressing the results of past gender and racial discrimination'. Section 27 uses the same phrase, and stipulates that responsible authorities, in issuing general authorizations and licences, must take into account the need to redress the results of past gender and racial discrimination. The same obligation applies for proposed allocation schedules (section 45). It is the Department of Water and Sanitation's (DWS) competence to allocate these water rights. The catchment management agencies are therefore the regional water authorities that play only a subsidiary role in the allocation of water. These allocation schedules must show that water is allocated to people who are entitled to a licence based on redress. Section 79(4) is directed at catchment management agencies, and states that 'in performing its functions a catchment management agency must … be mindful of the constitutional imperative to redress the results of past racial and gender discrimination'. This is thus a general obligation, which relates to all activities carried out by a catchment management agency.

In South Africa, the use of water can be authorized in four ways.

1. It can be a licensed water use (section 27 NWA).
2. It can be permitted in Schedule 1 of the NWA, which permits every 'reasonable domestic use'.
3. There can be a general authorization for the water use (section 39 NWA).
4. A water use can be the continuation of an existing lawful water use (sections 32–33 NWA).[107]

Section 27 states the requirements for a water licence. According to this provision, a responsible authority must take into account *all* relevant

[107] See Kidd, M. 'Water rights and permitting: a South African approach' in *Handbook of Water Law and Policy*, 38–50.

factors, and it then mentions eleven factors that should be taken into account. Among these factors are existing lawful water uses, the need to redress past racial discrimination, efficient and beneficial use of water in the public interest, the socio-economic impact of the water use or the refusal to authorize the water use and the likely effect on water resources and other water users. These requirements apply to all applications for licences. Issuing authorities generally provide these licences for a specified duration, such as for 20 or 30 years.

General authorizations are for water uses in a specific area, either for all of the water users or only for designated ones. The authorizations can reduce the administrative burden, and have the effect that people who otherwise would not be ready to apply for a licence still have an authorization to use water. This can be an instrument to help achieve equity, by making it easier for resource-poor farmers to have lawful access to water.[108]

Existing lawful water uses are water uses that the responsible authorities do not explicitly authorize, but that are still legally valid. They are generally continuations of water uses that were lawful before the NWA came into force. There are two forms of existing lawful water uses. The first is described in section 32(1)(a) NWA. These are water uses that have taken place in the two years before the promulgation of the NWA. The second, described in section 32(1)(b), are declared as such by the responsible authority, following the procedure of section 33. Existing lawful water uses can be carried on without a licence, until the competent authority starts a compulsory licensing procedure.

This compulsory licensing procedure is part of a mechanism for re-allocation of water rights, which is used to implement the WAR. The WAR is a responsibility of the Department of Water and Sanitation. The aim of this reform is to redistribute water rights to help improve the economic position of HDIs. All of these redistributed water uses should be licensed. The legal basis for this is section 43 NWA. When it is desired that water uses be allocated fairly and equitably within a water-stressed area, all people interested in using water may apply for a water licence. When all relevant applications have been received, the responsible authority has to make a water allocation plan according to section 45 NWA. This water allocation plan should take into account various factors, such as licences already issued, entitlements on the basis of redress, current existing lawful water uses and the general criteria of section 27.

---

108 See <https://www.dwa.gov.za/war/generalauthorisations.aspx>.

According to Article 49, these licences are also subject to review. That means that after a certain amount of time, the responsible authority can review the licence and is authorized to change the conditions of the licences and the amount of water allocated. Further, Article 54 provides the responsible authority with the capacity to withdraw licences if the licensee fails to comply with provisions of the licence or the NWA in general.

9.3 Re-allocation in Practice: Experiences from the Inkomati-Usuthu Catchment

The research for this section has been carried out in close collaboration with the Inkomati-Usuthu Catchment Management Agency (IUCMA) in Mpumalanga, South Africa in June 2016. Catchment management agencies are governmental authorities responsible for the co-ordination of activities of water users in their catchment area, and the management of water quality.[109] The IUCMA thus has a legal obligation to take into account the issue of redress. This is an open norm, and therefore gives the IUCMA a margin of appreciation as to how to implement it.

9.4 Stakeholder Involvement

Public participation also plays a role relating to redress. In this context, we use the example of the irrigation boards. These boards are collectives of various stakeholders, mostly farmers, who are concerned with the irrigation of their land. They are active in small sub-sections of the catchment area, such as smaller rivers or sections of larger rivers. One of the boards in the Inkomati catchment area is the White River Irrigation Board. Its members are mostly black farmers, but the board members themselves are still older white males. According to Debbie Turner, the secretary of the board, this is mostly because everything is working properly and effectively, and therefore black members do not feel the urge to manage the board, which is time and labour intensive. The board also deals with illegal abstraction, and says that this is not an urgent problem in the area. TsGro is another relevant institution for public participation. It is closely tied to the Irrigation Board of the region of the Malalane Sugar Plant. This company actively helps land reform beneficiaries to make effective use of their allocated water. This is done

---

[109] National Water Act, section 80.

through education and providing financial mechanisms, infrastructure and maintenance on existing water infrastructure.

9.5 Competent Authorities and Policy Instruments: The Verification Procedure and Permits

The Department of Water and Sanitation (DWS), a department of the national government, is responsible for the allocation of water and licensing of water uses. The end responsibility for water allocation thus lies with DWS, but the IUCMA has an important role in water allocation as well. It is currently involved with the verification procedures in its catchment management area. Verification means that current water uses are registered, in order to learn how the water is being used, by whom and in which quantities. This is an important first step in the WAR, as it makes it easier to provide water licences when there is knowledge about the exact amount of water actually used. It should be noted that the IUCMA currently has no delegated powers to carry out this verification, but does so regardless because, in its opinion, the DWS' previous verifications were not successful.[110]

As the Inkomati catchment area is very dry, agricultural land needs to be irrigated. The water for irrigation has to be allocated to the prospective water users, as every water use, with the exception of existing lawful water uses, has to be authorized beforehand by the responsible authorities. These land reform beneficiaries are therefore not allowed to take the water themselves. Next to irrigation, these small-scale emerging farmers need to have knowledge of irrigation and effective water management. Capacity building is therefore of utmost importance in a dry area like the Inkomati area.

9.6 Discussion

Redress and public participation are essential obligations for the IUCMA. The open norm on redress gives the IUCMA much freedom to choose the way in which it implements this obligation. The agency does this by supporting DWS with its allocation reform. By verifying existing water uses, without having the legal competence to do this, the IUCMA is an important contributor to the water allocation reform. South Africa's sophisticated water licensing law, including its compulsory licensing system, gives issuing authorities the capacity to distribute water rights in

---

110 As stated by Johan Boshoff.

an equitable way that takes into account all relevant interests. However, from our field research and the interviews, we learned that a sophisticated legal system is not automatically easy to implement. The verification procedure causes many problems, as not all historic water users are eager to participate in the verification process. It is difficult to get all the necessary information and it takes a long time before results can be shown. This hampers trust in the legal system by those who are dependent on the reform for their basic needs and water to make a living.

10 EVALUATION OF THE ALLOCATION MECHANISMS

The allocation of water use rights can be researched and designed from different perspectives. The general administrative law approach emphasizes the importance of criteria such as an appropriate degree of transparency on the availability of the right, the applicable allocation procedure, the deadlines for applications and the allocation criteria. From an economic point of view, the competition for available water use rights is very important too. There has been an academic debate framed under the name of scarce public rights, and the Advocate-General of Administrative Law in the Netherlands has summarized the relevant Dutch and European case law.[111] The subject of this opinion is the degree to which allocation procedures for scarce public rights should give interested persons the opportunity to compete for these scarce public rights. The opinion focuses therefore on the economic perspective combined with requirements following from administrative law and good governance. However, this approach does not lead to a preference for a specific allocation procedure. It also states that scarce public rights cannot be allocated for an indefinite amount of time, as in such cases there is not enough room for competition.[112] The highest Dutch court on general

[111] ABRvS 25 May 2016, ECLI:NL:RVS:2016:1421.

[112] Drahmann, A. *'Transparante en eerlijke verdeling van schaarse besluiten: een onderzoek naar de toegevoegde waarde van een transparantieverplichting bij de verdeling van schaarse besluiten in Nederland'* (PhD thesis, Leiden University 2015), 214; Johan Wolswinkel, *'De verdeling van schaarse publiekrechtelijke rechten. Op zoek naar algemene regels van verdelingsrecht'*, (Boom juridische uitgevers 2013), 305.

administrative law and environmental law has drawn from the Advocate-General's opinion in a subsequent judgment.[113] This approach of the Advocate-General and the court is quite a minimal yardstick, though, as it only aims to guarantee a minimum degree of market competition. Another, broader standard that is better suited to the allocation of water use rights is the sustainable, balanced and equitable use of natural resources.[114] This is derived from the EU Water Framework Directive and is more appropriate to allocation than a purely economic approach because water use also has many social, ecological and environmental implications. A better system is therefore one that takes into account relevant social, environmental and economic aspects, and balances them in an appropriate way.

In this chapter, we used a method to assess water allocation that allows us to look at these values and interests and how they have been balanced. The method used consists of ten elements and their mutual relationship and connectivity necessary to improve water governance. We highlighted aspects concerning values and principles, including the recognition of the right to water and human and ecological basic needs and the way citizens are involved in decision-making when it comes to the allocation of water use rights. After that, we turned to the policy instruments used in several countries, the ways in which the allocation of water use rights can be enforced and how conflicts can be avoided or solved. We paid special attention to the sometimes large difference between 'the law in the books' and the 'law in practice'. To get ideas about how a re-allocation of water use rights can be designed and achieved, we looked at the South African approach, learning that designing the legal system to achieve a more fair allocation is indeed one such idea, but implementing it in practice is much more difficult to achieve.

A comparison of the approaches in the different countries reveals that they all have allocation policies in place that are based on shared values, at least to a certain extent. We can see a shift from a mainly economic approach, serving primarily irrigation and agriculture, industry, shipping and electricity, towards a more balanced approach aiming at sustainability and equal access. All countries researched acknowledge the need to ensure water for basic human needs, sometimes by explicitly recognizing the right to water (Kenya, South Africa), sometimes indirectly by means of ordinary water legislation (the Netherlands) or by turning toward

[113] Administrative Jurisdiction Division of the Dutch Council of State 2 November 2016, ECLI:NL:RVS:2016:2927.

[114] van Rijswick (n 15).

religion or philosophies that reflect shared values (Indonesia, China). A minimum amount of water is also available for small-scale agriculture in some of the countries. For example, in Indonesia and China we see that basic human needs are combined with non-commercial use such as small-scale agriculture. It is also interesting to see that a basic need for ecosystems, and therefore for future generations, is acknowledged by allocating a certain amount of water for necessary environmental flows or needs (Kenya, China, the Netherlands, South Africa). Depending on the water availability in a country, this may differ from a general reserve for ecological interests (Kenya, South Africa) to specific legal arrangements that are applicable only in times of drought (the Netherlands, China). Also, the transboundary nature of many watercourses is reflected in the water allocation mechanisms, sometimes by an explicit reserve due to international obligations (Kenya), and sometimes only by separate water treaties (South Africa, the Netherlands, China).

Despite this attention to more sustainable and equitable water allocation, practice shows that the economic interests and existing water property or water use rights still play a dominant role, even if policies have been developed to tackle this inequality. A balancing between economic, social and ecological interests can be justified by the fact that enabling the population to make a living is also a strong shared value, although our research shows that this sometimes leads to an insufficient fulfilment of basic needs and a strong position for commercial water use. Furthermore, the transparency of how the different interests have been balanced can be improved.

To enable all interested parties and interests to be involved in the allocation process, we see that the focus on public participation is growing, following general insights that stakeholder involvement improves water governance.[115] This is sometimes organized by means of water user associations and irrigation boards. However, it becomes clear

---

[115] Akhmouch, A. and Clavreul, D. 'Stakeholder Engagement for Inclusive Water Governance: "Practicing What We Preach" with the OECD Water Governance Initiative' (2016) 8 *Water*, 204; Fitzmaurice, M. 'Note on the participation of civil society in environmental matters. Case study: the 1998 Aarhus Convention on access to information, public participation in decision-making and access to justice in environmental matters' (2010) 1 *Human Rights and International Legal Discourse*, 47; Spyke, N. 'Public Participation in Environmental Decision making at the New Millennium: Structuring New Spheres of Public Influence' (1999) 26 *Boston College Environmental Affairs Law Review*, 263; Arnstein, S. 'A ladder of citizen participation' (1969) 35 *Journal of the American Planning Association*, 216.

that the most vulnerable groups often still lack effective participation possibilities, which may decrease the legitimacy of the allocation mechanisms used.

To come to an allocation mechanism, all countries use a mix of instruments, starting with more general planning (all countries), sometimes followed by more specific allocation plans or guidelines (China). Priority setting for designated water uses takes place by means of priority setting in plans, legislation or guidelines (Kenya, China, the Netherlands, South Africa). A system of permits or licences is used in all countries, although not for all water uses. The same goes for payment for water use, which is sometimes free, for example for basic human needs or for small scale or even large scale agriculture. This is sometimes based on solidarity or beneficiary arguments.

Redress of unequal water use by re-allocation asks for additional instruments, such as verification procedures and compensation mechanisms. It appears that re-allocation in practice is a sensitive and politically difficult job as it asks for the cooperation of traditional water use right holders.

Enforcement, conflict prevention and resolution mechanisms appear to be weak in all countries, although improvements can be seen. The argument of the assessment method we used – that all elements of water governance should be addressed and should be connected – appears to be sound as the case studies show. Enforcement is receiving more attention in those countries in which water governance is based on shared values and participation. Policy instruments that enable enforcement help achieve sustainable and equitable water allocation. Without thinking about solutions for potential conflicts in the future when designing the allocation mechanism, we see problems shifting from the policy design phase towards the implementation and enforcement phases. This is shown by some more general conclusions in the countries we researched.

The Brantas Basin in Indonesia is a water rich basin and water allocation is performing moderately to well. The current situation of water-related legislation in Indonesia strongly influences policy instruments and enforcement. The main problem regarding stakeholder involvement is the relation between 'higher' and 'lower' stakeholders, in this case mainly industry and small-scale farmers. This is because priority for water allocation during the dry season is given first to industry, as this beneficiary pays for its water while the farmers do not. This creates an even bigger gap where the participation of farmer groups is also ignored to some extent. To resolve this issue, the authorities have attempted to introduce a new concept of 'the beneficiary-pays'. However, this effort needs understanding and cooperation from the relevant farmers. Integrated water resources management and the role of a central water management authority, PJT-I, has had a positive effect on water allocation in the Brantas Basin. It uses policy instruments as planning, a priority system with attention to basic needs and non-commercial use, a permit system and a financing system to recover costs for water services.

Water resources allocation mechanisms in Kenya do have some severe weaknesses. Improvements are therefore necessary, especially in the field of stakeholder involvement, decision-making processes, regulations, enforcement and finances.

A number of allocation and permitting systems are in place in Dutch water law, namely, plans, permits, general rules and a statutory list of priorities. In the case of the plans, neither the criteria on which distribution is based, nor the procedure (how distribution takes place), nor how the various interests involved are weighed up against each other is clear. The publication of plans does mean that information on shortage and distribution is made known, but it is questionable whether this meets transparency requirements. As it is, the granting of permits on a 'first come, first served' basis and permanently until the available water use rights have been fully utilized protects existing rights and prevents a more sustainable and equitable water use as well as the creation of a competitive market place where new entrants can apply.

Based on the Chinese philosophy of the 'unity of heaven and men' and the 'harmonious society', the Chinese Government is rebalancing its social development by slowing down economic growth and focusing more on environmental protection. During this transition process, the roles of both government and society are changing. Public participation regarding water allocation, although strongly promoted in recent times, is still difficult to implement in practice owing to the lack of such a tradition. Water allocation mechanisms are also still in their infancy. Cooperation and collaboration between different stakeholders can be improved in order to better align water resource management across scales and levels as the water resource management structure is fragmented at all levels, in the horizontal, vertical and the interactive dimensions.

South Africa has a sophisticated legal system on water allocation. The factors mentioned in section 27 of the Water Act, on the issue of licences, are comprehensive and include ecological, environmental, economic and social aspects. These criteria also apply to the compulsory licensing procedure, which is carried out primarily to make water distribution more equitable. In the compulsory licensing procedure, every interested person is invited to send in an application, and the deliberation on the licences and the issue of the licences only take place when all applications have been received. The South African legal framework provides, at least in theory, sufficient room for competition. It has a compulsory licensing mechanism, which offers the possibility of re-opening the market for water use rights in favour of new interested parties. Therefore, it does not protect existing rights, but rather creates an opportunity for a new distribution of rights. In addition, when looking at possibilities for achieving a sustainable, equitable and balanced water allocation, the South African legislation performs well. In the compulsory licensing procedure, the responsible authority must balance competing interests, and take into account many social, ecological and economic factors. In this way, it looks for the best allocation from the perspective of sustainability, equity and economic goals.

Obviously, there are huge differences in the political, social, economic and historical situation of the countries researched in this chapter. The most important one is that South Africa has to deal with its history of apartheid. Therefore, the principle and legal obligation to redress the results of past discrimination does not translate one-on-one from South Africa to the other countries. At the same time, all countries have to cope with existing rights and the need to create a more sustainable and equitable water use. Therefore, the comparison with South Africa can still be useful when redress is perceived as an instrument to empower disadvantaged people or to re-open certain water use for newcomers. The compulsory licensing procedure also reflects the awareness that a redistribution of water rights should entail a balancing of social, environmental and economic interests. This can be helpful for new applications for water use rights.

To conclude: all countries studied here show a development towards more sustainable and equitable water use and the allocation of water use rights. Furthermore, every country has to find solutions for dealing with existing rights and the re-allocation of water use rights to achieve this sustainable and equitable water use. Improving participation and a further development of a mix of policy instruments that is based on shared values and that enables a transparent balancing of interests may be helpful. Finally, taking the possibilities for enforcement and conflict solution into account when designing allocation policies seems to be a step to which all countries should give more attention.

7. The role of private property rights in Australia's and Canada's modern water allocation regimes

Sharon Mascher and Deborah Curran

1 INTRODUCTION

The commonly told 'natural history of property rights'[1] goes something like this: When a resource is plentiful, articulated property rights are not necessary – the resource is 'unowned, unmanaged and open to all'.[2] Then, as the resource becomes less plentiful, 'somewhat diffuse common-property arrangements develop.' However, when the resource becomes scarce enough, and demand is high, 'full-blown individualized property rights'[3] are needed to create incentives to conserve and manage the resource and to avoid conflict and waste.[4]

Based on this natural history of property rights, as demand increases and water resource becomes scarcer, one might expect to see the articulation of comprehensive individualized property rights to allocate access to the resource.[5] However, there are several reasons why resort to a full-blown private property regime may not offer a first-best solution to

---

[1] Rose, C. 'Energy and Efficiency in the Realignment of Common-Law Water Rights' (1990) 19(2) *Journal of Legal Studies*, 261, 263.

[2] Ibid.

[3] Ibid.

[4] Rose, C. 'The Comedy of the Commons: Custom, Commerce and Public Property' (1986) 53 *University of Chicago Law Review*, 711, 715–16; Hardin, G. 'The Tragedy of the Commons' (1968) 162 *Science*, 1243; Demsetz, H. 'Toward a Theory of Property Rights' (1967) 57 *American Econ. Rev. Papers and Proc*, 347. See also Harris, E. 'The Evolution of Water Rights in the Nineteenth Century: The Role of Climate and Asset Type' (2013) 53 *Natural Resources Journal*, 217 for a discussion of the drivers of scarcity in the context of water.

[5] There are many 'counterexamples' of this progression, where individualized exclusive property rights give way to a commons, see Smith, H.E.

questions of water allocation in the face of scarcity. First, water is essential to all life and ecosystems and individually held private property rights will not ensure these ecological functions are provided for. Second, water moves freely and is 'constantly changing in form, quantity, and location' making it difficult to exclusively possess.[6] In addition, the quantity of water available within the system at any given time is dependent on drought, precipitation and human uses – with variability and scarcity increasingly exacerbated by climate change. Entitlements to a fixed amount of water may no longer be possible in many watersheds. Reflecting these differences, commentators such as Joseph Sax have noted, '[t]he roots of private property in water have simply never been deep enough to vest in water users a compensable right to diminish lakes and rivers or to destroy the marine life within them … The rights of use in water, however long standing, should never be confused with more personal, more fully owned property.'[7]

Modern water allocation regimes, therefore, must balance the desire to provide security to water users and incentivize efficiency in the face of scarcity with the need to manage water in a manner that incorporates public values for ecological health. In so doing, these regimes create rights that possess characteristics of private property while incorporating mechanisms to allow for adaptive management of the water resource. Using modern Australian and Canadian water law, this chapter explores the question of whether the result is resort to a regime of private rights.

The chapter starts by defining the characteristics of private property rights used in this chapter. It then turns to consider whether the rights created under Australian and Canadian water allocation regimes bear these characteristics. The chapter concludes by offering observations and conclusions as to consistencies and divergences in practice within these jurisdictions.

'Exclusion Versus Governance: Two Strategies for Delineating Property Rights' (2002) XXXI *Journal of Legal Studies*, 453, 457–8.

6 Saxter, S.R. 'The Fluid Nature of Property Rights in Water' (2010) *Duke Environmental Law & Policy Forum*, 56.

7 Sax, J.L. 'The Limits of Private Rights in Public Waters' (1989) 19 *Environmental Law*, 473, 482.

2 DEFINING THE CHARACTERISTICS OF PRIVATE PROPERTY IN THE CONTEXT OF WATER

As Henry E. Smith has observed, while a great deal has been written about the evolution of property rights, 'it is often unclear what is expected to evolve.'[8] Terry Anderson and P.J. Hill also note that what is meant by 'more private property' is often not specified in commentary on the evolution of property rights.[9] Presumably, this is because opinions as to how property rights should be described and ultimately defined remain divergent. Indeed, very little consensus can be found as to the definition of private property, much less its characteristics.[10]

Early efforts to define private property did focus on the standard incidents or characteristics of property.[11] In his classic treatment of ownership, Professor Tony Honoré identified eleven such characteristics:[12]

> Ownership comprises the right to possess, the right to use, the right to manage, the right to income of the thing, the right to capital, the right to security, the right to incidents of transmissibility and absence of term, the duty to prevent harm, liability to execution, and the incident of residuary.

Of course, different commentators describe these characteristics in different ways and whether any specific characteristic is essential remains a matter of disagreement.[13] While it is therefore difficult to point with certainty to an 'irreducible core'[14] of characteristics necessary for a private property right to exist, the literature and case law do suggest that

---

8 Smith (n 5), 453.

9 Anderson T. and Hill, P.J. 'The Evolution of Property Rights: A Study of the American West' (1975) 18 *The Journal of Law and Economics*, 163, 165.

10 Zellmer, S. and Harder, J. 'Unbundling Property in Water' (2007–2008) 59 *Alabama Law Review*, 679, 700.

11 Ibid., 699.

12 Honoré, T. 'Ownership' in Honoré, T. *Making Law Bind* (Clarendon Press, 1987), 165.

13 See for example: Merrill, T.W. 'Property and the Right to Exclude' (1998) 77 *Nebraska Law Review*, 730; Gray, K. 'Property in Thin Air' (1991) 50 *Cambridge Law Journal*, 252; Cohen, F. 'Dialogue on Private Property' (1954) 9 *Rutgers Law Review*, 357; Bentham, J. 'Idea of a Complete Law', in Hart, H.L.A. (ed.), *Of Laws in General,* (Athlone 1970), 77; Harris, W. *Property and Justice* (Clarendon Press 1996); Penner, J.E. *The Idea of Property in Law* (Clarendon Press 1997). For more modern conceptualizations of property, see: Waldron, J. 'What is Private Property?' (1985) 5 *Oxford Journal of Legal Studies*, 313, 327; Zellmer and Harder (n 10), 700.

14 Ziff, B. *Principles of Property Law* 3rd edn (Carswell 2000), 5.

certain characteristics at least point to the existence of a property right in the sense that 'the more of these characteristics attach to particular rights, the more likely that they amount to a property right.'[15] Zellmer and Harder, for example, point to the rights of exclusive possession or control and alienation as key characteristics of property.[16] Meanwhile, Merrill describes property as something that is irrevocable, exclusive and relating to a discrete marketable asset[17] and Saxter identifies the classical real property characteristics as exclusivity, alienability and utility.[18]

The courts have also weighed in the question of what constitutes property. In his well known description of property, Lord Wilberforce in *National Provincial Bank Ltd v Ainsworth* stated the right must be 'definable, identifiable by third parties, capable in its nature of assumption by third parties, and have some degree of permanence or stability.'[19] Meanwhile, in considering whether bore licences issued pursuant to the New South Wales Water Act 1912 amounted to a species of property, for Justices Hayne, Kiefel and Bell of the Australian High Court '[t]hat the entitlements attaching to the licences could be traded or used as security amply demonstrates that to be so.'[20]

From these various conceptualizations in the literature and case law, four key characteristics which point to the existence of private property in water emerge: exclusivity, duration, security and transferability.[21] This chapter will focus on the existence of these four characteristics to consider the role assigned to private property rights in Australia's and Canada's modern water allocation regimes.

---

15 McKenzie, M. 'Water Rights in New South Wales' (2009) *Sydney Law Review*, 442, 452.

16 Zellmer and Harder (n 10), 685.

17 Merrill, T. 'The Landscape of Constitutional Law' (2000) 86 *Virginia Law Review*, 885, 952–60 and 969–81.

18 Saxter (n 6), 56.

19 *National Provincial Bank Ltd v Ainsworth* [1965] AC 1175 at 1247–1248.

20 *ICM Agriculture Pty Ltd v Commonwealth* (2009) 240 CLR 140 at para 147 (*ICM*). For a review of the relevant Australian case law, see McKenzie (n 15), 449–452.

21 McKenzie (n 15), 452.

3 THE AUSTRALIAN WATER REGIMES

3.1 Introduction

The colonies of Australia originally relied on the common law rules relating to water received from England. However, it quickly became apparent that these common law rules were not well suited to the hot and arid climate of Australia.[22] Towards the end of the nineteenth century, Australian states and territories responded by passing legislation directed at the regulation and control of the water resource.

In most states, the resulting legislation vested the right to the use, flow and control of surface water and ground water in the state[23] or the Crown[24] to allow for state-based regulation and allocation of water.[25] The taking of water was prohibited without a licence, statutory authorization or exemption.[26] Rights were created to allow for the taking of water in the form of licence. Licences were issued for a fixed term[27] and attached to specified land.[28] Gradually, annual volumetric limits were introduced along with specified purposes for which the water could be used.[29] Later

---

22 Gardner, A., Bartlett, R. and Gray, J. *Water Resources Law* (LexisNexis Butterworths 2009), 177–81.

23 Water Act 2000 (Queensland), s 19.

24 See for example: Water Management Act 2000 (New South Wales) ('WMA'), s 392; and Water Act 1989 (Vic), s 7.

25 Gardner, Bartlett and Gray (n 22), 191. Drawing an analogy with the High Court's reasoning in *Yanner v Eaton* (1999) 201 CLR 251, it is possible to conclude that these declarations assert imperium, rather than dominium, over the water resource (at para 29). With this assertion of imperium, commentators suggest that the declaration provisions 'tacitly convert common property in water into public property', Carruthers, P. and Mascher, S. 'The Story of Water Management in Australia: Balancing Public and Private Property Rights to Achieve a Sustainable Future' (2001) 1 *Property Law Review*, 104.

26 See for example: WMA 2000 (NSW), ss 60A–60E; and Water Act 1989 (Vic), ss 33E and 289.

27 Water Act 1912 (NSW), s 12(1)(b). These licences could be renewed from time to time (s 14), amended (s 13AA) or cancelled with notice if not used within three years (s 13F).

28 Gray, J. 'Legal Approaches to the Ownership Management and Regulation of Water from Riparian Rights to Commodification' (2006) 1 *Transforming Cultures eJournal*, 64, 82. See also Water Act 1912 (NSW), ss 16 and 20F.

29 Gray (n 28) 82–83; *ICM* (n 20) at para 60.

amendments allowed volumetric allocations to be varied.[30] Some jurisdictions introduced basic trading, albeit on a temporary basis.[31]

Over time, the drawbacks in this regulatory framework emerged. With the bulk of Australia's population and agricultural activities located in the places where water resources are most limited,[32] many water resources became over-allocated. This meant that licensed extractions were equal to, or greater than, the amount of water that could be taken without adverse impacts on water ecosystems and other users.[33] The impacts of climate change will make the situation worse with predictions of increased temperatures, reduced rainfall, increased rainfall variability and evaporation, and a significantly increased frequency and severity of drought[34] punctuated by extreme weather events, such as flooding.[35]

Recognizing the need for national water reform, the Council of Australian Governments (CoAG)[36] entered into two major agreements: the 1994 Water Reform Framework Agreement (1994 CoAG Agreement); and the 2004 National Water Initiative (NWI).[37] Together these agreements recognize the 'national imperative' to increase 'the efficiency of Australia's water use' while at the same time ensuring 'the health of river

---

[30] *ICM* (n 20) at para 61; Water Act 1912 (NSW), at ss 20X and 22B(4).

[31] Tan, P. *Legal Issues Relating to Water Use: Issues Paper No 1* (Murray Darling Basin Commission Project 2004), 12.

[32] Chartres, C. and Williams, J. 'Can Australia Overcome its Water Scarcity Problems?' (2006) 1 *Journal of Developments in Sustainable Agriculture*, 17.

[33] Burchmore, J. 'The Development of New Water Legislation for NSW – a Policy Perspective' (2000) 17 *EPLJ* 309.

[34] Department of Climate Change and Energy Efficiency, *Water Resources*. Available at: <http://www.climatechange.gov.au/climate-change/impacts/water-resources.aspx> accessed 15 June 2011. See also: Intergovernmental Panel on Climate Change (IPCC), *Fourth Assessment Report: Climate Change 2007: Working Group II: Impacts, Adaptation and Vulnerability* (Cambridge University Press 2007), Chapter 11.4 and 11.6 ('IPCC').

[35] IPCC (n 34), Chapter 11.6.

[36] The CoAG is an intergovernmental forum, made up of the Prime Minister, State Premiers, Territory Chief Ministers, and the President of the Australian Local Government Association.

[37] CoAG Water Reform Framework Agreement, 25 February 1994, Attachment A of the Communiqué to the CoAG Meeting (1994 CoAG Agreement), <http://www.environment.gov.au/system/files/resources/6caa5879-8ebc-46ab-8f97-4219b8ffdd98/files/policyframework.pdf> accessed 17 January 2017; Council of Australian Governments, *Intergovernmental Agreement on a National Water Initiative* 2004 (NWI) available at <http://www.environment.gov.au/topics/water/australian-government-water-leadership/national-water-initiative> accessed 17 January 2017.

and groundwater systems by establishing clear pathways to return all systems to environmentally sustainable levels of extraction.'[38] To incentivize efficient use, the 1994 CoAG Agreement endorsed the creation of a water market 'backed by separation of water property rights from land title, and clear specification of entitlements in terms of ownership, volume, reliability, transferability and, if appropriate, quality.'[39] The NWI affirmed this approach, again recognizing the role of water markets and the need for secure water access entitlements – being entitlements which specify the 'essential characteristics' of the water product, which are exclusive and can be traded, which can only be cancelled at Ministerial or agency discretion where the responsibilities or obligations of the entitlement holder have clearly been breached and which, subject to emergencies, may only be varied through mutual agreement[40] – to allow the market to function. However, the CoAG Agreements did not suggest simply turning water management over to the markets and property rights. Rather, the NWI in particular endorsed the use of mechanisms designed to ensure a return to healthy water systems and to allow for adaptive management of the water resource.[41] To this end, the NWI made provision for a comprehensive and mandatory statutory water-planning framework; for direct allocations to the environment; and for share-based, rather than volumetric, water entitlements.[42]

The CoAG Agreements therefore position Australian jurisdictions to adopt a unique approach to water allocation and management – one which draws on the strengths of markets and private property-like water rights to incentivize efficiency but positions those rights within a

---

[38] NWI (n 37) preamble at cl 5.

[39] 1994 CoAG Agreement (n 37) at cl 4(a).

[40] To achieve this, the NWI directed that water access entitlements should: 'specify the essential characteristics of the water product'; 'be exclusive'; 'be able to be traded, given, bequeathed or leased'; 'be enforceable and enforced'; 'only be able to be cancelled at Ministerial or agency discretion where the responsibilities or obligations of the entitlement holder have clearly been breached'; and, subject to emergencies, 'be able to be varied … when mutually agreed upon.' NWI (n 37) at cll. 23, 31 and 31. See also, Fisher, D. 'Markets, Water Rights and Sustainable Development' (2006) 23 *Environmental Planning Law Journal*, 100.

[41] For a full discussion of the adaptive management features in the CoAG Agreements and the resulting New South Wales Water Management Act 2000, See Curran, D. and Mascher, S. 'Adaptation in Water Law: Evaluating Australian (New South Wales) and Canadian (British Columbia) Law Reform Initiatives' (2016) 12(1) *McGill Journal of Sustainable Development Law*, 93.

[42] NWI (n 37) at cl 25(ii).

framework that allows for flexibility and adaptive management to deliver an efficient and sustainable water allocation system. Many Australian jurisdictions have responded by passing modern water legislation following the framework established in the CoAG Agreements. The question of whether the resulting statutory regimes create rights that possess the four characteristics of property identified above – exclusivity, security, duration and transferability – will be explored in context of the New South Wales (NSW) Water Management Act 2000 (WMA).[43]

3.2 Adaptive Exclusivity and Security in New South Wales (NSW)

The WMA, amended to align with the NWI in 2004, is a complex and sophisticated piece of legislation. In broad overview, the WMA vests 'State water rights'[44] in the Crown and requires a person who takes water to do so in accordance with one of several categories of water access licences or a statutory authorization.[45] In keeping with the CoAG Agreements, the Act provides for the creation of perpetual share-based water access licences and the development of a water market. It also requires the creation of statutory water management plans containing rules to ensure water dependent ecosystems are protected and restored, pursuant to which licensing and management decisions are made, and the allocation of water to the environment.[46]

Water access licences remain the primary method by which water is allocated to users other than the environment. And, at first blush, these water access licences have the appearance of private property rights. The extraction component of a water access licence entitles its holder to take water at specified times, rates and/or in specified circumstances from specified areas or locations[47] – manifesting the characteristics of exclusivity and security. Unlike the fixed term statutory allocations of earlier water legislation, water access licences endure indefinitely, absent a breach of conditions or some other form of non-compliance – creating

---

43 For a detailed discussion of these characteristics as they relate to the WMA, see McKenzie (n 15). The impacts that the Commonwealth's Water Act 2007 has on these characteristics is not considered here.

44 This term is defined in WMA s 392.

45 WMA s 60A. Statutory authorization is provided for domestic and stock rights (s 52), harvestable rights (ss 53 and 54) and native title rights (s 55).

46 WMA, ch. 2.

47 WMA s 56(1)(b).

certainty of duration.[48] And, the WMA also authorizes the transfer of water access licences – providing for transferability.[49]

However, upon closer examination it becomes clear that the water access entitlements created by the WMA do not create what commentators have called a 'pure perpetual water right'[50] – a right to exclusively take a specified volume of water in perpetuity. Such a right would be inconsistent with the WMA's object of ensuring that water sources, and their associated ecosystems, are protected, enhanced and restored.[51] It would also be inconsistent with the WMA's guiding principle of adaptive management. In keeping with the NWI, the WMA resolves this tension by creating secure rights in a *share* of the available consumptive pool in the specific water resource rather than a secure right to a specified volumetric allocation of water.[52] The volumetric allocation associated with that share is determined in accordance with the water plan and its environmental rules, as amended from time to time, and the 'available water determinations' made under the plan.[53] In this way, the volumetric allocation associated with the right to a share can fluctuate depending on annual flow without interfering with the right itself. That the risk associated with fluctuations in the available water is born by the holder of a water access licence is made clear in the WMA's compensation provisions, which provide that compensation is not available if 'the reduction in water allocations is for the purpose of restoring water to the environment because of natural reductions in inflow to the water source, including but not limited to changes resulting from climate change,

---

48 The WMA provides that an access licence ceases to be in force when cancellation of the licence is recorded in the Access Register (s 69). Access licences issued for specific uses are cancelled when the specific use ceases. If an access licence is cancelled/suspended, compensation is not payable (s 87A).

49 Consent is required for transfers involving a prescribed category of water access licences, falling within prescribed circumstances or involving trades outside the state (WMA ss 71M and 71U). Other types of dealings associated with access licences (e.g. assignments or consolidations) require prior approval of the Minister (WMA ss 71P and 71T).

50 See Pagan, P. and Crase, L. 'Property Right Effects on the Adaptive Management of Australian Water' (2005) 12 *Australasian Journal of Environmental Management*, 77, 85. See also McKenzie (n 15), 454.

51 WMA, ss 3 and 5(2)(h).

52 WMA s 56. See also McKenzie (n 15), 444. For a discussion of the effects of strengthened property rights on adaptive capacity, see also Pagan and Crase (n 50), 82–85.

53 WMA, ss 8(1)(a), 45 and 59.

drought or bushfires.'[54] In times of water shortage, the priorities assigned to various categories of licences through water management plans or the Act further diminish security, most significantly for lower priority licences.[55]

As a result, while the water access licence itself remains secure, to accommodate adaptive management of the water resource, the guarantee of a known volumetric allocation of water is compromised. So too is exclusivity, representing only an exclusive right to a share of the water available in the system rather than an exclusive right 'in the strict sense'[56] to a specified volume of water. In this same way, duration is affected as, while perpetual, the right is to a perpetual share of an unknown quantity of the available resource. Security and ultimately certainty of duration are also undermined by the power given the Minister to compulsorily acquire a water access licence when it is in the public interest to do so. In these circumstances, however, the Act provides that the person from whom the licence is acquired is entitled to compensation from the state for the fair market value of the licence.[57]

So, while a key object of the CoAG Agreements was the creation of security in water entitlement to facilitate a market-based approach and incentivize efficient use, the resulting water access licences in the WMA do not clearly amount to a form of private property. The Australian courts have made clear that '[r]egulatory machinery is not antithetical to the concept of property',[58] recognizing that fishing rights that 'confer only a defeasible interest' because of their vulnerability to changes to a management plan, remained property because they were assignable and 'changed

---

54 WMA s 87AA(3)(c). The risk is shared if the available consumptive pool is reduced due to a change in state government policy (s 87AA(5)) or to provide additional water to the environment because of more accurate scientific knowledge that demonstrates that the amount previously allocated to the environment was inadequate (s 87AA(6)). The National Water Commission and others have noted the practical difficulties associated with attributing a reduction in water allocation to climate change, government policy or new knowledge, see Bennett, M. and Gardner, A. *Groundwater Regulation in a Drying South West*, University of Western Australia Faculty of Law Research Paper, No 2014-42, 67–8.

55 WMA ss 49A, 57, 58 and 60(3). In the event of severe water shortage the Minister holds the power to further vary these priorities and allow basic domestic purposes and the environment priority over any other water users.

56 McKenzie (n 15), 454.

57 WMA s 79.

58 *Western Mining Corporation Ltd v Commonwealth* (1994) 50 FCR 305 at 335.

hands at substantial prices'.[59] The Australian courts have even been willing to recognize that fixed term volumetric licences created under the Water Act 1912 (NSW) may be a species of property.[60] However, while the water access licences created under the WMA possess features of the characteristics of exclusivity, security, duration and transferability, the limitations – particularly in relation to exclusivity and security – may suggest they do not possess enough of these characteristics to be considered private property rights.[61] This, however, was a compromise necessary to ensure these rights do not undermine the 'public good uses of water'[62] or restrict the ability for adaptive management while still allowing enough certainty to support a functioning water market.

4 WESTERN CANADIAN WATER LAW REGIMES

4.1 Introduction

Similar to Australia, the roots of water law in Canada are based on riparian rights, which still exist in different forms in approximately half of Canada from the province of Ontario eastward.[63] As colonial settlement occurred in the western provinces, in order to accommodate security of water use for agricultural and mining within a more arid climate, the mid-1800s to early 1900s saw the creation of water management regimes that prohibited the taking of water unless allocated under licence.[64] Based on examples from the western United States and Australia,[65] these developments culminated in the enactment of the BC

---

59 *Minister for Primary Industry and Energy v Davey* (1993) 47 FCR 151 at 175 and 171.

60 *ICM* (n 20) at paras 81–4.

61 See McKenzie (n 15), 462–3 for the conclusion that the answer to this question is uncertain.

62 Pagan and Crase (n 50), 82.

63 Percy, D.R. *The Framework of Water Rights Legislation in Canada* (Canadian Institute of Resources Law 1988).

64 Percy (n 63), 5–7; Lucas, A.R. *Security of Title in Canadian Water Rights* (Canadian Institute of Resources Law 1990), 11–16.

65 Percy (n 63), 7.

Water Act in 1909[66] and the Alberta Water Resources Act in 1931,[67] regimes that are largely still in force today.[68]

The foundation of these water laws is the principle of prior allocation. Similar to the principle of prior appropriation and 'first in time, first in right' found in the western United States,[69] the premise of prior allocation is that older licences take precedence over more junior or recent licences issued by the provincial government.[70] This priority, coupled with water licences that had no end date, provided security and duration of use as long as there was any water to be taken and it was used according to the purpose for which the government issued the licence and any conditions contained in the licence.[71] The attachment of a water licence to land, a work or undertaking,[72] such as a dam, provided an element of exclusivity, with transferability only present in the ability to sell the land, work or undertaking to which the water licence adhered,[73] or apply to the provincial government.[74] As long as a licensee adhered to the legislation or regulations, licence conditions or orders of water officials,[75] there was no ability to rescind or alter licences.

Since the 1990s, over-allocation and changing hydrological conditions have challenged the security of water rights under these prior allocation regimes in some watersheds in BC and Alberta.[76] Similar to Australia, some watersheds are overallocated,[77] and pursuant to treaty and aboriginal rights Indigenous communities are increasingly contesting the

---

[66] SBC 1909 ch. 48.

[67] SA 1931 ch. 71.

[68] Curran, D. 'Leaks in the System: Environmental Flows, Aboriginal Rights and the Modernization Imperative for Water Law in British Columbia' 50(2) *UBC Law Review* (forthcoming 2017).

[69] Wilkinson, C.F. 'Aldo Leopold and Western Water Law: Thinking Perpendicular to the Prior Appropriation Doctrine' (1989) 24 *Land and Water L Rev*, 1; Tarlock, D.A. 'Prior Appropriation: Rule, Principle or Rhetoric?' (2000) 76 *NDL Rev*, 881.

[70] Water Act, s 30, Water Sustainability Act ('WSA'), s 22.

[71] Water Act, ss 49 and 51; WSA, ss 1–2, 20(3)(c), 30 and 94.

[72] Water Act, s 45; WSA, s 20(1)–(2).

[73] Water Act, s 80; WSA, s 25(1).

[74] Water Act, s 81; WSA, s 27.

[75] Water Act, s 18; WSA, s 8.

[76] Sandford, R. *Restoring the Flow: Confronting the Worlds Water Woes* (Rocky Mountain Books 2009).

[77] Curran and Mascher (n 41), s 4.1; Alberta Environment, 'Approved Water Management Plan for the South Saskatchewan River Basin' August 2006 <http://aep.alberta.ca/water/programs-and-services/river-management-frameworks/

validity of water management decisions.[78] Changing hydrology without reform to the water management regimes weakens the security and exclusivity attached to water entitlements in western Canada.

In response, both provincial governments in Alberta and BC have undertaken law reform in the past 20 years to build some flexibility into the water rights regime but still maintain the system of licence priority based on date. The changes have focused on permitting watershed planning and the ability to change water rights to address low flows that threaten ecological values. While the range of responses to water scarcity varies significantly in the two provinces, the treatment of water licences under the Alberta regime has more indicia of private property by creating a water market and requiring compensation to be paid to affected licensees for administrative orders aimed at increasing environmental flows.

4.2 Adaptive Exclusivity and Security in Alberta and British Columbia

It is important to examine the water law reforms in the context of western water law in the US and the approach to land use in Canada. Although some courts view water rights as a form of private property rights in the western United States,[79] and private water trusts take responsibility for public ecological values in some states,[80] the Canadian treatment is less clear. Starting with the robustness of private property rights in private land regimes, Canadian jurisdictions allow governments to restrict virtually all use of land by, for example, zoning regulation without compensating the rights holder. This 'no compensation' principle is codified in the BC Local Government Act[81] and the Alberta Municipal Government Act,[82] which state that no compensation will be paid for changes in the value of land caused by decisions made pursuant to a land use bylaw or

---

south-saskatchewan-river-basin-approved-water-management-plan/documents/SSRB-ApprovedWaterManagementPlan-2006.pdf> accessed 17 January 2017.

[78] See, for example, *Chief Sharleen Gale and Fort Nelson First Nation v Assistant Regional Water Manager* 2012-WAT-013(c) (BC EAB 2015); *Tsuu T'ina Nation v Alberta*, 2010 ABCA 137; *Sak'uz First Nation and Stellat'en First Nation v Rio Tinto Alcan Inc.* 2015 BCCA 154.

[79] See, for example, *Tulare Lake Basin Water Storage District v United States*, 49 Fed Cl 313 (2001).

[80] King, M.A. 'Getting Our Feet Wet: An Introduction to Water Trusts' (2004) 28 *Harvard Environmental Law Review*, 495.

[81] RSBC 2015, ch. 1, s 458.

[82] RSA 2000, ch. M-26 at s 621.

permitting function. It is only when regulation takes away virtually all incidents of private ownership that the regulation will be found to be improper.

These 'regulatory takings' or challenges to land use decisions as expropriations of private property rights are few and far between in Canada. Although we read about successfully argued cases in the US courts, in Canada a court has never found land use regulation to result in a regulatory expropriation.[83] Courts have awarded compensation for loss of mineral rights upon the creation of a park[84] or for the removal of all economic viability including goodwill.[85] It is within this permissive regulatory context for private property that we examine the enactment of the Alberta Water Act in 1996 and BC Water Sustainability Act (WSA) in 2016, and their distinctive features of watershed planning, review of licences, compensation, and water markets.

Both provincial water laws enable the creation of water management plans (WMPs) that may affect water rights and decision-making in watersheds.[86] While not mandatory, WMPs in both provinces can change how water is used under an existing licence, and to mandate how decisions about water licensing are made.[87]

The recent water law reforms also permit adapting licences due to changing ecological conditions, but treat the consequences of those adaptations differently. Under the WSA, licences will be subject to WMPs that can reduce water diversions,[88] and to having their terms and conditions reviewed and changed anytime after 30 years from when the WSA comes into force.[89] At the same time, the WSA makes it clear that when licence entitlements change the provincial government will not pay compensation for those changes. The WSA clarifies that there will not be any compensation paid and no legal proceedings allowed relating to '... loss or damages arising from an effect on ... rights under a licence ... or rights and privileges in relation to land and resources.'[90] The WSA

---

[83] See *Mariner Real Estate Ltd v Nova Scotia (Attorney General)* 1999 CanLII 7241 (NSCA) and *Canadian Pacific Railway Co v Vancouver (City)* [2006] 1 SCR 227, 2006 SCC 5.

[84] *R v Tener*, [1985] 1 SCR 533; *Casamiro Resource Corp v British Columbia (Attorney General)* 1991 CanLII 211 (BCCA).

[85] *Manitoba Fisheries Ltd v The Queen* [1979] 1 SCR 101.

[86] Water Act ss 9–10; WSA Part 3 Division 4.

[87] Water Act s 11(3); WSA ss 76–83.

[88] WSA s 79.

[89] WSA s 23.

[90] WSA s 121(1)(a)–(f).

notes that the activities that could attract this alleged loss or damage include a change in the precedence of water rights, a restriction on the exercise of rights, or a change or imposition of new terms and conditions on a licence.[91] There are two exceptions to this no compensation rule. Cabinet may make regulations respecting the payment of compensation,[92] and if a water sustainability plan submitted to the minster recommends a significant change to a licence or drilling authorization, the plan must contain a detailed proposal recommending responsibility for compensating the licensee.[93] In contrast, the Alberta Water Act requires that compensation be paid for changes to water rights, subject to a contrary intention expressed in regulations, when the Director amends, suspends or cancels licences for conservation purposes.[94] The changes to licences are authorized under various sections of the Water Act if an adverse effect on the aquatic environment occurs and it was not reasonably foreseeable at the time the licence was issued.[95]

The Alberta law reforms explicitly provide for a limited market in water, whereas the BC WSA retains the connection of each water licence to land or an undertaking. In Alberta, where there is an approved water management plan or government order, parties may apply to have a water licence transferred where there is no impact on the environment or rights of others.[96] This allows licensees to transfer and third parties to acquire a water licence where a basin is closed to further allocation or the third party desires a more senior licence to provide greater security of rights.[97] There is no similar mechanism in BC. Acquiring the right to use a specific volume of water already licenced for use requires the purchase of land, a work or undertaking to which the licence is appurtenant.

In summary, while both the Water Act and WSA take similar approaches to regulating water rights, they differ in several ways that make water rights under the Alberta regime appear to more closely resemble private property rights. However, given the significant administrative discretion under each law, one could not conclude that water rights in western Canada are, indeed, private property rights. Under WMPs in both provinces, provincial government can adopt aspects of WMPs into

---

91 WSA s 121(1)(g)–(l).

92 WSA s 134.

93 WSA s 74(2)–(3).

94 Water Act s 158(1).

95 See, for example, s 54(2).

96 Water Act, ss 81–82.

97 Bankes, N. 'Policy Proposals for Reviewing Alberta's Water (Re)Allocation System' (2010) 20(2) *Journal of Environmental Law and Practice*, 81, 82.

law and change water entitlements and provincial decision-making. While water managers have extensive powers to make orders under both laws, orders to cease or cut back on taking water for ecological reasons mandate compensation under the Alberta Water Act, whereas it is clear that under the WSA no compensation is payable and licences may be reviewed every 30 years for adjustment. Finally, although heavily regulated, the water market under the Alberta Water Act provides for the transfer of water rights separate from the land or works to which they were attached. These qualities make the Alberta water law regime more like private property as security, exclusivity, and transferability are all explicitly safeguarded. In contrast, the BC regime weakens security and exclusivity by permitting administrative action that changes water rights without compensation. Both water law regimes are responding and take similar approaches to conflicts in water use and scarcity. However, the end result, largely due to different compensation rules, adaptation of licence entitlements and ability to transfer water, diverges when evaluated for characteristics of private property.

5 EVOLVING PRIVATE PROPERTY IN WATER IN AUSTRALIA AND WESTERN CANADA

Ultimately, the security, exclusivity and transferability of water rights – whether they are private property or usufructuary rights – depends on a reliable quantity of water being available to be taken, with the security and exclusivity of property rights in water undermined by changing ecological conditions. While it looks as if law reform that addresses changing environmental flows creates uncertainty, it is, in fact, attempting to address the legal uncertainty created by ecological change and to avoid the inevitable conflicts in water use in the long term. The evolving Australian regime best reflects the uncertainty of water rights in high-demand watersheds – as hydrology changes, the regime allocates water based on a fluctuating percentage above the environmental flow baseline. The 'property' in the water right, if any, is to a percentage of available flow. The western Canadian law reforms continue to base water rights on fixed volume entitlements that may change over the very long term, while increasing administrative flexibility to make orders and take the environment into account during licencing decisions. This administrative flexibility decreases security and exclusivity with Alberta and NSW bolstering this weakness through water markets and compensation provisions.

It is instructive to note that law reform in three different jurisdictions motivated by water scarcity and conflicts among water users contains WMP, environmental flow, and licence adaptation approaches that are aimed at incorporating public values for ecological health into water management. This attention to public values may mean that water rights will look less and less like private property as hydrological uncertainty mandates innovative water governance, administrative intervention on behalf of the environment and the sharing of water to preserve watershed economies.[98] These modern water law regimes maintain some of the incidents of private property while acknowledging that adaptation will be necessary and the public values that water provides cannot be constrained solely within a regime of private rights.

---

[98] For a detailed case study in this regard see Holly Doremus and A Dan Tarlock. *Water War in the Klamath Basin: Macho Law, Combat Biology, and Dirty Politics* (Island Press 2008).

8. Recovery of costs for water uses at the different levels of water law

Antti Belinskij[1]

1 INTRODUCTION

The Dublin Statement on Water and Sustainable Development reacted to the scarcity and misuse of freshwater in 1992 by declaring that a failure to recognize the economic value of water has led to its wasteful and environmentally damaging uses.[2] Still, it is largely assumed that public authorities are responsible for distributing water to different user groups with low tariffs, which increases water consumption and deficits as well as raising the costs of water abstraction. From an economic viewpoint, water should be treated as an economic commodity and priced in a way that avoids its wasteful use.[3] The promotion of efficient water use is also well in line with the European Union (EU) environmental policy objectives to contribute to the prudent and rational utilization of natural resources.[4]

References to economic viewpoints are not uncommon in international, regional or national water law. At the international level, the 1992 United Nations Economic Commission for Europe Convention on the Protection and Use of Transboundary Watercourses and International Lakes (ECE

[1] This research was funded by the Strategic Research Council's Winland project (#303628, #13303623).

[2] The International Conference on Water and the Environment (ICWE, Dublin 26–31 January 1992), *Dublin Statement on Water and Sustainable Development* (31 January 1992), principle 4.

[3] Jihad C. Elnaboulsi, 'An Incentive Pricing Policy for Sustainable Water Use' (2009) 42 *Environ Resource Econ*, 451, 452.

[4] Consolidated Version of the Treaty on the Functioning of the European Union [2012] OJ C326/47, art 191; Commission, 'A Blueprint to safeguard Europe's water resources' (Communication) COM (2012) 673 final, 10. On the resource efficiency objectives of the EU see Commission, 'Roadmap to a resource efficient Europe' (Communication) COM (2011), 571 final.

Water Convention)[5] includes the polluter pays and the best available technology principles.[6] In Finland, the Water Services Act (119/2001) requires the recovery of all costs of a water utility[7] while the Water Act (587/2011) states that all infringements resulting from a water resources management project must be minimized and any losses of benefits compensated.[8]

The clearest demand for the use of economic instruments in water management is presented in the EU Water Framework Directive (WFD).[9] The WFD requires the EU Member States to take account of the principle of recovery of the costs of water services in accordance with the polluter pays principle and ensure that water pricing policies provide adequate incentives to use water resources efficiently.[10] Despite this, the pricing of water has been a source of controversy between the European Commission and many EU Member States.[11] In case C-525/12 (Water Services Case) of the European Court of Justice (ECJ), the Commission claimed that a wide range of water uses from hydroelectric power generation to personal consumption must be priced, whereas Germany, supported by five other Member States, insisted on a narrower interpretation according to which the water pricing obligation covered only water or wastewater services in a bilateral relationship.[12]

---

[5] Convention on the Protection and Use of Transboundary Watercourses and International Lakes (adopted 17 March 1992, entered into force 6 October 1996) 1936 UNTS 269.

[6] ECE Water Convention, art 2(5).

[7] Water Services Act, s 18(1).

[8] Water Act, ch 2 s 7 and ch 13, s 1.

[9] Directive 2000/60/EC of the European Parliament and of the Council of 23 October 2000 establishing a framework for Community action in the field of water policy [2000] OJ L327/1. On the innovativeness of the recovery of costs in the Water Framework Directive see Case C-525/12 *Commission v Germany* [2014] ECLI:EU:C:2014:449, Opinion of AG Jääskinen, para 49; and R. Brouwer, 'Payments for Ecosystem Services: Making Money Talk' (Vrije Universiteit, Amsterdam, 15 April 2010), 21 <http://www.ivm.vu.nl/en/Images/101760_Oratietekst_for_Pedel_final_280510_tbv_Internet_tcm234-163494.pdf> accessed 5 January 2017; Dominic Moran and Sabrina Dunn, 'The economic value of water use: Implications for implementing the Water Framework Directive in Scotland' (2008) 87 *Journal of Environmental Management*, 484, 484.

[10] Water Framework Directive, art 9(1).

[11] See Commission, 'Environment: Commission asks Belgium, Denmark, Finland and Sweden to recover costs of all water services' (Press Release, 27 October 2011) IP/11/1264.

[12] Case C-525/12 *Commission v Germany* [2014] ECLI:EU:C:2014:2202.

This chapter discusses how the principle of cost recovery is reflected in international, EU and Finnish water law. The key questions are: (1) What kind of legal requirements are there for the recovery of costs for water uses at the different levels of law? and (2) What kind of measures must be taken to meet those requirements? While the recovery of costs has widely been dealt with in the context of EU law, this is not the case with either international or Finnish law.

The chapter focuses on legal provisions that regulate water uses, such as the abstraction, impoundment, storage, treatment and distribution of water, as well as waste-water collection, treatment and discharge. It is divided into four sections. Following this introductory section, Section 2 discusses the principle of cost recovery at the different levels of law, while Section 3 investigates the instruments for cost recovery. Section 4 presents the conclusions.

2 PRINCIPLE OF THE RECOVERY OF COSTS

2.1 Theoretical Underpinnings

The principle of cost recovery is strongly linked to the polluter pays principle. The ECE Water Convention defines the polluter pays principle to mean that the costs of pollution prevention, control and reduction measures are borne by the polluter (article 2, paragraph 5). According to the OECD, the polluter should bear these costs on the basis of either the damage done to society or an unacceptable level of pollution.[13] However, the idea of the recovery of costs for water uses is not only to prevent pollution but also to limit the quantitative use of water due to its burden on ecosystems. While the polluter pays principle directs an operator to pay environmental protection costs, the principle of cost recovery aims to cover all the costs that result from water uses despite best efforts to prevent them.

Following the EU Water Framework Directive the costs of water uses can be divided into financial, environmental and resource costs.[14] *The financial costs* include the costs of providing and administering water

---

[13] OECD, 'Polluter-Pays-Principle' (Glossary of Statistical Terms) <https://stats.oecd.org/glossary/detail.asp?ID=2074> accessed 10 November 2016. On the polluter pays principle in the context of the Water Framework Directive see William Howarth, 'Cost Recovery for Water Services and the Polluter Pays Principle' (2009) 10 *ERA Forum*, 565, 584–6.

[14] Water Framework Directive, art 9(1).

uses and can be divided into capital, operation and maintenance costs whereas *the environmental costs* are the marginal external costs caused by the damages imposed on the environment, ecosystems and other users of the environment. *The resource costs* relate to the scarcity of water associated with water consumption or water quality depletion and appear as foregone opportunities when the resource is depleted over its natural rate of recharge or recovery.[15] The environmental and resource costs are manifold and may occur in different situations of, for example, pollution, flooding or competing water uses.[16]

The recovery of environmental and resource costs is much more difficult compared to recovering financial costs. Water uses can damage the status of water by impairing its natural rate of recharge or by causing eutrophication, salinization, desiccation, loss of biological diversity, and morphological changes.[17] Also, the resource costs may be considered to include either all opportunity costs or merely the costs of unsustainable water use. While the financial costs are best recovered at the scale of water service suppliers, the assessment of environmental and resource costs should be based on the river basin or sub-basin scale depending on where these costs occur.[18]

---

[15] Commission, 'Pricing policies for enhancing the sustainability of water resources' COM (2000) 477 final, 10; Drafting Group ECO2, 'Assessment of Environmental and Resource Costs in the Water Framework Directive' (Information sheet, Common Implementation Strategy, Working Group 2B, June 2004), 3–4. On the financial costs see Case C-280/00 *Altmark Trans GmbH, Regierungspräsidium Magdeburg v Nahverkehrsgesellschaft Altmark GmbH* (2003), para 93; Herwig Unnerstall, 'The Principle of Full Cost Recovery in the EU-Water Framework Directive – Genesis and Content' (2007) 19:1 *JEL*, 29, 37; and Working Group 2.6 – WATECO, 'Economics and the Environment – The Implementation Challenge of the Water Framework Directive' (Common implementation strategy for the water framework Directive, Guidance Document No 1, European Communities 2003), 118. On the environmental and resource costs see Howarth (n 13), 581–3; Moran and Dunn (n 9), 486.

[16] United Nations Economic Commission for Europe (UNECE), 'Recommendations on Payments for Ecosystem Services in Integrated Water Resources Management' (ECE/Mp.WAT/22, United Nations 2007), 3–4.

[17] Roy Brouwer, 'The Concept of Environmental and Resource Costs; Lessons learned from ECO2' (RIZA Working Paper 2004.112x: Environmental and Resource Costs and the Water Framework Directive – An overview of European practices), 3, 5.

[18] See Commission 'Pricing policies' (n 15), 17; Howarth (n 13), 568.

In practice, water users only pay operation and maintenance costs in full in many European countries while public organs subsidize investment costs.[19] The environmental and resource costs remain largely unaccounted for although these external costs have been partly included in the financial costs by environmental protection requirements such as the requirement of best available technology.[20] It can be argued that the more external costs remain unaccounted for, the more inefficient water use patterns are.[21] One example of the inadequately recovered environmental and resource costs for water uses is a situation where the general public pays the treatment costs of drinking water that is contaminated by agriculture and industry.[22]

It is suggested that the concept of ecosystem services should be utilized in the assessment of environmental and resource costs. In EU law, the Environmental Liability Directive (2004/35/EC) refers to natural resources services, meaning the functions performed by a natural resource for the benefit of another natural resource or the public.[23] Water related ecosystem services include, for example, flood control, water quality improvements, clean drinking water and groundwater recharge as well as cultural services such as recreational benefits.[24] The idea is that the ecosystem services that provide water should have an economic value.[25]

All in all, the principle of cost recovery for water uses aims to allocate the costs to the person responsible for causing them, but a lot of

---

[19] Ibid., 9–10, 105. According to the EEA report, many water companies struggle to renew or replace ageing water infrastructure due to low profit margins. Some water companies are therefore still dependent on government subsidies in relation to necessary capital investments.

[20] Commission, 'Pricing policies' (n 15), 10; and ECO2 (n 15), 3–4. According to the ECO2, a distinction can be made between the *environmental costs*, which are the costs that occur due to water uses, and the *environmental protection costs* that prevent, avoid, abate or mitigate pollution, and other environmental damages.

[21] See Moran and Dunn (n 9), 486.

[22] See European Environment Agency (EEA), 'Towards Efficient Use of Water Resources in Europe' No 1/2012 EEA Report 32–3.

[23] Directive 2004/35/CE of the European Parliament and of the Council of 21 April 2004 on environmental liability in regard to the prevention and remedying of environmental damage [2004] OJ L143/56, art 2(13).

[24] UNECE, 'Recommendations on Payments for Ecosystem Services' (n 16), 2.

[25] See Brouwer (n 9), 3–6.

confusion surrounds its implementation.[26] Next, the question of whether the different levels of water law include concrete legal obligations concerning the recovery of costs for water uses is discussed.

2.2 International Water Law

Neither the United Nations Watercourses Convention (UN Watercourses Convention)[27] nor the ECE Water Convention,[28] which are the two global water law agreements in force, explicitly includes the principle of cost recovery. However, the two Conventions contain principles and rules such as polluter pays, no-harm, equitable and reasonable utilization, cooperation and compensation that are related to the recovery of financial, environmental and resource costs for water uses.

Perhaps the clearest example of an economic approach is the requirement for the best available technology in the ECE Water Convention. Accordingly, States must implement relevant measures, including economic and financial measures to ensure, for example, that the limits for discharges of hazardous substances are based on the use of best available technology and that the ecosystem approach is promoted as a part of sustainable water-resources management.[29]

The ECE Water Convention defines the polluter pays principle to mean that the costs of pollution prevention, control and reduction measures are borne by the polluter.[30] Also, the UN 1992 Rio Declaration on Environment and Development endeavours to promote the internalization of environmental costs and the use of economic instruments for that purpose,[31] while the UN Agenda 21 Action Plan seeks to enhance the

---

[26] See Erik Gawel, 'Article 9 of the EU Water Framework Directive: Do We Really Need to Calculate Environmental and Resource Costs?' (2014) 11 *Journal for European Environmental & Planning Law*, 249, 255–6.

[27] Convention on the Law of the Non-Navigational Uses of International Watercourses (adopted 21 May 1997, entered into force 17 August 2014) UN Doc A/51/869.

[28] Convention on the Protection and Use of Transboundary Watercourses and International Lakes (adopted 17 March 1992, entered into force 6 October 1996) 1936 UNTS 269.

[29] ECE Water Convention, art 3. Best available technology is defined in the ECE Water Convention, annex I, to mean 'the latest stage of development processes, facilities or methods of operation which indicate the practical suitability of a particular measure for limiting discharges, emissions and waste.'

[30] ECE Water Convention, art 2(5).

[31] The Rio Declaration on Environment and Development (1992 United Nations Conference on Environment and Development), principle 16.

polluter pays and natural-resource-user-pays principles in environmental law and regulation.[32] Nevertheless, it is still open to discussion whether the polluter pays principle is part of customary international law.[33]

The no-harm rule established by the two global water Conventions requires States to take all appropriate measures to prevent, control and reduce the causing of significant harm to other watercourse States when utilizing an international watercourse. States also need to pay attention to the protection of ecosystems.[34] The no-harm rule provides a due diligence obligation for a State to conduct itself in a manner that is generally considered appropriate and proportional in relation to the degree of risk of transboundary harm.[35] It is an obligation of best efforts instead of an obligation of result.[36] Therefore, the no-harm rule mitigates but does not totally prevent transboundary environmental and resource costs from occurring.

The principle of equitable and reasonable utilization aims to reduce transboundary costs of water uses as well. The principle requires that an international watercourse is used and developed with a view to attaining optimal and sustainable utilization of water resources, taking into account the interests of all the watercourse States concerned.[37] The factors relevant to the equitable and reasonable utilization include, for example,

---

[32] Agenda 21 Action Plan (1992 United Nations Conference on Environment and Development) paras 8.27–8.30. On the application of the polluter pays principle in water pollution prevention and control see para 18.40.

[33] International Law Association (ILA), 'Water Resources Law' (Berlin Conference 2004, Fourth Report), 30–31. In the ECE Water Convention the polluter pays principle is regarded as having primarily a national scope. UNECE, 'Guide to Implementing the Water Convention' (ECE/MP.WAT/39 United Nations 2013), 29.

[34] UN Watercourses Convention, arts 7, 21; ECE Water Convention, art 2. In the ECE Water Convention, art 1, the transboundary impact refers to any *significant adverse effect* on the environment resulting from a change in the conditions of transboundary waters while the UN Watercourses Convention, art 7, refers to all appropriate measures to prevent the causing of *significant harm.*

[35] UNECE, 'Guide to Implementing the Water Convention' (n 33), 19. See also Alistair Rieu-Clarke, Ruby Moynihan and Bjørn-Oliver Magsig, *UN Watercourses Convention: User's Guide* (IHP-HELP Centre for Water Law 2012), 117; and Attila Tanzi and Alexandros Kolliopoulos, 'The No-Harm Rule' in Attila Tanzi, Owen McIntyre, Alexandros Kolliopoulos, Alistair Rieu-Clarke and Rémy Kinna (eds), *The UNECE Water Convention on the Protection and Use of Transboundary Watercourses and International Lakes: Its Contribution to International Water Cooperation* (Brill 2015) 133, 140–41.

[36] UNECE, 'Guide to Implementing the Water Convention' (n 33), 117.

[37] UN Watercourses Convention, art 5; ECE Water Convention, art 2(2).

the economy of water use and the costs of measures taken to that effect.[38] The principle of equitable and reasonable utilization calls for the weighing and balancing of different interests of watercourse States, but in much the same way as the no-harm rule it does not prevent decisions that cause transboundary environmental and resource costs.[39]

In order to internalize part of the environmental costs into financial costs, the ECE Water Convention requires the prior licensing of waste-water discharges as well as the application of environmental impact assessment.[40] Also, a State must take measures to prevent water pollution at the source, and apply low and non-waste technology.[41] Stricter requirements must be imposed when the quality of the receiving water or the ecosystem so requires.[42] The UN Watercourses Convention, for its part, specifically demands the protection and preservation of international watercourses and the prevention, reduction and control of pollution.[43]

2.3 EU Water Law

The EU Water Framework Directive (WFD) requires the recovery of financial, environmental and resource costs of water services.[44] In general, the Directive is a far-reaching and significant piece of EU environmental legislation. It introduces a holistic river basin district approach to the management and protection of water resources through river basin management plans and programmes of measures.[45] The overall objectives are 1) to achieve a good status of waters and 2) to

---

[38] UN Watercourses Convention, art 6.

[39] See, on the principle of equitable and reasonable utilization in more detail, e.g. Attila Tanzi and Maurizio Arcari, *The United Nations Convention on the Law of International Watercourses* (Kluwer 2001), 95–142; and Owen McIntyre, 'The Principle of Equitable and Reasonable Utilisation' in Attila Tanzi, Owen McIntyre, Alexandros Kolliopoulos, Alistair Rieu-Clarke and Rémy Kinna (eds), *The UNECE Water Convention on the Protection and Use of Transboundary Watercourses and International Lakes: Its Contribution to International Water Cooperation* (Brill 2015), 146.

[40] ECE Water Convention, art 3(1); see UNECE, 'Guide to Implementing the Water Convention' (n 33), 28–9.

[41] ECE Water Convention, arts 2(3), 3(1).

[42] Ibid., art 3(1).

[43] UN Watercourses Convention, arts 20–21. See UNECE, 'Guide to Implementing the Water Convention' (n 33), 20.

[44] Directive 2000/60/EC of the European Parliament and of the Council of 23 October 2000 establishing a framework for Community action in the field of water policy [2000] OJ L327/1, art 9.

[45] Water Framework Directive, arts 1, 11, 13.

prevent the deterioration of water status.[46] The recovery of costs of water services provides one mechanism for the achievement of the objectives of the Directive.[47]

The WFD states that Member States must take account of the principle of recovery of the costs of water services in accordance with the polluter pays principle.[48] To assess the costs of water services, each Member State must undertake an economic analysis of water use for each river basin district.[49] Furthermore, water pricing policies must provide adequate incentives for water users to use water resources efficiently and the different water uses, disaggregated into at least industry, households and agriculture, must contribute adequately to the recovery of costs.[50]

A Member State may, however, in recovering the costs of water services have regard to the social, environmental and economic effects of the recovery as well as the geographic and climatic conditions.[51] A Member State may also derogate from the recovery of costs of a given water-use activity where this is in accordance with established practices and does not compromise the purposes and the achievement of the objectives of the Directive.[52] Overall, the obligation to recover the costs of water services is formulated rather weakly as a principle that must be taken into account and that includes exceptions which complicate its implementation.[53]

---

46 Water Framework Directive, art 4(1).

47 See Howarth (n 13), 565–6; and Petra E. Lindhout, 'A Wider Notion of the Scope of Water Services in EU Water Law: Boosting payment for water-related ecosystem services to ensure sustainable water management?' (2012) 8:3 *Utrecht Law Review*, 86, 87–8 <http://doi.org/10.18352/ulr.207> accessed 8 January 2017. See also Moran and Dunn (n 9), 484. On the difficulties to define the status of waters see Henrik Josefsson and Lasse Baaner, 'The Water Framework Directive – A Directive for the Twenty-First Century?' (2011) 23:3 *JEL*, 463; and on the legal status of the environmental objectives of the Directive J.J.H. van Kempen, 'Countering the Obscurity of Obligations in European Environmental Law: An Analysis of Article 4 of the European Water Framework Directive' (2012) 24:3 *JEL* 499.

48 On the relationship between the EU and Member States' legal systems see Merita Huomo-Kettunen, 'Heterarchial Constitutional Structures in the European Legal Space' (2013) 6:1 *European Journal of Legal Studies*, 47.

49 Water Framework Directive, arts 9(1), 5(1). See Howarth (n 13), 570–571.

50 Water Framework Directive, art 9(1).

51 Ibid.

52 Ibid., art 9(4).

53 Ibid., art 9(1).

It has been unclear what water uses are included in the concept of water services in the WFD and thus in the sphere of the recovery of costs.[54] According to Article 2(38)–(39) of the WFD, water services means 'all services which provide, for households, public institutions or any economic activity: (a) abstraction, impoundment, storage, treatment and distribution of surface water or groundwater, (b) waste-water collection and treatment facilities which subsequently discharge into surface water.' Water uses, for their part, are defined more widely to mean 'water services together with any other activity ... having a significant impact on the status of water.'[55] The innovativeness of the WFD – at least from the economic viewpoint – is indeed heavily related to the sphere of water uses under the recovery of costs obligation.

The key question is whether the concept of water services includes only water supply and wastewater services provided in a bilateral relationship (narrow interpretation) or also own use such as raw water abstraction for, for example, agricultural, energy or industrial purposes (broad interpretation). The European Commission has especially been concerned about inadequate water pricing and low cost recovery in the agricultural sector in Southern European countries where agriculture is the largest consumer of scarce water resources.[56] The traditional narrow interpretation of water services as services between a service provider and consumer basically requires no new cost recovery mechanisms since the costs of water supply and wastewater services are largely recovered by water utilities' charges. In comparison, the broad interpretation would mean that, basically, all water uses for e.g. irrigated agriculture, industry and hydropower in addition to traditional water services should be priced.

---

54 See Erik Gawel, 'Cost Recovery for Water Services in the EU' (2015) 50:1 *Intereconomics*, 40, 40–41.

55 On the different views on water services in legal literature see Howarth (n 13), 571; Petra E. Lindhout, 'A Wider Notion of the Scope of Water Services in EU Water Law: Boosting payment for water-related ecosystem services to ensure sustainable water management?' (2012) 8:3 *Utrecht Law Review* 86, 87–8 <http://doi.org/10.18352/ulr.207> accessed 8 January 2017; and Moritz Reese, 'Cost Recovery and Water Pricing in Water Services and Water Uses in Germany' (2013) 10:4 *Journal for European Environmental & Planning Law*, 355, 361–4.

56 Commission, 'Pricing policies' (n 15), 2–3. According to the EEA report, the levels of cost recovery of irrigation water pricing range from 20 per cent to 80 per cent in Mediterranean EU countries. In northern EU Member States, there are no separate water tariffs for irrigation because agriculture relies mostly on rain water. European Environment Agency (EEA), 'Assessment of cost recovery through water pricing' No 16/2013 *EEA Technical report* 10, 55.

It would also mean that market-based mechanisms based on the value of ecosystem services largely directed our water uses.[57]

The ECJ Case C-525/12 (Water Services Case) dealt especially with the scope of water services and water pricing. The Commission asked the Court to declare that Germany had failed to fulfil its obligations under Articles 2(38) and 9 the Water Framework Directive since it had excluded certain services from the concept of water services. These excluded services included water impoundment for the purposes of hydropower, navigation and flood protection as well as water abstraction for irrigation, industrial purposes and personal consumption. The Court dismissed the action of the Commission but instead of clarifying the scope of water services and cost recovery, it discussed the possible failure to meet the obligations of the Directive.[58]

The ECJ stated that although the various activities listed in Article 2(38) of the WFD 'may have an impact on the state of bodies of water and are therefore liable to undermine the achievement of the objectives pursued by that directive, it cannot be inferred therefrom that, in any event, the absence of pricing for such activities will necessarily jeopardize the attainment of those objectives.' Especially, the Court referred to the possibility of derogation from the recovery of costs for given water-use activities according to Article 9(4) of the WFD. According to the ECJ, the objectives of the Directive do not necessarily imply that all activities referred to in the definition of water services are subject to the principle of cost recovery. Therefore, the fact that Germany did not make some of those activities subject to the principle of recovery of costs 'does not establish by itself, in the absence of any other ground of complaint, that the Member State has thereby failed to fulfil its obligations under Articles 2(38) and 9' of the WFD.[59]

The ECJ strongly emphasized teleological interpretation based on the objectives of the WFD in the Water Services Case.[60] This is understandable since the wording of the water services provisions in the WFD is

---

[57] Brouwer (n 9), 21–2. Brouwer points out that he is convinced that the broad interpretation of water services would enhance the sustainability of water use.

[58] See Gawel, 'Cost Recovery for Water Services in the EU' (n 54), 41–2; and Erik Gawel, 'ECJ on Cost Recovery for Water Services under Article 9 of the Water Framework Directive: Camera Locuta Causa non Finita' (2015) 12 *Journal for European Environmental & Planning*, 71.

[59] Water Services Case (n 12) paras 56–9.

[60] Ibid., paras 43–5. On the teleological interpretation becoming more frequent in the field of environmental law see Douglas Fisher, *Legal Reasoning*

unclear, and the systematic interpretation did not provide much clarity on the concept of water services.[61] The ECJ stated that the WFD is a framework directive and that the EU legislature's aim was to allow some discretion on measures to apply the principle of cost recovery. While the legislature promoted the pricing of costs, it did not intend to extend the pricing to all services associated with water use since water-pricing practices varied notably between the Member States. Hence, the Court concluded that there is no general requirement to price all activities related to water use in EU law.[62]

Overall, the ECJ appears to support the broad interpretation of the concept of water services on the basis of teleological interpretation. One may conclude that the Court dismissed the action of the Commission only because the Commission failed to prove that Germany had jeopardized the achievement of these objectives by not making some water use activities subject to the principle of cost recovery.[63] Nevertheless, the judgment fails to clarify the precise content of water services.[64]

The Water Services Case left a lot of discretion to the Member States in applying the recovery of costs in practice since Member State freedoms are only limited by the objectives of the Directive.[65] The key

---

in Environmental Law: A Study of Structure, Form and Language (Edward Elgar Publishing 2013), 32–5.

61 Art 4(1) of the Internal Market Directive (Directive 2006/123/EC of the European Parliament and of the Council of 12 December 2006 on services in the internal market [2006] OJ L376/36) defines services as economic activities that are normally provided for remuneration, whereas in art 2(13) of the Environmental Liability Directive (Directive 2004/35/CE of the European Parliament and of the Council of 21 April 2004 on environmental liability with regard to the prevention and remedying of environmental damage [2004] OJ L143/56) natural resources services mean the functions performed by a natural resource for the benefit of another natural resource or the public.

62 Water Services Case (n 12) paras 47–52.

63 See also Commission, 'The Water Framework Directive and the Floods Directive: Actions towards the "good status" of EU water and to reduce flood risks' (Communication) COM (2015) 120 final 11. The Commission states that in the light of the judgment the recovery of costs potentially applies to a wide range of water services.

64 See Gawel, 'Cost Recovery for Water Services in the EU' (n 54), 41–3. Gawel points out that the judgment actually invites the Commission to take renewed action on the grounds of a Member State's substantiated failure to fulfil the obligations of the Directive.

65 See Petra Lindhout, *Cost recovery as a policy instrument to achieve sustainable and equitable water use in Europe and the Netherlands* (Uitgeverij 2VM 2015), 114. Lindhout's take is that '[t]he Court judgment positions cost

objective of the Directive is to maintain and improve the quality of the aquatic environment while the control of water quantity is an ancillary measure in securing the quality objectives.[66] If a Member State does not apply the recovery of costs for some water use activities, it nevertheless must explain (in river basin management plans), on the basis of Article 9(2) and Article 9(4) of the WFD, which other measures are in place to achieve the objectives of the Directive.[67]

2.4 Finnish Water Law

In Finland, the definition of water services in the Decree on Water Resources Management (1040/2006, Section 7 a) follows closely the Water Framework Directive but includes one difference in emphasis. According to the Decree, water services include surface water and groundwater abstraction, impoundment, storage, treatment and distribution as well as wastewater collection and treatment *that are provided as services*.[68] While the Finnish definition refers to a bilateral relationship between a service provider and user, it can be interpreted in conformity with EU law to include also own water uses.[69] The more important issue is how the recovery of costs and the polluter pays principle are transposed into substantive legislation.

---

recovery as a practically voluntary tool to be used by Member States at least offering the Member States a huge margin of appreciation.' See also Gawel, 'ECJ on Cost Recovery for Water Services' (n 58), 71–9.

[66] See EEA, 'Assessment of cost recovery' (n 56), 100–101. See also Unnerstall (n 15), 38; Howarth (n 13), 581–2; and Water Services Case (n 12) para 55.

[67] Commission, 'The Water Framework Directive and the Floods Directive' (n 63), 10.

[68] On the Finnish approach to the recovery of costs see *Commission v Germany* (n 9) paras 36–8. The Finnish Government supported the German reading of the water services concept and was of the opinion that the Commission's overly broad interpretation would upset the balance between the different water management instruments of the Water Framework Directive.

[69] On the interpretative effect of EU law see Alina Kaczorowska-Ireland, *European Union Law* (4th edn, Routledge 2016), 342–3; and Case C-282/10, *Maribel Dominguez v Centre informatique du Centre Ouest Atlantique, Préfet de la région Centre* [2012] ECLI:EU:C:2012:33, para 24. Kaczorowska-Ireland points out that the interpretative effect is based on art 4(3) TEU that introduces e.g. the principle of sincere cooperation between the EU and the Member States and the Member States' duty to ensure fulfilment of the obligations arising out of the Treaties or resulting from the acts of the EU institutions.

The principle of cost recovery is included in the Finnish Water Services Act (119/2001) that applies to the services of water utilities; the conducting, treatment and supply of water for household use; and the disposal and treatment of wastewater.[70] The Act requires that the charges for water and wastewater services cover all the investments in the building, repair and maintenance of the services' infrastructure as well as other costs of a water utility.[71] Thus, the Water Services Act covers the recovery of the financial costs of water utilities but it does not mention environmental and resource costs.[72]

The recovery of the environmental and resource costs of water uses is regulated by the Water Act (527/2011) and the Environmental Protection Act (527/2014). The Water Act is based on the principle of sustainable use of water resources while the Environmental Protection Act enacts that the polluter must prevent and mitigate pollution as far as possible.[73]

Both Acts require the permitting of projects that affect the water environment. According to the Water Act, all management projects of water resources that may cause harmful consequences to a water body or its aquatic environment are subject to a water permit whereas the Environmental Protection Act requires an environmental permit for activities that may cause pollution of a water body.[74] The thresholds for the application of water and environmental permits are low, and all the various activities listed in the definition of water services in the WFD require a water or environmental permit excluding water uses with very little or no impact on the environment.[75] A water permit is granted if the weight of the benefits gained from the project by public or private interests is considerable in comparison to the losses incurred for these interests while an environmental permit requires that the activity does not result in significant environmental pollution or risk thereof.[76]

---

[70] Water Services Act, ss 2–3.

[71] Water Services Act, s 18. See also Finnish Government Bill HE 218/2013, 51.

[72] See Antti Belinskij, 'Vesivarojen käytön kustannusten kattaminen EU:n ja Suomen oikeudessa' (2015) 3–4/2015 Lakimies 387, 400–401.

[73] Water Act, ch 1 s 1; Environmental Protection Act, s 7.

[74] Water Act, ch 2, ss 2–3; Environmental Protection Act, s 27.

[75] On water and environmental permitting in Finland see Pekka Vihervuori, 'Vesitaloushankkeet' in Kari Kuusiniemi, Ari Ekroos, Anne Kumpula and Pekka Vihervuori (eds), *Ympäristöoikeus* (SanomaPro 2013), 951; and Anne Kumpula, 'Ympäristönsuojelulaki' (in the same book), 1325.

[76] Water Act, ch 3 s 4; Environmental Protection Act s 42.

The Water Act and the Environmental Protection Act require responsible parties to minimize the harmful impacts of water resource management and wastewater discharge projects and the latter obligates operators to use the best available technology.[77] The permit decisions must include necessary regulations in these regards.[78]

However, it is obvious that water resource management and wastewater discharge projects may result in environmental and resource costs. In the following section, the instruments by which these costs are recovered at the different levels of water law are discussed.

3 INSTRUMENTS FOR THE RECOVERY OF COSTS

3.1 Theoretical Underpinnings

The use of economic instruments such as water pricing or water markets is one option for the allocation of water resources between competing uses. Traditionally, these allocations have been managed through a public or government allocation mechanism such as water and environmental permits.[79] However, the traditional regulatory mechanisms may not consider either the economic efficiency or the recovery of costs of water uses.[80] While it is difficult, if not impossible, to allocate water solely through economic measures,[81] the economic instruments may provide an

77 Water Act, ch 2 s 7; Environmental Protection Act, ss 7–8. See also Kumpula (n 75), 1341; and Erkki Hollo, *Maankäyttö- ja vesioikeus* (Talentum 2006), 484–5.

78 Water Act, ch 1 s 10; Environmental Protection Act s 43.

79 EEA, 'Towards Efficient Use of Water Resources in Europe' (n 22), 29. See also Ariel Dinar, Mark W. Rosegrant and Ruth Meinzen-Dick, 'Water Allocation Mechanisms: Principles and Examples' (World Bank Policy Research Working Paper No. 1779, June 1997), 6–17. Dinar *et al.* point out that water allocation may also be managed by user-based allocation where water users or their associations allocate water. On the Australian Water Market see Vicki Waye and Christina Son, 'Regulating the Australian Water Market' (2010) 22:3 *JEL*, 431.

80 EEA, 'Towards Efficient Use of Water Resources in Europe' (n 22), 29.

81 Ibid., 29.

effective addition to the regulatory mechanisms to enhance the achievement of environmental policy objectives in the management of water resources.[82]

In order to make use of economic instruments for the allocation of water resources, water must have an economic value. The economic value should provide incentives for water users towards rational and efficient uses of scarce water resources. To maximize the social benefits, water should be reallocated to uses with high social values. Water has direct values in relation to extractive and consumptive uses and indirect values regarding recreational and aesthetic uses.[83] Water pricing aims to raise consumer awareness of the value of water and combine the environmental and economic benefits of water use.[84] It may offer an efficient information system for water users if it reflects the true financial, environmental and resource costs of water production and consumption and wastewater discharges. In this way, water pricing could correct market failures and ensure the application of the polluter pays principle in the management of water resources.[85]

The level of water prices is of the utmost importance. When the prices are too low water pollution and the demand for water are excessive. In that case water and wastewater payments may legitimize environmental damages resulting from water uses. If the prices are too high many beneficial water use activities are prevented altogether.[86]

There are different ways to price water, from taxes to charges, levies and tariffs.[87] To provide incentives for the efficient use of water, water pricing should be based on a metered consumption instead of fixed rates. It should also take into account regional conditions and the characteristics of different water uses. The concrete examples on water pricing in the EU Member States in addition to water utilities' charges include charging for wastewater discharges and, to a lesser extent, for water abstractions and impoundments. Raw water abstraction charges have

---

[82] Commission, 'Bringing our needs and responsibilities together – Integrating environmental issues with economic policy' (Communication) COM (2000), 576 final, 3.

[83] Moran and Dunn (n 9), 484–5.

[84] Commission, 'A Blueprint to safeguard Europe's water resources' (n 4), 10.

[85] EEA 'Assessment of cost recovery' (n 56), 8.

[86] EEA, 'Towards Efficient Use of Water Resources in Europe' (n 22), 31. See also Howarth (n 13), 584.

[87] EEA 'Assessment of cost recovery' (n 56), 8.

often provided no incentives for efficient water use, whereas water tariffs and point-source pollution taxes and fees may be more useful for the purpose.[88]

Water pricing affects the sustainability of water use but its effect is limited to rare conditions where the short run marginal costs and the price elasticity of water uses are both high. The elasticity of water uses in relation to water pricing has been rather low, especially in agriculture.[89]

3.2 International Water Law

Perhaps the liveliest discussion on water pricing in international law deals with the human right to water. It is stated that water must be affordable for everyone and that States must adopt the necessary measures, such as free or low-cost water or income supplements, to guarantee its affordability.[90] The need to cover the costs of water services through water pricing is recognized but progress towards the full recovery of costs is often seen to threaten the affordability of water for vulnerable and marginalized groups.[91]

Within the regime of the ECE Water Convention it is suggested that the polluter pays principle could be implemented through fines and taxes in addition to permit procedures. Accordingly, the economic measures for the cost internalization of pollution activities should meet three criteria. They should: (1) be proportional in relation to the pollution, (2) have an appropriate level of financial significance to create an incentive to invest in pollution prevention and control and (3) have exemption clauses for those who invest significantly in pollution prevention. In order to remedy

---

88 EEA, 'Assessment of cost recovery' (n 56), 7–14, 22–3, 53, 64, 71–8, 88–91; Brouwer (n 9), 22–5. According to the EEA, the block rates for households may mean that the luxurious use of water e.g. for car washing and swimming pools costs more than those uses that are necessary e.g. for drinking and laundry. Brouwer brings forward the idea of payable environmental permits and their trade.

89 See EEA, 'Towards Efficient Use of Water Resources in Europe' (n 22), 30; EEA 'Assessment of cost recovery' (n 56), 20–24; and Brouwer (n 9), 25.

90 Committee on Economic, Social and Cultural Rights, 'General Comment No. 15: The Right to Water (Arts. 11 and 12 of the Covenant)' (E/C.12/2002/11, 2003) paras 25–7.

91 UNECE and WHO, 'No One Left Behind: Good practices to ensure equitable access to water and sanitation in the Pan-European region' (UN and WHO 2012), 2.

accidental pollution incidents, the polluter should also have financial guarantees such as environmental insurances.[92]

Also, the payments for ecosystem services have been discussed in the context of the ECE Water Convention. According to the UNECE, an economic analysis associated to these payments should involve four steps from the identification of (1) water-related ecosystem services and (2) major stakeholders, to the assessment of (3) the net benefits of changes in water-related ecosystem and (4) the distribution of costs and benefits among major stakeholders. The idea is to establish an effective market that compensates for the changes in land use and management practises enhancing ecosystem services. For example, public or private drinking water providers could compensate farmers for the non-application of nitrates or for the sustenance of the natural filtration of water.[93] However, it is difficult to see how this kind of payment scheme, where a farmer who has deteriorated ecosystems services could be paid for their reconstruction, enhances the implementation of the polluter pays principle.

In a situation where environmental and resource costs do occur in a transboundary setting there are some applicable but rather general international water law provisions on compensations. According to the UN Watercourses Convention, where significant harm is caused to another State, a State has to discuss the question of compensation where appropriate[94] while the International Law Association's (ILA) Helsinki Rules state that the substantial injury caused by water pollution in the territory of another State must be compensated.[95] In relation to the environmental costs, the ECE Water Convention requires that States take all appropriate measures to ensure the restoration of ecosystems where necessary.[96] The concepts of significant harm, where appropriate, and

[92] UNECE, 'Guide to Implementing the Water Convention' (n 33), 30.

[93] UNECE, 'Recommendations on Payments for Ecosystem Services' (n 16), 4–17. According to the UNECE, the schemes for the payments of ecosystem services may include public, private and trading schemes where e.g. a municipality purchases ecosystem services (public scheme), private entities act as buyers and sellers (private scheme) or water rights can be exchanged, sold or leased (trading scheme). The financial arrangements for these schemes may include, for example, direct compensations, customer-charged payments and tax-based contributions.

[94] UN Watercourses Convention, art 7.

[95] International Law Association (ILA), 'The Helsinki Rules on the Uses of the Waters of International Rivers' (International Law Association, fifty-second conference, Helsinki, August 1966), art XI.

[96] ECE Water Convention, art 2. On the relationship between the UN Watercourses Convention and ECE Water Convention on compensation see Attila

substantial injury are very ambiguous, and in general compensation is not seen as an absolute obligation of strict liability but rather as an obligation to hold consultations between different States on the equitable balancing of water use interests.[97] The threshold of compensation is high and compensations have not been discussed often in the framework on international water law.[98]

3.3 EU Water Law

Article 9(1) of the Water Framework Directive requires the Member States to ensure that water-pricing policies provide adequate incentives for users to use water resources efficiently, and thereby contribute to the environmental objectives of the Directive. Thus, the WFD includes a clear obligation to apply water pricing that concretizes a general requirement for the recovery of costs of water services.[99] Water pricing is regarded as a novel and innovative measure to guarantee the application of the polluter pays principle in relation to water uses.[100] According to the European Commission, appropriate water pricing is of the utmost importance in the development of sustainable water policies.[101] The Member States had to implement water-pricing policies by 2010 but it is still unclear how this requirement is fulfilled.[102]

Tanzi, 'The Economic Commission for Europe Water Convention and the United Nations Watercourses Convention: An analysis of their harmonized contribution to international water law' (ECE/MP.WAT/42 Water Series No 6, United Nations 2015), 34–8.

97 Tanzi (n 96), 34–8. The ILA Helsinki Rules (n 95), art 5, clarify the approach of international water law to compensation by referring to the 'practicability of compensation to one or more of the co-basin States as means of adjusting conflicts among uses' as one of the relevant factors for the equitable and reasonable utilization. See also Tanzi and Arcari (n 39), 163–72.

98 ILA, 'Water Resources Law' (n 33), 22, 27, 53. The ILA Berlin Rules, arts 21 and 70, link the duty to compensate especially with severe circumstances where persons or communities are displaced by or suffer other damages due to a water programme, plan, project or activity.

99 See Howarth (n 13), 571–2. On the complex drafting history of provisions on water pricing see Maria Kaika and Ben Page, 'The EU Water Framework Directive: Part 1. European Policy-Making and the Changing Topography of Lobbying' (2003) 13 *European Environment*, 314.

100 *Commission v Germany* (n 9) para 49.

101 Commission, 'Pricing policies' (n 15), 2, 16.

102 See EEA, 'Assessment of cost recovery' (n 56), 7–9; and Commission, 'A Blueprint to safeguard Europe's water resources' (n 4).

Water pricing should affect water users' behaviour in a way that supports the achievement of the environmental objectives of the WFD.[103] It should be based on water quantities used or pollution produced.[104] In the light of Article 9(1) of the WFD, the different water use sectors, disaggregated into at least industry, households and agriculture, must contribute to the costs of water services corresponding to the degree to which their activities give rise to cost increases.[105]

The European Court of Justice dealt with the question of the incentive pricing in the Case C-686/15 where it considered whether Article 9 of the WFD allows water pricing to include not only a variable component calculated according to the volume of water consumed, but also a fixed component covering charges for the connection of buildings to the water supply works. The ECJ stated, referring to the framework nature of the Directive, that the methods enabling the incentive pricing are left to the discretion of Member States. Furthermore, while the setting of the price according to the volume of water consumed constitutes one of these methods, the Member States may also adopt other water-pricing methods which, for example, enable the recovery of the costs of the availability of water distribution services. Thus, the ECJ ruled that the WFD does not preclude national legislation where the price of water services includes also a fixed component.[106]

The WFD includes two derogations from the water pricing requirements. First, the Member States may have regard for the social, environmental and economic effects of the recovery of the costs as well as the geographic and climatic conditions of the regions affected. Second, it might be decided, in accordance with established practices, not to apply water pricing for a given water use activity where this does not compromise the purposes and the achievement of the objective of the Directive.[107] These derogations enable specific water management concepts and instruments in different Member States although the use of national level discretion is guided by the principle of pricing water services.[108]

---

[103] See EEA 2013, 8–9.

[104] Commission, 'Pricing policies' (n 15), 2, 16.

[105] Howarth (n 13), 572–3.

[106] Case C-686/15, *Vodoopskrba i odvodnja d.o.o. v Zeljka Klafurić* [2016] ECLI:EU:C:2016:927, paras 16–28.

[107] Water Framework Directive, art 9(1,4).

[108] See Erik Gawel 'Article 9 of the EU Water Framework Directive: Do We Really Need to Calculate Environmental and Resource Costs?' (2014) 11 *Journal for European Environmental & Planning Law*, 249, 256.

Water pricing policies driven by the European Commission include some conflicting objectives. On the one hand, water pricing should be as simple and transparent as possible to provide incentives for sustainable water uses. The gains in efficiency achieved through a water pricing system should clearly exceed its establishment and management costs. On the other hand, water pricing should recover the costs of each water use sector, set different unit prices on them, and include an implementation plan that ensures its affordability and acceptability, especially for users that have not previously paid the full costs for water uses.[109]

Also, the relationship between water pricing and other measures to recover the costs of water uses is not straightforward in the WFD. The Directive requires water pricing to be one of the basic measures that is included in the programme of measures of a river basin district but there are also other measures such as regulatory provisions for the internalization of the costs of water uses.[110] Also, the ECJ pointed out in the Water Services Case that 'measures for the recovery of the costs for water services are one of the instruments available to the Member States for qualitative management of water in order to achieve rational water use' giving some leeway to the Member States in the interpretation of the requirement for water pricing.[111]

3.4 Finnish Water Law

In Finland, the Water Services Act, which regulates water utilities' water supply and wastewater services, includes provisions on water pricing. Accordingly, the charges for water services must cover all the costs of a water utility as well as be reasonable and equitable for consumers. The need to promote the sparing use of water or wastewater reductions and the exceptional quantity or quality of wastewater can be taken into account in the charges.[112] A water utility must collect a charge for the use of water services based on the quantity and quality of water and wastewater. In addition, it may collect connection and other fixed charges for the water services.[113]

In addition to the water pricing in the Water Services Act, Finnish law contains provisions on compensation for the loss of benefit resulting from operations performed under water and environmental permits. The Water

[109] See Commission, 'Pricing policies' (n 15), 16–19.
[110] Water Framework Directive, art 11.
[111] Water Services Case (n 12) para 55.
[112] Water Services Act, s 18.
[113] Water Services Act, s 19.

Act and the Environmental Protection Act require that any loss of benefit or damage to fish stock and fishing caused by a water resources management project or wastewater discharge is compensated by the party responsible for the project.[114] The permit authority must order compensations *ex officio* when granting a permit under the Water Act or Environmental Protection Act.[115]

The Water Act and Environmental Protection Act cover the compensations for the loss of benefit to property owners and other water users. The compensations are lump sum monetary compensations, unless there are specific reasons requiring their payment at intervals, and the loss of benefit is calculated on the basis of water use and wastewater allowances in the permits.[116] Thus, the environmental and resource costs that are not related to the direct loss of benefit to natural or legal persons are not compensated and the compensations provide no incentives for the permit holders to use water resources efficiently or reduce water pollution caused by wastewater.[117]

Finland had a specific water protection charge in use until 2000. If an operator paid a yearly charge, it was possible to derogate from the water protection requirements of the old Water Act (264/1961, Chapter 10, Section 27). The charge was abandoned since it was at odds with the requirement of the best available technology.[118]

The Water Framework Directive requires Finland to report on the recovery of costs and possible derogations from it in the river basin management plans (Article 9(2) and Article 9(4)). Currently, these plans only consider the recovery of the costs of water utilities' services. Own water uses have so far been excluded from reporting.[119]

---

114 Water Act, ch 13, ss 1, 14; Environmental Protection Act, s 125.

115 Water Act, ch 13, s 7; Environmental Protection Act, s 125. On the compensations in more detail see Vihervuori (n 75), 1037–1039; and Kumpula (n 75), 1490–96

116 Water Act, ch 13, ss 1, 13; Environmental Protection Act s 125.

117 See Johanna Ketola, 'Kustannusten kattamisen periaatteen toteutuminen Suomen vesilainsäädännössä' in *Ympäristöoikeudellisia tutkielmia*, (Suomen Ympäristöoikeustieteen Seura 2001), 153, 222–3.

118 1961 Water Act, ch 10 s 27. See Finnish Government Bill HE 84/1999 vp.

119 See eg Milla Mäenpää ja Sini Tolonen (eds), 'Kooste vesienhoitoalueiden vesienhoitosuunnitelmista vuoteen 2015' (Suomen ympäristö 23/2011, Suomen ympäristökeskus 2011), 43–4; and Pertti Manninen and Juho Kotanen (eds), 'Vesien tila hyväksi yhdessä: Ehdotus Vuoksen vesienhoitoalueen vesienhoitosuunnitelmaksi vuosiksi 2016–2021' (Etelä-Savon ELY-keskus, Kaakkois-Suomen ELY-keskus, Pohjois-Karjalan ELY-keskus ja Pohjois-Savon ELY-keskus 2014), 75–7.

4 CONCLUSIONS

The manifestations of the principle of cost recovery for water uses differ significantly at the international, EU and Finnish levels of water law. While international law does not recognize such a principle and Finnish law relates it to water utilities' water and wastewater services, the EU Water Framework Directive requires water pricing to recover the financial, environmental and resource costs for a wide range of water uses including own uses. It is understandable from the viewpoint of the objectives of the WFD that the European Court of Justice interpretation includes a very far reaching water pricing obligation. However, the interpretation is somewhat surprising considering that the wording of the cost recovery provisions in the WFD is unclear; the interpretation is not largely supported by international law and it was opposed by many Member States.

The main idea of the recovery of costs for water uses is to prevent, mitigate and compensate environmental and resource costs. International and Finnish water law mitigates the environmental and resource costs in transboundary and national settings but does not prevent them from occurring. At the international level, compensations are usually discussed only when States negotiate on the equitability of the uses of international waters or serious transboundary damages have taken place. Finnish law secures the recovery of the financial costs of water utilities' services and compensates the loss of benefit resulting from water and environmental permits but does not recover the environmental and resource costs that are not related to the loss of benefit to natural or legal persons. The water pricing provisions in the WFD target particularly the recovery of these traditionally unaccounted environmental and resource costs in line with the polluter pays principle.

On the one hand, it is difficult for the European Commission to control the national implementation of the principle of cost recovery set in the WFD. First, it is still unclear what the environmental and resource costs mean according to the WFD. Second, the obligation to water pricing is ambiguous because of the wording of the WFD and because the ECJ in the Water Services Case linked the recovery of costs with the objectives of the Directive and referred to the possibilities to derogate from the water pricing obligations. Third, in addition to water pricing there are other measures, such as river basin management plans as well as water and environmental permitting, that are to be used to achieve the objectives of the Directive. On the other hand, while the judgment of the ECJ in the Water Services Case left a lot of discretion to the Member States, it nevertheless enables the European Commission to demand broad reporting on the recovery of costs for water uses in the river basin management plans. In the future, when a Member State leaves many water uses outside its water pricing policies and does not achieve the environmental objectives of the WFD, the Commission will probably initiate another application to ask the ECJ to declare an infringement of the WFD.

It is a good question whether Finland must make changes to its water pricing policies to implement the requirements of the WFD since, in Finland, water pricing covers only water utilities' financial costs. The Member States have a wide discretion in the recovery of costs for water uses and the quantitative use of water does not appear to jeopardize the attainment of the objectives of the WFD in Finland since only 2 per cent of the total water flow is utilized for water uses.[120] Nevertheless, Finland must at least widen its reporting on the recovery of costs in the river basin management plans to cover all significant water uses.

Water pricing and other economic instruments alone are insufficient methods for the achievement of water protection objectives. The value assigned to water can provide information to water users on the financial, environmental and resource costs of the uses. However, it is difficult to set this value correctly in such a way that it reflects the true environmental and resources costs and for it to steer towards the efficient uses of water without a very complicated water pricing system. Also, the price elasticity of water demand varies a lot between different water uses and is especially low in agricultural irrigation. Low wastewater prices may at worst legitimize the damages caused for other water users with no positive effects to the environmental objectives.[121] Therefore, the prevention of the harmful impacts of water uses requires a strong regulatory system that can then be complemented through water pricing policies.

Finally, it is necessary to ask what the polluter pays principle means in relation to the recovery of costs at the different levels of water law. In international water law, the approach to the polluter pays principle is somewhat dismissive in the sense that, for example, the ILA states that it cannot be concluded to be part of customary international law. Also, in Finland, many environmental and resource costs remain unaccounted for.

---

[120] On the amount of water flow used in Finland see Heikki Pajula and Juha-Pekka Triipponen (eds), '*A survey of the irrigation in Finland – case study Southwestern Finland*' (Finnish Environment Institute 2003), 7.

[121] See also Howarth (n 13), 584.

After all, the broad water pricing obligation of the WFD represents, despite its shortcomings, the first serious attempt to implement the polluter pays principle in Finnish water policies.

PART III

Transboundary water management

9. Tensions and relationships in international water law

Tuomas Kuokkanen[1]

1 INTRODUCTION

Clyde Eagleton has noted that when a river 'flows through two states a fertile source of dispute is found'.[2] While many water-related conflicts and disputes confirm this, in many instances the reverse is also a possibility: water can be a source for cooperation between states. Indeed, there are various interests and relationships relating to water. For example, some countries might have a more beneficial geographical position than others in relation to watercourses. Often there can be both short-term and long-term interests related to water. While water is usually a useful element, it can also be dangerous, as in cases of floods. Furthermore, water serves as a natural resource subject to human consumption and, at the same time, plays an important role in the ecosystem.[3]

This chapter discusses water from the perspective of how international law has sought to resolve, regulate and manage various problems and relationships relating to its protection and utilization. In order to better understand water-related issues and the relationships between them, the chapter distinguishes four dimensions: subjective, spatial, temporal and material.[4] The subjective dimension includes, first of all, states and

[1] The author has participated in the ENJUSTESS-project of the Sustainable Governance of Aquatic Resources Programme, financed by the Academy of Finland (263403).

[2] Clyde Eagleton, *International Government* (revised edn, The Ronald Press Company 1948), 65.

[3] Tuomas Kuokkanen, 'International Law and Water' in *International Environmental Law-making and Diplomacy Review* 2004 (University of Eastern Finland – UNEP Course Series 1 2005), 167.

[4] See the analysis by Sir Gerald Fitzmaurice of the term 'jurisdiction': Gerald Fitzmaurice, 'The Law and Procedure of the International Court of

international organizations, but also covers various other actors such as non-governmental organizations. The spatial dimension refers to areas and territories, in particular, adjacent to boundary waters or international watercourses. While the temporal dimension refers to issues involving time, the material dimension refers to various substantive interests involved in water-related issues. Each of these dimensions may be subject to water-related tensions, benefits, synergies, risks, problems and conflicts and may have a number of links and relationships with the other dimensions.

Having discussed four dimensional aspects of water, the chapter explores different tools and methods provided by international law to deal with water-related issues. To this end, the chapter discusses three broad themes: dispute settlement, regulations and management. Examples from water-related cases are used in order to discuss different dimensions and relationships. However, the objective is not to engage in presenting case studies, but to highlight points relevant to the chapter's theme. Furthermore, this does not aim to be an exhaustive inquiry into all the issues relating to international watercourses. Rather, its purpose is to understand water-related dimensions and relationships and set out the legal tools that may be used to deal with them.

2 DIMENSIONS RELATING TO WATER

2.1 The Spatial Dimension: The Tension Between International and National Jurisdictions

Boundary waters refer to such waters as rivers, lakes, and canals, parts of which are situated in different countries. They are called boundary waters either because they form a boundary between states or because they run through one. For instance, in many cases, state boundaries have been drawn to follow rivers or a watershed for easy recognition.[5] As Berber notes, 'water which is today in the territory of one state and therefore a part of its state territory will flow tomorrow into the territory of another state and become part of that state's territory'.[6] This fungible quality of

Justice', 34 *British Yearbook of International Law 1958*, 8–9. See also Veijo Heiskanen, 'Ménaga à trois? Jurisdiction, Admissibility and Competence in Investment Treaty Arbitration' *ISCID Review* (2013), 1–16.

[5] Tuomas Kuokkanen, *International Law and the Environment. Variations on a Theme* (Kluwer Law International 2002), 107.

[6] F.J. Berber, *Rivers in International Law* (Stevens & Sons 1959), 4.

water has been particularly important in relation to safeguarding the freedom of navigation.

Boundary waters are also called international watercourses because they are already, by definition, 'international'. Conversely, states do not have an interest in regulating internal waters that do not affect their boundary waters. For instance, the crux of the case concerning the diversion of water from the river Meuse, which related to the use of the canal known as Zuid-Willemsvaart, was the conclusion that the two parties had accepted limitations on their sovereignty only within the area governed by the treaty between them. Outside this area, the parties were free to take any action provided that it did not violate the treaty.[7]

The sphere of international jurisdiction has gradually expanded in international water law even though states have been reluctant to accept interference in their domestic jurisdiction. In order to regulate various other interests and concerns, states began to conclude treaties relating to watercourses in the nineteenth and twentieth centuries. Moreover, as early as the nineteenth century, a number of international organizations were established to secure the international administration of boundary waters.[8] Through the establishment of joint bodies and regulations, international jurisdiction began to develop. Nevertheless, a tension remained between national and international jurisdiction.[9]

The River Oder Case is a good example of the tension between international and domestic jurisdiction.[10] Pursuant to the Treaty of

---

[7] See *The Diversion of Water from the Meuse* (Netherlands *v.* Belgium), Judgment, 1937 PCIJ Series A/B No. 70, 26 ('As regards [canals which are situated in Dutch or Belgian territory and do not leave that territory], each of the two States is at liberty, in its own territory, to modify them, to enlarge them, to transform them, to fill them in and even to increase the volume of water in them from new sources, provided that the diversion of water at the treaty feeder and the volume of water to be discharged therefrom to maintain the normal level and flow in the Zuid-Willemsvaart is not affected').

[8] Kuokkanen (n 5), 165–6.

[9] See *Brownlie's Principle of Public International Law* (8th edn by James Crawford, Oxford University Press 2012), 337 ('Conceivably a body of water could be 'internationalized', that is, given a status entirely distinct from the territorial sovereignty and jurisdiction of any state, on the basis of treaty or custom, general or regional').

[10] River Oder Case (Great Britain, Czechoslovakia, Denmark, France, Germany, and Sweden v. Poland). The judgment of the Permanent Court of Justice is reproduced below from 1929 PCIJ Series A, No. 23, 5–32. See also Johan Lammers, *Pollution of International Watercourses* (Martinus Nijhoff Publishers 1984), 505–7.

Versailles concluded after World War I, the Oder, Ultava, Elbe, Niemen and Danube were declared international rivers. Article 341 of the Treaty provided for the Oder to be placed under the administration of an international commission comprising representatives of Poland, Prussia, Great Britain, the Czechoslovak State, France, Denmark and Sweden. Soon after the International Commission had commenced its work, a problem arose as to whether the Commission had jurisdiction over two tributaries of the Oder, the Warthe and the Netze, situated in Polish territory. At one of the sessions of the Commission, the Polish delegate argued that the Warthe and the Netze should be internationalized only up to the Polish frontier, while the other six delegates took the view that internationalization should be extended up to the navigable portions of tributaries situated in Polish territory. After the conciliation proposal put forward by the Advisory and Technical Committee for Communications and Transit of the League of Nations was rejected, the issue was submitted by a Special Agreement to the Permanent Court of International Justice. The Agreement was concluded by Great Britain, Czechoslovakia, Denmark, France, Germany, Sweden and Denmark on the one side and Poland on the other. Pursuant to the Special Agreement the following questions were submitted to the Permanent Court of Justice: 'Does the jurisdiction of the International Commission of the Oder extend, under the provisions of the Treaty of Versailles, to the sections of the tributaries of the Oder, Warthe (Warta) and Netze (Notec) which are situated in Polish territory, and, if so, what is the principle laid down which must be adopted for the purpose of determining the upstream limits of the Commission's jurisdiction?'[11]

In its judgment of 10 September 1929, the Court was of the opinion that the jurisdiction of the International Commission of the Oder extended to the sections of the Warthe and the Netze which were situated in the Polish territory. From a water law point of view, there has been particular interest in the part of the Court's judgment in which it considered the principles governing international fluvial law in general.[12] The Court opined as follows:

> [W]hen consideration is given to the manner in which States have regarded the concrete situations arising out of the fact that a single waterway traverses or separates the territory of more than one State, and the possibility of fulfilling the requirements of justice and the considerations of utility which

---

[11] River Oder Case (n 10), 6.

[12] See editorial note in Tuomas Kuokkanen (ed.), *Seminal Cases of International Environmental Law* (Edita 1999), 47.

> this fact places in relief, it is at once seen that a solution of the problem has been sought not in the idea of a right of passage in favour of upstream States, but in that of a community of interest of riparian States. This community of interest in a navigable river becomes the basis of a common legal right, the essential features of which are the perfect equality of all riparian States in the use of the whole course of the river and the exclusion of any preferential privilege of any one riparian State in relation to the others.[13]

As water pollution problems emerged, the process of internationalization was pushed further. For instance, in order to effectively protect joint watercourses, it is not sufficient to regulate only the boundary area, but rather extension of the geographical scope of such regulation to cover the whole drainage basin is required. Moreover, the need to mitigate climate change and adapt to its effects has extended international jurisdiction even further.

2.2 The Subjective Dimension: The Tension Between Upstream and Downstream Countries

The subjective dimension refers to various actors operating in respect of international watercourse issues: states, international organizations, non-governmental organizations, corporations, indigenous peoples and individuals. While states and international organizations are formal subjects of international law, other actors also play a role in international water governance.

States are the traditional subjects of international law. The idea of international law as law between sovereign states dates back to the 1648 Treaty of Westphalia, which established that states were masters within their territory and equal in their relations with other sovereigns.[14] Classical international law focused, in particular, on conflicts between sovereign states.

Sovereignty comprises two aspects. On the one hand, territorial sovereignty allows a state to use its territory, and on the other hand territorial integrity protects a state against harm from another state. That said, it is understandable that states pursue their own national interests. In the context of watercourses, this has in many instances led to tensions and conflicts between upstream and downstream states. One classical example is the conflict between the United States and Mexico which

---

[13] River Oder Case (n 10), 27.

[14] David Kennedy, 'International Law and the Nineteenth Century: History of an Illusion' (1996) 65 *Nordic Journal of International Law*, 403.

arose during the latter part of the nineteenth century concerning the use of the waters of the Rio Grande.

The Rio Grande, called the Rio Bravo in Mexico, has its source in the State of Colorado, United States and empties into the Gulf of Mexico. After flowing for about one-third of its length, it becomes a boundary water between the United States and Mexico. In 1895, Mr Matías Romero, the Mexican Minister to the United States sent a letter to Mr Richard Olney, the United States Secretary of State, in which he protested that the diversions of water from the Rio Grande by farmers and ranchers in Colorado and New Mexico reduced the water supply available to the Mexican communities in the vicinity of Ciudad Juarez. The diversions of water from the upper Rio Grande occurred at a point where the river was fully within the territory of the United States. The Minister emphasized that because Ciudad Juarez and the adjacent region downriver were situated in a dry zone they were obliged to depend upon irrigation from the Rio Grande. According to Mr Romero, the people of Ciudad Juarez had used the water of the Rio Grande for more than 300 years until the diversions of water increased on the United States' side of the border. In addition, he argued that, pursuant to the boundary conventions of 1848 and 1884, one country could not build any works that obstructed the navigation of the river without the consent of the other.[15]

The Secretary of State of the United States referred the issue to Judson Harmon, the Attorney General, for his legal opinion. The Attorney General submitted his opinion on 12 December 1895.[16] In his opinion, Harmon stated that he had not been able to find any support for the Mexican claim in the doctrine of international servitudes and was convinced of the supremacy of absolute sovereignty over international servitudes. Harmon noted that '[t]he fundamental principle of international law is the absolute sovereignty of every nation, as against all others, within its own territory'. The Attorney General concluded his examination by stating that international law imposed no liability or obligation upon the United States.

---

15 See Kuokkanen (n 12), 7–8.

16 The text of the official opinion by Attorney-General Harmon is reproduced from *Official Opinions of the Attorneys-General of the United States, Advising the President and Heads of Departments in Relation to Their Official Duties*, (ed. E.C. Brandenburg, Washington, 1898, vol. XXI), 274–83.

The doctrine propounded by Judson Harmon has become known as the *Harmon doctrine*.[17] Under it, a state wields absolute sovereignty with regard to that part of a river that lies within its territory. The state so situated is free to divert and use the river in any way it finds appropriate without liability to the state downstream. While a state could give, upon consideration of comity, consent to restrain its sovereignty, this is not inconsistent with the absolute nature of sovereignty, but is a reflection of it. Today, the Harmon doctrine is mainly of historical value.[18] Furthermore, the US State Department abandoned it after the Columbia River controversy with Canada where Canada was the upstream country.[19] Nevertheless, tensions between upstream and downstream countries remain in relationships between riparian states.

The first international waterway administration was established in 1804 to deal with the navigation of the Rhine. In 1821, a river commission was established to oversee navigation of the Elbe. The European Danube Commission, established by the 1856 Treaty of Paris, consisted not only

---

[17] For discussion of this doctrine, see J. Austin, 'Canadian-United States Practice and Theory Respecting the International Law of International Rivers: A Study of the History and Influence of the Harmon Doctrine' (1995) XXXVII *The Canadian Bar Review*, 393–443; Stephen C. McCaffrey, 'The Harmon Doctrine One Hundred Years Later: Buried Not Praised' (1996) 36 *Natural Resources Journal*, 965–1007; J. Simsarian, 'The Diversion of Waters Affecting the United States and Mexico' (1939), 27 *Texas Law Review*, 27–61; Kuokkanen (n 5), 9–24.

[18] Discussing the case of *G.J. Bier v. Mines de Potasse*, Antonio Cassese notes as follows: 'The Court held that in the last decade a general principle had evolved whereby the upstream users of an international river are no longer entitled to the unrestricted use of (the waters) of such a river, and are bound, when taking decisions concerning its use, to take reasonable account of the interests of other users in downstream areas'. Antonio Cassese, *International Law* (2nd edn, Oxford University Press 2005), 490.

[19] US Senate, Diversion of Water from Lake Michigan, Hearings before a Subcommittee of the Committee on Public Works, Letter of 29 July 1958, quoted by Johan Lammers in Lammers (n 10), 274–5. The letter stated as follows: 'The Department of State believes that a reiteration of the Harmon doctrine by any branch of the US Government would not be in the interests of this country or in line with the progressive development of international law during the last sixty years. The position which is being taken by the United States in its negotiations with Canada regarding the development of the Columbia-Kootenay river system is that there is no basis under either the Boundary Waters Treaty of 1909 or customary international law for any contention that Canada has a legal right unilaterally to divert within its own territory certain waters of this system, if such diversion may cause material injury to power and navigation interests on the Columbia river in the United States.'

of representatives of riparian but also of non-riparian states. Using the model of the Danube administration, the International Commission for the Navigation of the Congo was established in 1885. These early international commissions paved the way for later international organizations and bodies. As water pollution problems emerged, riparian states also began to establish joint bodies for the protection of international watercourses against pollution. For instance, the International Commission for the Protection of the Mosel against Pollution was established in 1961[20] and the International Commission for the Protection of the Rhine against Pollution in 1963.[21]

More recently, institutional arrangements for water bodies have developed in relation to the establishment of dynamic processes and frameworks which are characterized by being referred to as regimes. As it has been found that non-governmental organizations, the private sector, scientific experts, individuals, and indigenous peoples[22] can all make a contribution to the management of water, what was historically the province of state-only-governance has been broadened to encompass other actors. The Commission on Global Governance described this development very well: 'Governance is the sum of the many ways individuals and institutions, public and private, manage their common affairs. It is a continuing process through which conflicting or diverse interests may be accommodated and co-operative action may be taken.'[23]

2.3 The Temporal Dimension: The Tension Between Short-term and Long-term Interests

Many water-related issues have time-related dimensions, which may relate to past, current or future circumstances. For example, in many instances states are required to provide prior and timely notification on certain issues. According to the precautionary principle, lack of full

[20] Protocol Concerning the Constitution of an International Commission for the Protection of the Mosel against Pollution, Paris, 20 December 1961, 940 *United Nations Treaty Series*, 211.

[21] Agreement Concerning the International Commission for the Protection of the Rhine against Pollution, Bern, 29 April 1963, 994 *United Nations Treaty Series*, 3.

[22] See Dinah Shelton, 'Water Rights of Indigenous Peoples and Local Communities', in Laurence Boisson de Chazournes, Christina Leb and Mara Tignino (eds), *International Law and Freshwater. The Multiple Challenges* (Edward Elgar Publishing 2013), 69–93.

[23] Our Global Neighbourhood: The Report of the Commission on Global Governance (1995), 2.

scientific certainty shall not be used as a reason for postponing measures to prevent environmental degradation. Furthermore, the developmental and environmental needs of present and future generations have to be met pursuant to the doctrine of sustainable development.

The case concerning the Gabcíkovo-Nagymaros Project,[24] in which the International Court of Justice gave a judgment on a protracted dispute between Hungary and Slovakia over the construction and operation of dams on the river Danube, involved several time-related issues. The case concerned a treaty concluded in 1977 by Hungary and Czechoslovakia for the building of dam structures for the production of electric power, flood control and improvement of navigation on the Danube. In 1989, Hungary suspended and subsequently abandoned completion of the project, arguing that it posed grave risks to the Hungarian environment and the water supply for Budapest. Slovakia, a successor country to Czechoslovakia, insisted that Hungary fulfil its treaty obligations. Nonetheless, Slovakia planned and subsequently put into operation an alternative project only on Slovak territory, the operation of which affected Hungary's access to water from the Danube.

With regard to the temporal dimension of the case, the Court found that Hungary was not entitled to suspend and subsequently, in 1989, abandon work on the Nagyrnaros Project and on that part of the Gabcikovo Project for which responsibility was attributed to it under the 1977 Treaty and related instruments.[25] The Court also found that Czechoslovakia was entitled to proceed, in November 1991, to plan an alternative project, the so-called Variant C, but that it was not entitled to put that Variant into operation from October 1992.

The Court did not consider that the new developments that had taken place in respect of environmental knowledge and environmental law could be said to have been completely unforeseen. What is more, the formulation of certain provisions, designed to accommodate change, in the treaty between Czechoslovakia and Hungary, made it possible for the parties to take account of such developments and to apply them when implementing those provisions. The changed circumstances advanced by Hungary were, in the Court's view, not of such a nature, either individually or collectively, that their effect radically transformed the extent of the obligations still to be performed in order to complete the project. Consequently, the Court held that the notification of termination by

---

[24] Case Concerning the Gabcíkovo-Nagymaros Project (Hungary/Slovakia), ICJ Reports 1997, 7.

[25] Ibid., para 59.

Hungary of 19 May 1992 did not have the legal effect of terminating the 1977 Treaty and related instruments.[26]

Discussing the development of environmental regulations, the Court noted that, throughout the ages, mankind has constantly interfered with nature, and that in the past this was often done without consideration of the effects upon the environment. The Court pointed out that new norms and standards have been developed over the last two decades on the basis of new scientific insights and a growing awareness of the risks for present and future generations. The Court took the view that such new norms have to be taken into consideration not only when States contemplate new activities but also when continuing with activities begun in the past. Therefore, the Court held that the parties should together look afresh at the effects on the environment of the operation of the Gabcikovo power plant.[27] Having found both states to be in breach of their legal obligations, it called on them to give effect to the treaty between them while taking account of the factual situation that had developed since 1989.

Overall, the reconciliation of short-term economic interests and long-term environmental concerns requires optimization. Moreover, timing issues involve both proactive and retroactive considerations to cover both *ex ante* and *ex post facto* situations. For instance, many extreme water-related events, such as floods, appear to be difficult to control.[28] Therefore, as opposed to aiming merely at prevention, there is a need to promote preparedness for and responses to the effects of such events.

2.4 The Material Dimension: The Tension Between Utilization and Protection

Water is a good example of a natural resource that has a double function: on the one hand, it serves as natural resource subject to human utilization, and on the other, it has an ecosystem function.[29] While water may be used for purposes such as navigation and hydroelectric power, it also

---

[26] Ibid., para 115.

[27] Ibid., para 140.

[28] See, e.g. *Water and Climate Change Adaptation. Policies to Navigate Uncharted Waters*, OECD Studies on Water (OECD 2013).

[29] See UNGA, 'The Future We Want', Outcome Document, UNGA Res. 66/288, 27 July 2012, para. 122 ('We recognize the key role that ecosystems play in maintaining water quantity and quality and support actions with the respective national boundaries to protect and sustainably manage these ecosystems').

has an important ecosystem role. This dual character reflects different interests and concerns that states have in relation to water.[30]

In *Spain v Council of the European Union*,[31] Spain brought an action for annulment of Council Decision 97/825/EC.[32] The main issue in the dispute was whether the Convention on cooperation for the protection and sustainable use of the river Danube approved by Decision 97/825/EC concerned the protection and improvement of the quality of the waters of the Danube or the use of those waters and their management in its quantitative aspects. Spain argued that the legal basis for the Council Decision was inappropriate, and it should have been adopted on the basis of Article 130s(2) of the EC treaty (which required unanimity as the decision concerned 'management of water resources'). The Spanish Government took the view that the Convention related exclusively to the management of the water resources of the catchment area of the Danube.[33]

The Council argued that the Decision was correctly based on Article 130s(1) of the EC Treaty as that was the correct legal basis for the adoption of measures designed to improve the quality of water. That legal basis allowed decisions to be made by majority voting. The Council further argued that the Convention 'aims to promote at international level measures designed to deal with a very specific regional environmental problem, namely the pollution of one of the longest rivers in Europe'. It also noted that the Convention 'also claims to preserve, protect and improve the quality of the environment and to encourage the prudent and rational use of the resources of the river Danube'.[34]

In its findings the Court noted that Article 130s(2) refers to measures which regulate the *quantitative* aspects of the use of water resources and not those concerning the improvement and protection of the *quality* of those resources. It also noted that 'the concept of management of water resources does not cover every measure concerned with water, but covers only measures concerning the regulation of the use of water and the management of water in its quantitative aspect'. Thereafter, the Court

---

30 Kuokkanen (n 3), 167.

31 Case C-36/98 *Kingdom of Spain v Council of the European Union* [2001] ECR I-00779.

32 Council Decision 97/825/EC of 24 November 1997 concerning the conclusion of the Convention on cooperation for the protection and sustainable use of the river Danube [1997] OJ L342/18.

33 Case C-36/98, *supra* n 31, para 15.

34 Ibid., para 19. France, Portugal, Finland and the Commission intervened in the case to support the Council.

examined in a detailed manner the aim and the content of the Convention. It came to the conclusion that the primary purpose of the Convention is 'the protection and improvement of the quality of the waters of the catchment area of the river Danube, although it also refers, albeit incidentally, to the use of those waters and management in its quantitative aspects'.[35] The Court also pointed out that if a Community measure 'pursues a twofold purpose or that it has a twofold component and if one of these is identifiable as the main or predominant purpose or component, whereas the other is merely incidental, the act must be based on a single legal basis'.[36] Thus, the Court came to the conclusion that the Council had used the correct legal basis and therefore dismissed the action brought by the Spanish Government.[37]

Overall, there are variations in respect of how the relationship between water utilization and water protection appears. First, the relationship may be neutral, in which case there is no need to make a distinction between utilization and protection because both can be regarded as reflections of territorial sovereignty. For example, states can rely on sovereignty, on the one hand to justify utilization of water recourses, and on the other hand, to argue in favour of water protection. Second, legal issues relating to utilization and protection can be treated separately so that they hardly interact. For instance, issues relating to hydroelectric power are regarded as falling under energy law, while water protection is deemed to be part of environmental law. Third, an attempt can be made to integrate both aspects into the same sustainable framework in order to find an optimal solution.[38]

---

[35] Ibid., para 74.

[36] Ibid., para 59.

[37] Ibid., paras 75–6.

[38] For discussion see, e.g. Ulrich Beyerlin and Thilo Marauhn, *International Environmental Law* (Hart Publishing 2011), 87 ('The relationship between the use of international freshwaters and their protection has become one of the most pressing issues in international policy-making. To name some of the causes of the current crises, firstly, environmental degradation and overexploitation reduce the available amount of freshwater; secondly, increasing agricultural and industrial demand, and pressure from population growth contribute to water scarcity; thirdly, the infrastructure for water delivery is deficient in many states. Conflict interests on the use of freshwater resources are not new. However, today, water scarcity has reached an unprecedented level, and some even fear that wars will be fought over water').

3 APPLYING TECHNIQUES OF INTERNATIONAL LAW TO MANAGE WATER-RELATED TENSIONS AND RELATIONSHIPS

3.1 Recourse to Dispute Settlement to Resolve Contentious Relationships

Since early water-related issues were of a bilateral and sporadic nature, it was sufficient to seek to solve them retroactively through the settlement of disputes. In fact, the existence of a dispute is a necessary prerequisite for recourse to judicial dispute settlement, as the Court of International Justice, in particular, does not deal with issues *in abstracto*.[39] In the *Nuclear Tests Cases* the International Court of Justice pointed out that judicial settlement may provide a path to international harmony in circumstances of conflict.[40]

Current methods of dispute settlement developed from the nineteenth-century arbitral tribunals through the Permanent Court of International Justice and then the International Court of Justice. The *Lake Lanoux* arbitration is a good example of an arbitral tribunal managing to settle a water-related case by applying the burden of proof and other judicial techniques.

With regard to the spatial dimension, Lake Lanoux is one of the largest Pyrenean lakes. It lies in French territory at an altitude of approximately 2174 metres. The lake empties through a single stream, the Font-Vive, which flows into the River Carol. After having flowed about 25 kilometres from Lake Lanoux, the Carol crosses the Spanish border and continues its course in Spain for approximately 6 kilometres before joining the River Sègre, which eventually flows into the Mediterranean. Thus, the River Carol is a boundary river between France and Spain. The border between Spain and France was determined by three treaties, signed at Bayonne on 1 December 1856, 14 April 1862 and 26 May

---

[39] Nuclear Tests Case, New Zealand *v.* France, Judgement, ICJ Reports 1974, paras 61–2; Nuclear Tests Case, Australia *v.* France, Judgement, ICJ Reports 1974, 253, paras 58–9.

[40] Nuclear Tests Case, New Zealand *v.* France, Judgement, ICJ Reports 1974, 457, para 61 ('While judicial settlement may provide a path to international harmony in circumstances of conflict, it is nonetheless true that the needless continuance of litigation is an obstacle to such harmony').

1866. These treaties were accompanied by an Additional Act, also concluded at Bayonne on 26 May 1866.[41]

As to the material dimension, France and Spain had different interests. While France sought to utilize hydroelectric power, Spain was concerned about the possible adverse effects of this. As early as 1917, the French authorities drew up plans to divert the waters of Lake Lanoux towards the River Ariège and from there towards the Atlantic in order to use a natural drop of about 800 metres between the lake and the river for the purposes of generating electric power. The Spanish Government objected that this plan would affect Spanish interests and requested that it should not be carried out without its consent. Thereafter, the issue of the use of the waters of Lake Lanoux was subject to an exchange of views and negotiations between the two countries. On 21 September 1950, l'Électricité de France applied to the French Ministry for Industry for permission to divert the waters of Lake Lanoux. According to the plan by Électricité de France, an equal quantity of water to be diverted from the lake would be returned via an underground tunnel to the River Carol. The plan was communicated to Spanish authorities in 1952, following which it was considered by the Mixed Commission of Engineers and a Special Mixed Commission. Spain, however, maintained its objection to the project.

As the two governments were not able to settle the dispute, they decided to submit the matter to arbitration pursuant to the Arbitration Treaty of 10 July 1929. By a *Compromis* signed in Madrid on 19 November 1956, the two parties referred to the tribunal the question of whether the French Government was committing a breach of the provisions of the 1866 Treaty of Bayonne and the 1866 Additional Act in carrying out, without prior agreement with Spain, works for the utilization of the waters of Lake Lanoux. The French Government argued that the project would not be injurious to any of the rights or interests envisaged in the Treaty of Bayonne of 26 May 1866 and in the Additional Act. However, the Spanish Government objected that the project would be injurious to Spain's interests and rights in view of the fact that it altered the natural conditions of the hydrographic basis of Lake Lanoux and made the restitution of the waters to the Carol dependent upon human will, which would result in one party being in a

---

[41] Boundary Treaty between France and Spain, Bayonne, 14 April 1862, 125 *Consolidated Treaty Series*, 455; Boundary Treaty between France and Spain, Bayonne, 8 December 1856, 116 *Consolidated Treaty Series*, 85; Treaty of Delimitation between France and Spain along with the Additional Act, Bayonne, 26 May 1866, 132 *Consolidated Treaty Series*, 359.

stronger position than the other in contradiction to the preservation of the equality of the two parties provided for under the Treaty of Bayonne of 26 May 1866 and the Additional Act. The Spanish Government further asserted that under the Additional Act the project required the prior agreement of the two governments because the scope of the matter affected a general interest. The tribunal gave its award on 16 November 1957, confirming that the French Government had not committed a breach of the provisions of the 1866 Treaty of Bayonne and the Additional Act. The tribunal rejected the Spanish argument as to the necessity of a prior agreement. It pointed out that according to the rules of good faith, the upstream state is under an obligation to take into consideration the various interests involved and to show that in this regard it is genuinely attempting to reconcile the interests of other riparian states with its own. As Spain was not able to submit evidence that any injury had been suffered, there was no need for the tribunal to consider what kind of injury amounted to serious injury.

Even though dispute settlement does not dominate the management of water-related issues,[42] this does not mean that it is no longer relevant. Indeed, dispute settlement is still important in respect of contentious water issues between riparian states. This is illustrated, for example, by the recent case of *Pulp Mills on the River Uruguay* between Argentina and Uruguay.[43]

3.2 Regulating Water Issues to Prevent Tensions and Conflicts

While the purpose of international adjudication is to resolve concrete disputes retroactively, the purpose of a regulatory approach is to prevent such disputes by agreeing abstract rules in advance. The Permanent Court

---

[42] See e.g. Patricia Birnie, Alan Boyle and Catherine Redgwell, *International Law and the Environment* (3rd edn, Oxford University Press 2009), 544 ('Unlike the delimitation of continental shelves, third-party settlement has not been widely used in river disputes and comparable judicial elaboration is lacking. The better solution given the greater complexity of the balancing process involved and the likelihood that the needs of states may change, is probably some form of common management designed to achieve equitable and optimum use of the watercourse system' (footnotes omitted).

[43] *Pulp Mills on the River Uruguay* (Argentina v. Uruguay), International Court of Justice, Judgment, 20 April 2010, ICJ Reports 2010, 14. See also Rosalyn Higgins, 'Natural Resources in the Case Law of the International Court', in Alan Boyle and David Freestone (eds), *International Law and Sustainable Development. Past Achievements and Future Challenges* (Oxford University Press 1999), 103–11.

of International Justice highlighted this distinction in the *Diversion of Water from the Meuse Case* by noting that the treaty of 1863 between the Netherlands and Belgium was 'an agreement freely concluded between two states to reconcile their practical interests with a view to improving an existing situation rather than to settle a legal dispute concerning mutually contested rights'.[44]

The same point is reflected in many boundary waters agreements. For instance, the preamble of the Boundary Waters Treaty between Canada and the United States, concluded in 1909, states that the purpose of the treaty is 'to prevent disputes regarding the use of boundary waters and to settle all questions which are now pending'.

H.A. Smith emphasized the need for a regulatory approach in his work on the economic uses of international rivers by pointing out that in many cases a river system can present complex issues because the use of its waters is demanded simultaneously for navigation, irrigation, electric power, and the supply of large cities. He asserted that the function of law is 'to provide rules for settling the possible conflict of interests'[45] by trying to strike an equitable balance between them.

Berber has also argued in favour of treaty-making, which he regards as representing the highest form of political wisdom. Noting the developing character of international water law, he stated that 'the conclusion of specific and specialized water treaties remains far and away the best solution'.[46] Berber characterizes treaties as follows: 'A treaty like every other legislative act is always a compromise between different interests and diplomacy is the act whereby this compromise is achieved.'[47] In order to regulate the various interests at issue, states concluded many watercourse treaties during the period from the beginning of the nineteenth century up to World War II. In exceptional cases states established a joint jurisdiction or agreed on common use with regard to a particular watercourse. More often, substantial regulations concerning the navigational and non-navigational uses of boundary waters were drawn up.[48]

---

44 *The Diversion of Water from the Meuse* (Netherlands v. Belgium), Permanent Court of International Justice, 1937, Series A/B, No. 70, 20.

45 Herbert Arthur Smith, *The Economic Uses of International Rivers* (P.S. King & Son 1931), 13.

46 Berber (n 6), 270.

47 Ibid., 271.

48 See *Brownlie's Principles of Public International Law* (n 9), 338 ('The early assumption that navigational uses enjoyed primacy is no longer accurate; irrigation, hydro-electricity generation, and industrial uses are now more prominent in many regions than navigation, fishing, and floating of timber, and domes

Turning to non-navigational uses of boundary waters, even before World War I, states had concluded a number of bilateral and multilateral treaties. While some of these treaties regulated the utilization of boundary waters in general terms, others regulated such traditional uses as irrigation, domestic use, fishing and the floating of timber. After the Industrial Revolution, it was recognized that regulations should be extended to cover modern uses of boundary waters. To this end, bilateral agreements were concluded in order to lay down detailed regulations in respect of, for example, the use of hydroelectric power, the size of dams to be constructed in boundary waters, and the volume of water to be diverted for mining or industrial purposes.

As the recognition of freshwater pollution problems increased, it became clear that there was a need to widen the scope of environmental protection agreements. Furthermore, it was discovered that there was a need to comprehensively regulate hydrological units. Thus, the scope of the process of internationalization was broadened from only regulating boundary waters to also controlling watercourses of international interest. Accordingly, regulations were issued to protect, for example, the Mosel, the Rhine, and the Great Lakes. These regulations set specific water quality objectives or emission limits, or alternatively established frameworks under which specific regulations could be set. Thus, emphasis was placed on waters crossing boundaries rather than on waters forming boundaries. To emphasize this aspect, international instruments began to refer to transboundary or international waters rather than to boundary or frontier waters.

In 1966, the International Law Association (ILA) adopted the Helsinki Rules on the Uses of the Waters of International Rivers as a statement of existing rules of international law. The codification and further development of rules on non-navigational uses of international watercourses was developed further through the work of the International Law Commission and the adoption and entry into force of the Convention on the Law of the Non-Navigational Uses of International Watercourses, which was adopted under the auspices of the United Nations in 1997.[49]

---

use is growing rapidly.'). See also *Brierly's Law of Nations. An Introduction to the Role of International Law in International Relations*, 7th edn (ed. Andrew Clapham, Oxford University Press 2012), 207–8.

[49] See Stephen C. McCaffrey, *The Law of International Watercourses* (2nd edn, Oxford University Press 2007). See also Patricia Wouters and Ruby Moynihan, 'Benefit Sharing in the UN Watercourses Convention and under International Water Law', in *International Water Law*, supra n 22, 321–35.

3.3 Managing Water-related Tensions and Relationships Under the Same Framework

While the regulatory approach strove to prevent conflicts and find harmony of interest, the managerial approach acknowledged different interests related to water without trying to put them into a straitjacket. This approach relied on the doctrine of sustainable development and on technical-oriented management. It represented contextual justice which sought to coordinate, reconcile and optimize short-term economic interests and long-term environmental concerns. However, the fusion of these aspects was not total in the sense that it did not resolve the tension between exploitation and protection. Rather, the reconciliation process brought different tensions and interests together under the same framework.

The principle of reasonable and equitable utilization began to be developed at the beginning of the twentieth century in relation to international watercourses. It involved the highlighting of the need to manage international watercourses through the optimization of long-term interests and short-term needs by taking into account all relevant factors and reaching a conclusion on the basis of the whole. Emphasizing this comprehensive approach, H.A. Smith noted that conflicts of interest should be appraised, holding that 'the only interest which can be allowed to dominate is that of the community as a whole'.[50]

According to Jerome Lipper, the principle of equitable utilization means 'the division of the waters in such a manner to permit the reasonable use of its waters by each of the riparian states'.[51] To put it differently, pursuant to equitable utilization, a riparian state cannot deprive another riparian state of the right to an equitable share of the natural resources of an international watercourse. What amounts to a reasonable and equitable share is, pursuant to the 1966 Helsinki Rules adopted by the ILA, 'to be determined in the light of all the relevant factors in each particular case'. The rules contain a non-exhaustive list of such factors. For example, the economic and social needs of each basin state as well as the avoidance of unnecessary waste in the utilization of waters of the basin shall be considered. The principle of equitable and reasonable utilization was subsequently codified in the 1997 Convention

---

[50] Smith, *supra* n 45, 143.

[51] Jerome Lipper, 'Equitable Utilization' in *The Law of International Drainage Basins*, (eds A.H. Garretson, R.D. Hayton and C.J. Olmstead, Oceana Publications 1967), 15–88, 43.

on the Law of Non-navigational Uses of International Watercourses. Article 5 of the Convention, which is a key provision, states as follows:

> Watercourse states shall in their respective territories utilize an international watercourse in an equitable and reasonable manner. In particular, an international watercourse shall be used and developed by watercourse States with a view to attaining optimal and sustainable utilization thereof and benefits therefrom consistent with adequate protection of the watercourse.[52]

Since the beginning of the 1990s, along with the emergence of the sustainable development approach, the concept of the sustainable use of international watercourses was generally applied to their management. This managerial approach is reflected in many regional conventions, such as the 1994 Convention on the Danube River,[53] the 1995 Mekong River Basin Agreement,[54] the 1999 Rhine Convention, and the 2000 Revised SADC Protocol on Shared Watercourses.[55]

An important issue relating to the managerial approach is that of gradual regime-building. This has allowed dynamic processes and frameworks which encompass policy-making and scientific expertise to begin to be developed in the water sector. A good example of this is the 1992

---

[52] In its commentary on the Law of the Non-navigational uses of International Watercourses, the International Law Commission determined the meaning of optimal utilization as follows: 'Attaining optimal utilization and benefits does not mean achieving the "maximum" use, the most technologically efficient use, or the most monetarily valuable use much less short-term gain at the cost of long-term loss. Nor does it imply that the State capable of making the most efficient use of a watercourse – whether economically, in terms of avoiding waste, or in any other sense – should have a superior claim to the use thereof. Rather, it implies attaining maximum possible benefits for all watercourse States and achieving the greatest possible satisfaction of all their needs, while minimizing the detriment to, or unmet needs of each.' See Draft Articles on the Law of the Non-Navigational Uses of International Watercourses and Commentaries thereto, adopted on the Second Reading by the International Law Commission at its Forty-Sixth Session, A/CN.4/L.493 (12 July 1994), 26.

[53] Convention on Co-operation for the Protection and Sustainable Use of the Danube River, Sofia, 29 June 1994, ECOLEX TRE-001207.

[54] Agreement on Co-operation for the Sustainable Development of the Mekong River Basin, Chiang Rai, Thailand, 5 April 1995.

[55] Revised Protocol on Shared Watercourses in the Southern African Development Community, Windhoek, 7 August 2007. See Ed Couzens, 'Water-related Conflict and Security in Southern Africa: The SADC Protocol on Shared Watercourses', in *International Environmental Law-making and Diplomacy Review* 2014 (University of Eastern Finland – UNEP Course Series 14 2015), 91–126.

UNECE Convention on the Protection and Use of Transboundary Watercourses and International Lakes.[56] The institutional framework of this Convention includes a meeting of the parties, a secretariat and a number of specialist subsidiary bodies or working groups. In addition, two protocols relating to the Convention have been adopted: the 1999 Water and Health Protocol[57] and the 2003 Civil Liability Protocol.[58] In 2003, parties amended the Convention to allow accession by United Nations member states outside the UNECE region.[59]

Furthermore, the managerial approach has been extended to extreme events such as floods and storms. This development is reflected in the emergence of the concept of water security which highlighted the emergence of a new relationship to water. While water is usually regarded as an object to be utilized or protected by humanity, in extreme events humanity should be protected from water. This movement from subject to object reflects the particular character of water security.

---

[56] Convention on the Protection and Use of Transboundary Watercourses and International Lakes, Helsinki, 17 March 1992, 31 *International Legal Materials 1992*, 1312. For discussion, see Annukka Lipponen, 'The UNECE Water Convention and the Support It Gives to the Management of Shared Waters: from Obligations to Practical Implementation' in *International Environmental Law-making and Diplomacy Review* 2014 (University of Eastern Finland – UNEP Course Series 14 2015), 49–63. See also Guide to Implementing the Water Convention, United Nations Economic Commission for Europe, United Nations, New York and Geneva 2013; Attila Tanzi, *The Economic Commission for Europe Water Convention and the United Nations Watercourse Convention. An analysis of their harmonized contribution to international water law*, United Nations Economic Commission for Europe, Water Series No. 6, United Nations, New York and Geneva 2015.

[57] Protocol on Water and Health to the 1992 Convention on the Protection and Use of Transboundary Watercourses and International Lakes, London, 17 June 1999, 38 *International Legal Materials 1999*, 1708.

[58] Protocol on Civil Liability and Compensation for Damage Caused by the Transboundary Effects of Industrial Accidents on Transboundary Waters, Kyiv, 21 May 2003, <https://www.unece.org/env/civil-liability/welcome.html> accessed 23 December 2016.

[59] The Meeting of the Parties adopted decision VI/3 on accession by non-United Nations Economic Commission for Europe countries, 30 November 2012, <https://www.unece.org/fileadmin/DAM/env/documents/2013/wat/ECE_MP.WAT_41.pdf> accessed 23 December 2016.

4 CONCLUSIONS

The above categorization of international water issues into four dimensions represents different ways of arranging such issues. Although these dimensions have been discussed separately, this does not mean that they are also functionally separate. On the contrary, in a number of instances they are interlinked. For example, the River Oder Case also concerned the subjective dimension, the Harmon doctrine the spatial dimension, the Gabcikovo Case the material dimension and *Spain v Council* the temporal dimension.

By reference to these four dimensions, this chapter has sought to emphasize that issues relating to watercourses should not be dealt with in a vacuum but are surrounded by a web of water-related tensions, conflicts and relationships. The use of different dimensions as a method of categorization makes it easier to determine where the centre of gravity lies, what possible trade-offs and benefits are available, what kind of conflicts of interest exist and whether any synergies or interlinkages can be identified.

International law offers tools to resolve and manage various water-related issues. The three approaches discussed in this chapter – dispute settlement, the regulatory approach and the managerial approach – reflect three contextually different but interlinked ways of dealing with such issues. However, while such tools exist at a general level, their functionality depends on the specific circumstances at hand. Consequently, it is not always possible to resolve every issue that may arise. Indeed, the tensions between upstream and downstream countries, between international jurisdiction and national jurisdiction, between short-term and long-term interests and between exploitation and protection are all issues that must be resolved within the specific contexts in which they arise.

10. The Columbia River Treaty between Canada and the United States of America – time for change?

Nigel Bankes

1 INTRODUCTION

The Columbia River Treaty (CRT or Treaty) between Canada and the United States was concluded in 1961[1] and entered into force in 1964 along with an accompanying Protocol.[2] The Treaty is widely recognized as an outstanding example of the cooperative development of an international river basin based upon the idea of sharing the benefits of that cooperation. The Treaty principally addresses two values, flood control and power. The Treaty has no set termination date but either party may give ten years' notice as of September 2014 to terminate the power sharing provisions of the Treaty. Additionally, the flood control provisions of the Treaty change automatically in 2024. This chapter considers what we can learn from the 50 years of practice under this Treaty. How has the Treaty been able to accommodate (or not) new values? What are the implications of the opportunity to terminate the power provisions of the Treaty and the automatic changes that will occur to the flood control provisions?

The chapter is divided into six parts in addition to this introduction. Part 2 describes the Columbia River Basin and the circumstances that led to the conclusion of the Treaty. Part 3 describes the basic provisions of

---

[1] 15 UST 1555 and also available online <http://www.ccrh.org/comm/river/docs/cotreaty.htm> (all urls in this chapter accessed 8 September 2016 unless stated otherwise).

[2] Exchange of Notes Constituting an Agreement Between Canada and the United States of America Concerning the Treaty Related to Co-Operative Development of the Water Resources of the Columbia River Basin Signed at Washington, on 17 January 1961, Washington, 22 January 1964, 15 UST 1555 (hereafter the Protocol).

the Treaty. Part 4 offers more details as to the provisions dealing with termination and changes to flood control operations. Part 5 examines some of the more significant changes that have occurred in the basin and more generally over the last five decades that might have implications for the Treaty. These include changing values, climate change, and the legal and political empowerment of indigenous communities. Part 6 describes how the US and Canada have each analysed the challenges facing the treaty regime and how each will approach the possibility of treaty termination, as well as possible negotiations with respect to a renewed or amended treaty. Part 7 offers some concluding observations.

2 THE COLUMBIA BASIN AND THE BACKGROUND TO THE TREATY

The Columbia River Basin covers 672,000 km² in Canada and the United States; 15 per cent of the basin lies in Canada, the balance in the United States.[3] Canada contributes disproportionately to the flow on the main stem (38 per cent of the average annual flow and as high as 50 per cent of peak flows). The river exhibits considerable seasonal and inter year variability and has experienced serious flooding in the past, most notably in 1948 when a flood destroyed the town of Vanport in Oregon and caused substantial damage in Trail, British Columbia.

The United States began to develop the hydro potential of the main stem and tributaries of the Columbia at the end of the nineteenth century, followed by significant dam building during the first 40 years of the twentieth century including the massive Grand Coulee Dam on the main

---

[3] Some of the details in this part of the chapter draw upon my previous publications on the Treaty (sole and co-authored) including, Nigel Bankes, 'The Columbia Basin and the Columbia River Treaty: Canadian Perspectives in the 1990s' working paper published by Northwest Water Law and Policy Project, Northwestern School of Law of Lewis and Clark College, 1996 <http://www.law.ucalgary.ca/system/files/Columbia+River+Treaty+Lewis+%2526+Clark+paper+Bankes9504-1.pdf>; Nigel Bankes and Barbara Cosens, 'The Future of the Columbia River Treaty', a working paper prepared for the Program on Water Issues, Munk School of Global Affairs, University of Toronto, 2012 <http://powi.ca/wp-content/uploads/2012/12/The-Future-of-the-Columbia-River-Treaty-2012.pdf>; Nigel Bankes and Barbara Cosens, 'Protocols for Adaptive Water Governance: The Future of the Columbia River', prepared for the Program on Water Issues of the Munk School of Global Affairs, University of Toronto, 2014 (Bankes and Cosens, 'Protocols') <http://powi.ca/wp-content/uploads/2014/10/Protocols-for-Adaptive-Water-Governance-Final-October-14-2014.pdf>.

stem just below the Canadian border, completed in 1941 and with a generating capacity of 6809 MW. This dam cut off the escapement of salmon to the entire upper Columbia in Canada. Notwithstanding these significant developments, both Canada and the US understood the importance of constructing additional upstream storage to provide enhanced flood control protection and also to make more efficient use of the installed generation on the main stem.

Accordingly, the two governments invited the International Joint Commission (IJC)[4] to examine the options for developing additional storage in Canada to provide power and flood control benefits. Headwater storage sites can be evacuated to provide flood control but the storage can also be used to smooth out the natural hydrograph and firm-up capacity at downstream dams which might otherwise be forced to spill water or operate at less than full head.[5] Having received the IJC's Report on the development options, the governments also asked the IJC for its advice as to how the two countries should share the costs and benefits associated with the development of upstream storage.[6] The two governments drew upon this advice in negotiating and concluding the Columbia River Treaty. The Treaty addresses developments on both the main stem of the Columbia and on one major transboundary tributary of the Columbia, the Kootenay River. Other significant tributaries of the Columbia including the Snake River (which lies entirely within the United States) and the Pend d'Oreille (which is also a transboundary river) were not the subject of cooperative development under the CRT, although the

---

[4] The IJC was established under the terms of the Boundary Waters Treaty, 1909, 3636 Stat. 2448 <http://www.ijc.org/en_/BWT>. The IJC has some decision-making functions with respect to the levels of boundary and transboundary waters but it also has a so-called reference or advisory jurisdiction which allows the two governments to ask the IJC for its advice in relation to a broad range of matters.

[5] For more detailed discussion and evaluations of the options *see* the classic work: John V. Krutilla, *Columbia River Development – A Study in International Cooperation* (Resources for the Future 1961). Another useful source is Neil Swainson, *Conflict over the Columbia: The Canadian Background to an Historic Treaty* (McGill-Queen's University Press 1979).

[6] IJC, Principles for determining and apportioning benefits from cooperative use of storage of waters and electrical interconnection within the Columbia River system (1959), available at <https://www.crt2014-2024review.gov/Files/IJC.pdf> and discussed in Nigel Bankes, 'The Flood Control Regime of the Columbia River Treaty: before and after 2024' (2012), 2 *Washington Journal of Environmental Law and Policy*, 1–74.

operation of facilities on those and other rivers may be implicated in the post-2024 flood control operations.

Historically the Columbia River was one of the most significant salmon rivers in the world with different species of salmon making use of the full length of the stream and spawning in the headwater lakes of Lake Windermere and Columbia Lake in British Columbia some 2000 km from the mouth of the river in Oregon.[7] Hydro developments on both the main stem and the tributaries have significantly reduced the access of salmon to spawning areas and degraded remaining accessible habitat and migration routes. As a result of the Grand Coulee dam (above), the only Columbia salmon returning to the Canadian side of the boundary are in the Okanagan system which joins the Columbia in the United States downstream (south) of the Grand Coulee Dam. The loss of the returning salmon also changes the nutrient cycling within the ecosystem with implications for high profile species such as bears and eagles. Main stem and tributary dams have also affected other fish species, most notably sturgeon, by creating barriers to movement both upstream and downstream.

The availability and abundance of salmon was an important nutritional and cultural resource for indigenous communities on both sides of what became the international boundary. Those communities have suffered greatly as a result of loss of access for salmon to spawning grounds as well as a loss of access to preferred harvesting sites as a result of flooding.[8] In general there was little if any consultation with indigenous communities as to the implications of hydro developments and storage facilities for those communities. More specifically, neither Canada nor the United States consulted with or involved the Tribes and First Nations in the negotiation of the Treaty.

The Columbia and its tributaries are a major source of electrical energy for both Canada and the United States. Hydro power is a particularly valuable source of electrical energy given that hydro systems offer the opportunity to store energy and can provide instantaneous availability. Hydro can be used for base load but it is particularly important for

---

[7] Chris Peery, 'The Effects of Dams and Flow Management on Columbia River Ecosystem Processes' in Barbara Cosens (ed.), *The Columbia River Treaty Revisited: Transboundary River Governance in the Face of Uncertainty* (Oregon State UP 2012) 138–47 (hereafter *Revisited*) and Carmen Thomas Morse, 'When Courts Run Regulated Rivers: The Effects of Scientific Uncertainty' in *Revisited*, 148–74.

[8] Mary Pearson, 'The River People and the Importance of Salmon' in *Revisited* (n 7), 70–83.

peaking purposes and to backstop more intermittent forms of energy such as wind and solar. While an important source of renewable low carbon energy, hydro developments are not without negative environmental and social consequences. In addition to the impacts on fisheries resources referenced above these negative consequences include: loss of productive valley bottom lands, dislocation of residents to make way for storage reservoirs, disruption of transportation routes, and seasonal draw-down of reservoirs interfering with recreational and non-recreational access as well as exposing unsightly mudflats with the potential for dust-related pollution.[9]

3 THE MAIN PROVISIONS OF THE TREATY

The recitals to the Treaty acknowledge that 'the greatest benefits to each country can be secured by cooperative measures for hydroelectric power generation and flood control' while making other benefits possible as well.[10] This statement confirms the two main values that underlie and drive the Treaty. The Treaty does not effect an apportionment of the flows of the Columbia. Flood control and power structures and operations may affect the timing of flows across the international boundary but the Treaty does not limit or apportion consumptive uses of water on either side of the boundary.[11] However, the Treaty does prohibit most non-consumptive diversions[12] and thus prohibited a major interbasin transfer from the Columbia to the Fraser which was under consideration in Canada. This was important to fishing interests in the United States because it reduced the prospect of Canada developing the hydro potential of the Fraser River, thus leaving the Fraser as a major salmon river on the west coast and salmon stocks in which the US had, and continues to have, a shared interest.

The power and flood control benefits of the Treaty were to be realized by the construction of three dams in Canada (Mica, Duncan and

---

9 For discussion see George Penfold, 'A Review of the Range of Impacts and Benefits of the Columbia River Treaty on Basin Communities, the Region and the Province', 5 December 2012, <http://blog.gov.bc.ca/columbiarivertreaty/files/2012/07/A-Review-of-the-Range-of-Impacts-and-Benefits-of-the-Columbia-River-Treaty6.pdf>

10 CRT, Preamble, para 4.

11 Indeed, CRT art XIII(1) expressly authorizes consumptive uses.

12 Art XIII did permit Canada to divert part of the flow of the Kootenay River to the Columbia River in stages but Canada has never exercised this option and it is inconceivable that it will now do so.

Keenleyside, the 'treaty dams') and authorization of another dam in the United States (Libby on the Kootenay River).[13] Once constructed, a specified amount of storage in the three Canadian dams was to be operated in accordance with a prescribed flood control operating plan (FCOP) designed to offer effective flood control,[14] and in accordance with other operating plans designed to maximize the power benefits available to both parties. Actual operations under the Treaty are carried out by the 'Entities' designated by each government.[15] The Canadian Entity for most purposes is British Columbia Hydro and Power Authority (BCH or BC Hydro), a Crown corporation of the province of British Columbia. While Canada is the party to the Treaty, all of the rights and obligations under the Treaty flow to the province as a result of an agreement between Canada and British Columbia[16] and the general constitutional order within Canada.[17] The US Entity is comprised of the Division Engineer, North Pacific Division of the Army Corps of Engineers (Corps) and the Administrator of the Bonneville Power Administration, US Department of Energy. Article XV of the Treaty establishes the Permanent Engineering Board (PEB) to oversee the operation of the Treaty and to 'assist in reconciling differences that may arise between the Entities'.[18]

Under the flood control regime that is in force until 2024, the US Entity effectively gets to prescribe releases and refill of a certain amount of storage space in the three Canadian reservoirs. This is referred to as

---

[13] CRT art IV(2). Treaty authorization for this dam was required because the reservoir behind Libby (Lake Koocanusa) floods back into Canadian territory. The balance of this section is an abridged version of Nigel Bankes, 'What Does the Future Hold for the Columbia River Treaty?' (2013) 59 *Rocky Mountain Mineral Law Institute* 7–1, 7–34.

[14] For the current version of the FCOP see US Entity, *Columbia River Treaty Flood Control Operating Plan* (May 2003) (FCOP 2003), <http://www.crt2014-2024review.gov/Files/FCOP2003.pdf>.

[15] For reviews of actual operations *see* James Barton and Kelvin Ketchum, 'The Columbia River Treaty Managing for Uncertainty' in *Revisited* (n 7), 43–9 and Anthony G White, 'The Columbia River: Operation under the 1964 Treaty' in *Revisited* (n 7), 50–60.

[16] Canada – B.C. Agreement, July 8, 1963, reprinted in Canada Dep't of External Affairs & Dep't of N. and Nat'l Res., *The Columbia River Treaty: Protocol and Related Documents* 100 (1964).

[17] Rights to public lands and water resources in each province accrue to the provincial government and not the federal government. See The Constitution Act, 1867, 30 and 31 Vict., ch. 3, s 109.

[18] CRT art XV(2)(c).

the assured flood control operation and the basic objective 'is to operate reservoirs to reduce to non-damaging levels the stages at all potential flood damage areas in Canada and the United States insofar as possible, and to regulate larger floods that cannot be controlled to non-damaging levels to the lowest possible level with the available storage space.'[19]

In addition, the United States is also entitled, on certain terms and conditions, to an 'on-call' operation of additional storage to provide further flood control protection. Article IV(2)(b) obliges Canada to 'operate any additional storage in the Columbia River Basin in Canada, when called upon by [the U.S. Entity] ... within the limits of existing facilities ... to meet flood control needs for the duration of the flood period for which the call is made.'[20] The 'additional storage' to which the paragraph refers includes at a minimum: (1) additional storage at the three Canadian treaty dams (i.e. beyond the space dedicated to treaty operations); (2) storage at other Canadian facilities already built at the time of the Treaty; and, perhaps, (3) any subsequently built storage. The United States has yet to trigger these provisions.

In return for the *assured* flood control operation through 2024, the United States agreed to pay Canada a onetime payment of $64.4 million. The United States pays for on-call storage during the initial term of the CRT in accordance with article VI(3) which requires the United States to pay $1.875 million for each of the first four on-call operations, and to deliver equivalent power 'lost by Canada' in respect of each of those four operations and any additional on-call operations.[21] As noted, there has been no on-call operation to date.

The power benefits of the Treaty are to be realized by having the Canadian Entity operate the storage reserved by the Treaty at the three Canadian treaty dams in order to optimize generation in both the United States and Canada. Optimization is based on a set of operating plans initially prepared six years in advance (the Assured Operating Plan, AOP) based on all possible flow scenarios and then adjusted on the basis of updated information in advance of the operating year by the Detailed Operating Plan (DOP) with a view to securing mutual benefit to both Entities. During the operating year the DOP may be further amended by so-called supplemental operating agreements to provide for flows or

[19] FCOP 2003 (n 14), 8, 16. Discussed in more detail in Bankes (n 6).
[20] CRT art IV(2)(b).
[21] CRT art VI(3).

levels to benefit fish, wildlife or recreational interests.[22] This is one significant way in which the Treaty has been able to respond to changing values and environmental concerns.[23]

In return for agreeing to operate in accordance with the operating plans, Canada is entitled to one half the downstream power benefits.[24] These benefits comprise both capacity benefits and energy benefits made possible by storage and releases from the Canadian treaty dams.[25] The capacity credit is based on the amount of incremental firm energy that the US base system is able to produce at times of critical stream flow by making use of the Canadian storage. The energy benefit represents the increase in average annual usable energy output made possible by the use of the Canadian storage. This calculation is based on an extended period of streamflow conditions (including both high and low flow conditions). The calculation is based on an *assumed* operation of US main stem dams 'in a manner that makes the most effective use of the improvement in stream flow resulting from the operation of the Canadian storage for hydroelectric power generation ...'[26] The Treaty does not *require* the United States to operate the facilities in this way, but it does require the United States to compensate Canada on the assumption that it does so.[27] The value of the annual downstream power benefits varies but currently averages around $215 million.[28] In general, US dam operators are not able to operate their facilities to optimize generation because of constraints imposed by various orders issued under the Endangered Species Act (ESA) to protect certain listed species including runs of salmon and sturgeon.[29]

---

22 Barton and Ketchum (n 15), 47–8; John Shurts, 'Rethinking the Columbia River Treaty' in *Revisited* (n 7), 192–248, esp. 217 noting that these arrangements 'are entirely voluntary and can always be refused'.

23 See Bankes (n 3), 66–72.

24 CRT art V(1).

25 CRT art VII.

26 CRT art III(1).

27 CRT art III(2).

28 British Columbia, Columbia River Treaty Review, FAQs <http://blog.gov.bc.ca/columbiarivertreaty/faqs/#faq6>.

29 16 USC 1531. For a summary of the relevant orders see US Army Corps of Engineers & Bonneville Power Administration, Columbia River Treaty 2014/2024 Review: Supplement Report and Executive Summary (2010) <http://www.crt2014-2024review.gov/SupplementalReport.aspx> and for further analysis and commentary, see Morse (n 7).

4 THE PROVISIONS DEALING WITH THE TERMINATION AND AUTOMATIC CHANGES TO THE FLOOD CONTROL PROVISIONS

As noted in the introduction, the Treaty does not contain any automatic sunset provision and thus, in principle, may continue in perpetuity. However, the Treaty may be terminated by either party with ten years' notice after 2014. These general statements with respect to duration and 'termination' however are subject to two caveats. The first is that termination by notice effectively only extends to the power provisions of the Treaty, and the second relates to the automatic changes to the flood control provisions in 2024.

As to the first, Article XIX makes it clear that notwithstanding 'termination' (evidently a misnomer), the flood control provisions of the Treaty (as automatically changed in 2024, see below) continue for so long as the relevant dams have a useful life.[30] Thus, as a practical matter, all that can be terminated by the notice provision are the power provisions of the Treaty. This more limited form of termination is of considerable interest to some constituencies in the United States who see this as a way of retaining flood control benefits while avoiding the power sharing provisions of the Treaty which are seen as unduly favourable to Canada, especially because they are calculated on the basis of assumed operations rather than actual operations which, as noted above, must (as a matter of US domestic law) observe obligations arising under ESA.[31]

As to the second, it is important to understand that regardless of any notice with respect to termination, the flood control provisions of the Treaty change automatically in 2024.[32] In the subsequent years the US is no longer entitled to the assured flood control operation and thus loses the ability to direct drawdown and fill operations for flood control purposes. In its place, the US is entitled to a 'called upon' operation.[33] Similar to the pre-2024 'on-call' operation, this operation allows the United States to require Canada to operate its facilities to provide the US with flood protection but only where 'all the related storage facilities' in the US prove inadequate to provide the requisite degree of protection (the effective use requirement).[34] The US and Canada have different views as

---

30 CRT art XIX(4).
31 See n 29 above.
32 CRT art IV(2)(b).
33 CRT art IV(3).
34 CRT, Protocol, art. 1(1).

to the interpretation and application of these provisions which they will need to resolve prior to 2024. In particular, the two parties disagree as to the required level of protection and as to the scope of the facilities of which the US must make effective use before calling upon Canada to provide protection.[35]

The Treaty and the Protocol to the Treaty express the required level of protection in terms of maximum desirable flows at one particular location on the main stem known as The Dalles. The US takes the position that flows should be managed to below 450 kcfs at The Dalles on the basis that this is the level of protection that the parties have agreed to adopt in successive FCOPs. Canada takes the position that this practice is not controlling (especially since it related to the assured flood control operation) and that the target prescribed by the Protocol to the Treaty is actually 600 kcfs – which means of course that there would be fewer occasions on which the US would be able to require a Canadian flood control operation.

As for the scope of the 'effective use' requirement, the US takes the view that this obligation only applies to a small number of facilities that are specifically licensed under US law for flood control purposes. Thus it would not apply to dams that are licensed for irrigation purposes or for power purposes. Canada takes the position that the US is only entitled to call upon Canadian storage once it has made effective use of all storage available that is capable of reducing flows at The Dalles. If correct, this might have considerable implications for the operation of US storage. It might mean, for example, that the US would have to draw down irrigation storage reservoirs to manage flows even though this might increase the risk of the storage failing to refill.

Whatever the trigger for, and the scope of, Canada's post-2024 obligations, the US has a duty to make Canada whole in economic terms,[36] but such calculations will be far from straightforward. Furthermore, there is some room for thinking that Canada should be compensated on a different basis (e.g. sharing the benefits conferred by the operation of Canadian storage – i.e. the avoided losses that the storage delivers).[37]

It is clear from the above that the treaty regime faces considerable change and uncertainty in the next few years. The next part of the chapter

---

35 I discuss the different views in greater detail in Bankes (n 6), 28–31.

36 CRT art VI(4).

37 This principle, sharing the benefits, is the principle that underlies both the power provisions and the pre-2024 flood control provisions of the Treaty. It is also the principle that governs the adoption of both DOP and supplementary

examines other pressures on the Treaty in light of changing values and other developments. Part 6 of the chapter returns to consider how each party has analysed the challenges facing the treaty regime and how it will approach possible termination and possible negotiations with respect to a renewed or amended treaty in light of these changing values.

5 CHANGES IN THE BASIN

Drawing on the work of the Universities Consortium on Columbia River Governance,[38] Professor Barbara Cosens has identified five major changes that have occurred within the Basin and which therefore should have some implications for any review or re-assessment of the Treaty:[39] changes in energy markets, climate change, declines in populations of anadromous fish populations, changing societal values in relation to the river and the empowerment of local populations espousing different values.

The power market of the Pacific Northwest did not evolve as the treaty negotiators had anticipated. Energy demand did not grow as fast as expected and the region never built the anticipated thermal generation to meet the load. As a result, hydro power continues to provide much of the base load within the system and is thus more valuable than might have been expected. The consequences of climate change within the Basin include increased temperatures resulting in increased vegetation demand and an overall drying effect; the form of precipitation is expected to change with some watersheds in the basin becoming rain-dominated while previously they were snow-dominated; the timing of runoff will change and will occur earlier in the spring. All of this signifies that built storage will assume greater importance as natural storage (snow) declines. Populations of salmon and steelhead trout in the basin have declined and thirteen populations are listed as threatened or endangered under ESA. As noted above, this has significant implications for the

---

operating plans i.e. the Entities should only move away from the AOP if both parties benefit (albeit perhaps not to the same degree). See Bankes (n 6) 29–31, 55–8.

[38] For background, see <http://www.columbiarivergovernance.org/>. The *Revisited* volume (n 7) is a product of the Consortium.

[39] Barbara Cosens, 'The Columbia River Treaty: an Opportunity for Modernization of Basin Governance' (2016) 27(1) *Colorado Natural Resources, Energy and Environmental Law Review*, 33–8. This section draws heavily on Professor Cosens' summary.

manner in which dams in the US part of the basin are operated. The changing values of society are reflected on both sides of the border in the adoption of significant environmental laws dealing with endangered species and environmental assessment as well as legislation dealing with access to information. Finally, both jurisdictions have accorded increased recognition to and protection of the rights of the indigenous peoples of the Basin, the Tribes and First Nations.

The opportunity to terminate the power provisions of the Treaty, combined with the automatic changes to the Treaty's flood control regime (all as outlined above) provide the parties, as well as stakeholders within the basin, with the occasion and perhaps the impetus to ask whether and if so how the Treaty might be amended in response to these changed circumstances or to clarify some of the uncertainties associated with the changes to the flood control regime. The next part of the chapter examines how each party has been considering these questions.

6 THE TREATY REVIEW PROCESSES IN CANADA AND THE US

Both Canada and the US have put in place procedures to allow each of them to come to a timely conclusion as to whether or not to recommend treaty termination and/or possible topics for negotiations on treaty amendments.[40] On the American side of the border this task has been undertaken by the US Entity (i.e. Administrator of the Bonneville Power Administration and Division Engineer of the Northwestern Division of the US Army Corps of Engineers);[41] and on the Canadian side of the border by the Province of British Columbia.[42] The US Entity forwarded its Regional Recommendation to the US Department of State in December 2013.[43] Following a long delay it was finally reported in October 2016 that the State Department had formalized the US negotiating

---

[40] This part of the chapter draws on a more detailed analysis in Bankes and Cosens, 'Protocols' (n 3) ch. 2.

[41] USACE and BPA, *Columbia River Treaty: 2014/2024 Review* <http://www.crt2014-2024review.gov/>.

[42] British Columbia, Columbia River Treaty Review <http://blog.gov.bc.ca/columbiarivertreaty/>.

[43] USACE and BPA, Regional Recommendation for the Future of the Columbia River Treaty after 2024, December 13, 2013 <http://www.crt2014-2024review.gov/Files/Regional%20Recommendation%20Final,%2013%20DEC%202013.pdf>.

position in a document known as Circular 175.[44] However the details of this document are not yet publicly available. As a result the following discussion is based on the Regional Recommendation.

The US Regional Recommendation consists of regional goals, general principles, detailed recommendations and a set of domestic matters to be addressed. The first goal is to include 'additional ecosystem operations to expand, enhance and complement' existing arrangements for taking into account ecosystem and ecological considerations.[45] The document explicitly recommends inclusion of ecosystem-based function as a third primary purpose of the Treaty to accompany flood control and power benefits. Second, the US Entity recommends amending the current formula for sharing the power benefits of the Treaty so as to make the sharing formula 'more equitable' and based on 'the more realistic measure of the power value of coordinated operations as compared to non-coordinated operations.'[46] Third, the US Entity recognizes the value of the flood control measures of the Treaty and that the nature of the flood control operation changes in 2024. As a result, the US Entity calls for a coordinated and resilient flood risk management plan to provide for public safety, developed in light of improved understanding of the implication of climate change. The Regional Recommendation summarizes these goals as follows:[47]

> ... the region's goal is for the United States and Canada to develop a modernized framework for the Treaty that ensures a more resilient and healthy ecosystem-based function throughout the Columbia River Basin while maintaining an acceptable level of flood risk and assuring reliable and economic hydropower benefits. Therefore, it is important to achieve a modernized framework for the Treaty that balances power production, flood

---

44 The reasons for the delay are not clear. Members of Congress from the Pacific Northwest repeatedly urged the Department to finalize the negotiating position and open negotiations with Canada. The most recent letter from the Pacific Northwest Congressional Delegation to Secretary of State Kerry (August 11, 2016) is available here <https://www.cantwell.senate.gov/imo/media/doc/2016.08.11%20to%20Dept%20of%20State%20(Kerry),%20Columbia%20RIver%20Treaty.pdf>. For information about the authorization of negotiations see 'Cantwell says U.S. ready to start talks with Canada on Columbia River Treaty' *The Columbia Basin Fish and Wildlife New Bulletin*, October 14, 2016 <http://www.cbbulletin.com/437746.aspx#> accessed 3 January 2017.

45 US Regional Recommendation (n 43), 1–2.

46 Ibid., 2.

47 Ibid., 2, footnotes omitted. The goals section of the paper also refers to other needs and authorized purposes including irrigation, municipal and industrial use, in-stream flows, navigation and recreation.

risk management, and ecosystem-based function as the primary purposes, while also recognizing and implementing all authorized purposes.

Accompanying these goals were nine general principles that should inform the modernization of the Treaty. These principles may be summarized as follows:[48] maximize shared benefits; ecosystem health should be treated as a shared cost and benefit; a reformed treaty should be of sufficient duration to provide stability but also be adaptable to changing conditions; operations should be based on the best available science; operations should observe all obligations under domestic law; cooperative arrangements to include Canada's non-treaty storage; the treaty should build in resilience to take account of climate change; costs and benefits under the treaty should be aligned; and ecosystem function requirements should be compatible with the sharing of power benefits.

Finally, in addition to the principles and goals, the Regional Recommendation also offered a series of more detailed recommendations under the headings of hydropower, flood risk management, ecosystem-based function, water supply, navigation, recreation and climate change.

The Province of British Columbia released its 'Decision'[49] in March 2014. This short document comprises a Preamble and a set of 14 principles. Its key message is that BC seeks to 'continue the Columbia River Treaty and seek improvements within the existing Treaty framework'. The document suggests that the fourteen principles will guide any changes to the Treaty that the Province may pursue. These principles can be summarized as follows: maximize benefits to both countries through coordination; ongoing impacts require compensation; benefits from coordinated operations should be shared equitably; treaty duration should provide certainty while allowing for adaptation; no called-upon operation without demonstration of effective use; coordinated flood risk management should cover US reservoirs as well as treaty dams; the parties should continue to apply ecosystem values to treaty operations; explore ecosystem improvements within and outside the treaty; domestic laws and constitutionally protected aboriginal rights may restrict the operation of Canadian facilities; strive for the improved coordination of the Libby Dam; upstream salmon migration should be the responsibility of each

---

48 This section is an abridged version of Bankes and Cosens, 'Protocols' (n 3), ch. 2.

49 British Columbia, Columbia River Treaty Review, B.C. Decision, <http://blog.gov.bc.ca/columbiarivertreaty/files/2012/07/Columbia-River-Treaty-Draft-BC-Recommendation.pdf>.

country regarding their respective infrastructure; climate change considerations should be incorporated in treaty planning and implementation; treaty negotiations should engage First Nations and communities; and non-treaty issues should be addressed in other fora.

While the differences between the two positions would seem to outweigh the similarities there is certainly some common ground. Thus, both agree that there is a need to balance certainty and adaptability; both agree that there are gains to be made through coordination and that such gains and benefits should be shared; both agree that the current regime is accommodating ecosystem values to some degree even if not expressly recognized in the treaty text; both make reference to the constraints of domestic laws in guiding their respective negotiating positions; and both agree that there is merit in exploring changes in the post-2024 flood control regime to enhance the level of protection beyond that provided for in the called-upon regime. Both recognize a need to involve Tribes/First Nations and affected communities during any negotiation process.

But the differences are substantial. The most significant differences would seem to be three: the treatment of ecosystem values and perhaps other values; the approach to sharing the benefits; and the assessment of what the Treaty requires for the post-2024 called-upon flood control regime. It is clear that the US Entity is proposing a major revision of the Treaty to expressly incorporate ecosystem-based function as a third primary purpose of the Treaty. It also appears that the US Entity seeks to establish the significance of other values within the Treaty (e.g. navigation, irrigation and perhaps recreation) although it less clear how these other values might be recognized. By contrast, British Columbia largely seems to take the view that ecosystem considerations can be accommodated in the current Treaty although it is committed to exploring mechanisms outside the Treaty as well as inside. British Columbia does not favour adding other values and interests to the Treaty.

The US Entity is firmly convinced that the current arrangements for sharing the downstream power benefits are unjust and seeks to have that position recognized. Furthermore, the Entity takes the view that any compensation payable to Canada for flood control measures taken by Canada should be based on the actual costs and losses incurred by Canada and should not be based on the idea of sharing the benefits (i.e. the US's avoided losses). Finally, and more generally, the health of the ecosystem, including any measures taken to restore anadromous populations to the upper Columbia in Canada, should be a shared cost (and benefit).

The government of British Columbia takes a different view on all of these issues. Thus it has released a report suggesting that the Treaty conferred benefits on the US that are not fully accounted for.[50] Furthermore, it calls for sharing the benefits that result from coordination rather than simply compensating for losses and it suggests that the party responsible for disrupting the passage of migrating fish should be solely responsible for the costs of any restoration measures within its jurisdiction and that such costs should not be a shared responsibility.

Finally, the two documents reflect the continued difference of opinion as to proper interpretation of the post-2024 flood control provisions of the Treaty as discussed above.

7 CONCLUSIONS

For more than half a century the Columbia River Treaty has delivered on its goals of providing flood control and enhanced power production for the communities of the Pacific Northwest. That is a significant achievement. But the Treaty is also under pressure and there is a significant body of opinion, especially in the United States, that considers that the Treaty should be significantly amended. Canada's position (or more particularly the position of British Columbia) is less ambitious. BC likely considers that incremental improvements can be made in actual operations under the Treaty and through collateral agreements addressing some of the more ambitious ecosystem goals without amending the Treaty itself and the balance of benefits and obligations embedded in the Treaty.

The Treaty does not contain an amendment procedure. Any amendment will require the consent of both parties. The Treaty does contain a termination provision but, as emphasized above, termination is something of a misnomer since all that such a notice would accomplish is termination of the power provisions of the Treaty. The flood control provisions change automatically in 2024 and those provisions will continue notwithstanding purported 'termination' for so long as the treaty dams can continue to provide flood protection. There is therefore nothing inevitable about a treaty amendment. The parties may not be able to reach an agreement and if they cannot the Treaty will continue with or without the power provisions (which are subject to unilateral termination). However, there are some common interests in improving the operation of the Treaty and good faith negotiations may well produce

[50] U.S. Benefits from the Columbia River Treaty – Past, Present and Future: A Province of British Columbia Perspective, June 25, 2013, available at <http://blog.gov.bc.ca/columbiarivertreaty/files/2012/07/US-Benefits-from-CRT-June-25-132.pdf>.

some common ground and some opportunities for give and take. For example, both parties surely have an interest in clarifying the terms of the post-2024 flood control obligations, and both have an interest in incorporating an ecosystem perspective. However, as revealed above, the parties likely have different views on how these results or interests should be achieved. For example, while the US seems to prefer comprehensive re-writing of the Treaty to incorporate ecosystem values on a par with the two original motivating purposes of the Treaty, British Columbia likely prefers a more incremental approach, emphasizing what can be achieved within the Treaty through an agreement or agreements between the Entities.

11. *Kansas v. Colorado*: state sovereignty and the equitable allocation of water

Itzchak E. Kornfeld

Sitting, as it were, as an international as well as a domestic tribunal, we apply Federal law, state law, and international law, as the exigencies of the particular case may demand ...[1]

1 INTRODUCTION

The United States Supreme Court has heard 137 cases under its Original Jurisdiction.[2] *Kansas v. Colorado* (Original No. 105), like many of its sister original jurisdiction disputes, dealt with an interstate water dispute. Indeed, the case involves a problem that has bedevilled the American west for over one hundred years: the apportionment of river water, mostly for irrigation purposes, in a desert to semi-desert environment.

In adjudicating inter-state water disputes, the Supreme Court's first legal duty and challenge – like that of all international courts and tribunals – is to balance the competing interests of state sovereignty.[3] *Kansas v. Colorado* demonstrates these competing interests in the guise of water allocation and the rough and tumble wrangling between the

[1] *Kansas v. Colorado*, 206 U.S. 46, 97 (1907). (Emphasis added).

[2] The Court was provided with the jurisdiction to hear disputes between the states of the Union, by the Constitution's Article III and the Judiciary Act of 1789. The Court's original jurisdiction provides, for our purposes, 'the Supreme Court shall have *original and exclusive jurisdiction of all controversies between two or more States ...*' Thus, the Court sits as a trial court or a *quasi* international court when it hears disputes between the states.

[3] See e.g., Itzchak E. Kornfeld, 'Is News of "Sovereignty's Death" Exaggerated?' (2012) 18 *ILSA J International & Comparative L*, 315; Noah Hall, 'Bilateral Breakdown: U.S.–Canada Pollution Disputes' (2006) 21 *Nat. Resources & Envt*, 18, 68.

litigants over water. The dispute also demonstrates the lengths to which states situated in the arid west will go for a greater share of the existing pool of water. For example, Kansas sued Colorado numerous times without success, claiming that the latter was depriving it of its 'rightful' share of river water. Another excellent example of this 'hunger' for water and the jealousy state sovereignty engenders is demonstrated by another of the Court's disputes, *Arizona v. California*, where California fought a decades-long rancorous dispute with Arizona, over 548 ac-ft (675 947 m^3) of water/day,[4] or 200 000 acre-feet (246 696 000 m^3) of water annually. But, this appetite for the allocation of water was and is not limited to the United States. It is also a current phenomenon in India's Sutlej and Krishna Rivers,[5] Bangladesh,[6] and South Africa.[7]

With regards to the issue of state sovereignty, it has been a constant in the United States' Supreme Court's transboundary dispute portfolio for almost one hundred years. Indeed, in 1931, Justice Oliver Wendell Holmes observed, in *New Jersey v. New York*[8] that the competing interests of state sovereignty must be balanced where an upper riparian has the physical power to cut off all the water within its jurisdiction, to the lower riparian. Moreover, he also observed that:

---

4 An acre-foot of water is the quantity of water required to cover an acre to a depth of one foot, which converts to 43 560 cubic feet. 548 ac-ft/day converts to 222 hectares/day. The following conversion rates were used: one acre = 0.405 hectares. That is: 178 566 567 gallons, or 678 552 122 litres.

5 K.S. Murty, 'River water sharing: India's problems, FRIEND 2002 – Regional Hydrology: Bridging the Gap between Research and Practice' (Proceedings of the Fourth International FRIEND Conference held at Cape Town, South Africa, March 2002). IAHS Publ. no. 274. 2002, <http://hydrologie.org/redbooks/a274/iahs_274_323.pdf> accessed 17 January 2017.

6 Robie I. Samanta Roy, *Trade and Environment Data Base, India-Bangladesh Water Dispute, ICE Case Number 78* (1997) <http://www1.american.edu/ted/ice/indobang.htm> accessed 17 January 2017 ('The problem of water resource allocation and sharing, primarily for irrigation purposes, has plagued relations between India and its neighbors … as well as the … water allocation issues associated with the Ganges river system.')

7 Synne Movik, *The Dynamics and Discourses of Water Allocation Reform in South Africa*, STEPS Working Paper 21, Brighton: STEPS Centre (2009), <http://steps-centre.org/wp-content/uploads/Reform_web_version.pdf> accessed 17 January 2017. ('The increasing perceptions of water scarcity have prompted many countries to engage in reforming their water legislations and systems of water use rights.')

8 283 U.S. 336 (1931).

> the exercise of such a power to the destruction of the interest of lower [riparian] States could not be tolerated. And on the other hand equally little could New Jersey be permitted to require New York to give up its power altogether in order that the river might come down to it undiminished. Both States have real and substantial interests in the River that must be reconciled as best they may be.[9]

The foregoing speaks not only to the Court's balancing of state sovereignty interests, but also to its consideration of the 'community of interests' between states, in the use of water from their transboundary river.

It is telling that after all these years, the notion of 'community of interest' is the current trend in water allocation and appears to be the future direction in which the law and practice in the apportionment of water is traveling.[10] In the modern era, the concept stems from the sense that a community of interest in a transboundary river's water is shaped by the fact that it is part of the commons or, a common property, and ought to be shared by the community.[11] It also captures the view of a watercourse as part of a wider hydrological system and suggests collective action when administering it. Similarly, the community of interest embraces a more exacting theory of the sovereign relationships of the states.

The paramount importance of the sovereignty issue[12] means that it casts a long shadow over the Court's adjudication. Consequently, the artifact of federalism is always in the background of the Court's adjudication of intra-state water disputes. Similarly, in a union of states, such as the European Union, sovereignty is also a key aspect for the members of the union. As Joseph H.H. Weiler has observed, 'Europe has charted its own brand of constitutional federalism.'[13]

---

9 Ibid., 342–3. (Emphasis added).

10 See generally, Alistair Rieu-Clarke, R. Moynihan and B. Magsig, 'UN Watercourses Convention – User's Guide' (2012), 103–105 <http://www.unwatercoursesconvention.org/documents/UNWC-Fact-Sheet-10-Theories-of-Resource-Allocation.pdf> accessed 17 January 2017.

11 Ibid.

12 The sovereignty issue plays out as follows inside a federal state: The U.S. Constitution's Bill of Rights' 10th Amendment states that: 'The powers not delegated to the United States by the Constitution, nor prohibited by it to the States, are reserved to the States respectively, or to the people.' Thus, there exists a continuous interplay between state and federal sovereignty, and in Congress the issue of states' rights versus federal rights.

13 Joseph H.H. Weiler, 'Federalism without Constitutionalism, Europe's Sonderweg' in Kalypso Nicolaidis and Robert Howse (eds), *The Federal Vision:*

2 THE AMERICAN WEST AND THE WATER PARADOX

The American West is an arid region. In order to understand the underpinnings of the *Kansas v. Colorado* dispute, some historical perspective is necessary. In 1893, Maj. John Wesley Powell, the explorer who opened up the West declared:

> When all the rivers are used, when all the creeks in the ravines, when all the brooks, when all the streams are used, when all the reservoirs along the streams are used, when all the canyon waters are taken up, when all the artisan waters are taken up, when all the wells are sunk or dug that can be dug in all this arid region, there is still not sufficient water to irrigate all this arid region ... *Not one more acre of land should be granted to individuals for irrigating purposes.*[14]

Powell underscored the fact that in the western United States water is life. However, for over 170 years both the federal and state governments have been attempting to fool Mother Nature. In some ways, they have succeeded. Nevertheless, given the continuing droughts over the past decade, the notion of control over water resources was at best illusory.

Returning to the earlier period, beginning in the 1840s the United States Government provided exclusive rights to people who were willing to farm the land in states like Kansas, Nebraska and Wyoming. Although most eastern politicians and government officials had never been west of St. Louis, they believed that eastern and western European farming methods were appropriate for the areas west of the 100th meridian, which represents the separation between the moist east and the arid west. However, this was and is not the case. The homesteaders, who flocked to the west on promises of rich soil and sufficient water resources were fooled by the federal government and local entrepreneurs. Many of these homesteaders were immigrants who were recruited to move west – and to extend the European hold on these lands – so that they might prosper and make their fortunes.[15] Of course, these people were entirely unaware of

---

Legitimacy and Levels of Governance in the United States and the European Union (Oxford University Press 2001), 70.

[14] Maj. John Wesley Powell speaking extemporaneously at the Second International Water Irrigation congress in Los Angeles, in mid-October 1893, quoted in Jonathan Waterman, *Running Dry: A Journey from Source to Sea Down the Colorado River* (National Geographic 2010) 242. (Emphasis added).

[15] The Homestead Act of 1862, Act of May 20, 1862, Public Law 37–64, enacted during the Civil War, provided, among other benefits, that any adult

the fact that the region encompassing Colorado, Kansas, Nebraska and Wyoming, among other states, was extremely arid, receiving approximately 38 cm (15 inches)–48 cm (19 inches) of rain per year,[16] in contrast to the 102 cm–152 cm (40–60 inches) average that fall from Virginia northwards,[17] and what many of these settlers were used to in Europe. Many of them perished from hunger or lost whatever money they had and returned east.[18]

When the western and southwestern states were settled, the settlers quickly learned that the farming methods they brought from Europe were utterly unsuitable because there was a paucity of water. Accordingly, in order to cultivate crops they needed irrigation and, just as was done in ancient Mesopotamia, water had to be channeled and diverted from a river to cultivate crops. When *Kansas v. Colorado* was filed in the Supreme Court, as is true today in these agrarian lands, water equalled money, in state revenue and revenue for the citizens who farmed the land and others who provided services to farmers, i.e. *water was and is the lifeblood of the economy*. Therefore, states were extremely jealous and combative of and about the apportionment of water from their trans-boundary river. Indeed, Kansas not only had water disputes with Colorado over water allocation, it also sued other states, including Nebraska, its neighbour to the north,[19] and Wyoming, its neighbour to its

---

citizen, or future citizen, i.e., European settlers, who had never borne arms against the United States, which excluded Confederate soldiers, would be granted plots of 160 acres of government surveyed land, if he or she paid a $10 fee, lived on the land for five years and 'improved' the parcel by constructing a dwelling building and cultivating the plot. After five years the original patent holder or filer, was entitled to a fee simple deed. Homestead Act (1862), Ourdocuments.gov (1995), <https://www.ourdocuments.gov/doc.php?flash=true&doc=31> accessed 17 January 2017.

16 United States Department of the Interior, United States Geological Survey, Nationalatalas.gov, *Colorado, Precipitation* (2011), <https://nationalmap.gov/small_scale/printable/images/pdf/precip/pageprecip_co3.pdf> accessed 17 January, 2017. ('The average annual precipitation for Colorado is 15.47 inches.')

17 The average rainfall in Virginia for the period 1895–1998 was 42.7 inches. See Bruce P. Hayden and Patrick J. Michaels, *Virginia's Climate*, University of Virginia Climatology Office (2000) <http://climate.virginia.edu/description.htm> accessed 17 January 2017.

18 See generally, Marc Reisner, *Cadillac Desert: The American West and Its Disappearing Water* (Penguin Publ., revised edn 1993) and James Malin, *Winter Wheat in the Golden Belt of Kansas* (University of Kansas Press, Lawrence Kansas 1944).

19 *Kansas v. Nebraska*, 538 U.S. 720 (2003).

northwest, seeking greater appropriative rights. Similarly, Wyoming sued Colorado, its neighbour to the south, over appropriative water rights.[20]

Today, as has been true for decades, the importance of water, as a resource in the western United States, is demonstrated by the fact that every southwestern and western state has a water court[21] and a state water engineer.[22] Indeed, one thing is certain: these types of disputes will continue to flare up in this region and around the world, because of global warming and climate change.

3 *KANSAS V. COLORADO*

3.1 Introduction

Kansas v. Colorado commenced in 1901[23] and concluded in 2009,[24] an epoch of 108 years. The dispute centered upon the allocation of the waters of the Arkansas River between irrigators on either side of these state's border. In the Court's 1907 decision[25] – the second in the dispute – Justice David Brewer was confronted with two different water norms: Colorado subscribed to the first in time, first in right, or the prior appropriation western water law doctrine, while Kansas maintained the common law riparian rights system of water allocation. Given these opposing doctrines, the Court established a Solomonic rule: the equitable – but not equal – allocation of water between the two states. The rule was developed in order to resolve this water allocation dispute in the semi-desert agricultural environment. Indeed, the rule was utilized as a means of pacific dispute resolution, i.e., use of the courts, as opposed to turning to armed conflict.

---

20 *Wyoming v. Colorado*, 259 U.S. 419 (1922).

21 See e.g., Colorado State Judicial Branch, *Water Courts* (2011), <http://www.courts. state.co.us/Courts/Water/Index.cfm> accessed 17 January 2017.

22 See generally, Kansas Dept. of Agriculture, Division of Water Resources (2011), <http://www.ksda.gov/dwr> accessed 17 January 1917. ('The Division of Water Resources administers 30 laws and responsibilities including the Kansas Water Appropriation Act which governs how water is allocated and used; statutes regulating the construction of dams, levees and other changes to streams; the state's four interstate river compacts; as well as coordinating the national flood insurance program in Kansas.')

23 185 U.S. 125 (1902).

24 556 U.S. 98 (2009).

25 206 U.S. 46 (1907).

Brewer also developed the concept of 'equality of right',[26] which refers to states' equal standing in their sovereign power and right.[27] In attempting to reconcile the common law, international law and dispute resolution, he observed that:

> For after all, the common law is but *the accumulated expressions of the various judicial tribunals in their efforts to ascertain what is right and just* between [parties] ... We have exercised that power in a variety of instances, determining in the several instances the justice of the dispute. *Nor is our jurisdiction ousted, even if, because Kansas and Colorado are States sovereign and independent in local matters, the relations between them depend in any respect upon principles of international law.* International law is no alien in this tribunal.[28]

The foregoing set the stage for many of the Court's subsequent equitable apportionment cases, including two of its most recent ones, 2010 *North Carolina v. South Carolina*, [29] and 2015 *Montana v. Wyoming.*[30] Indeed, there was an essential need for a structure to replace 'the waging of war and to negotiating treaties',[31] which would allow for the pacific resolution of disputes, while according the respect and dignity due to the sovereignty and equality of each state.[32] Moreover, some years later, in *Wyoming v. Colorado*[33] Justice Van Devanter hailed the doctrine of equitable allocation as 'a pioneer in its field'.[34]

The gravamen of Kansas' claim was that Colorado acting as *parens patria* and on behalf of a number of corporate irrigation companies, deprived and threatened to deprive the State of Kansas and its citizens of

---

[26] *Kansas v. Colorado*, 206 U.S. 46, 98. 'One cardinal rule underlying all the relations of the states to each other is that of equality of right. Each state stands on the same level with all the rest. It can impose its own legislation on no one of the others, and is bound to yield its own views to none ... the question of the extent and the limitations of the rights of the two states becomes a matter of justiciable dispute between them, and this Court is called upon to settle that dispute in such a way as will recognize the equal rights of both, and at the same time establish justice between them.'

[27] Ibid., 97.

[28] Ibid. (Emphasis added).

[29] No. 138 Original, *South Carolina v. North Carolina*, 558 U.S. 256 (2010).

[30] No. 137 Original, 577 U. S. ____ (2016).

[31] Kristin L. Myles, 'South Carolina v. North Carolina – Some Problems Arising in an East Coast Water Dispute' (2012) 12 *Wyoming Law Review*, 3, 5.

[32] Ibid.

[33] 259 U.S. 419 (1922).

[34] Ibid., 464.

the water that naturally and habitually flowed in the Arkansas river from Colorado to Kansas.[35] Nevertheless, the allocation formula that the Court inaugurated back at the turn of the last century was fashioned for an agrarian society,[36] and focused on apportioning the water for one purpose: irrigation.

3.2 *Kansas v. Colorado*: The Subject Matter of the Dispute

Colorado's Rocky Mountains are the source for the arid West's four major rivers, the Arkansas, the Colorado, the Platte and the Rio Grande. Without these rivers' watersheds, the American West would remain the arid region that America's natives discovered when they arrived on the continent at least 12 000 years ago. But for these four rivers, irrigation would not be possible. Consequently, any water projects that are developed along Colorado's transboundary rivers will likely affect water flowing into its downstream neighbor, Kansas.

The river at issue in this dispute was the Arkansas River,[37] which flows along the eastern flanks of the Rocky Mountains. Its headwaters are located near Leadville, Garfield County, Colorado, and its mouth is located at Napoleon, Arkansas.[38] 'Once proudly known as the "Nile of America"',[39] it is 2364 kilometers (1469 miles) long, making it the sixth

---

[35] 206 U.S. 46, 47–8 (1907).

[36] Professor Gabriel Eckstein has noted that 'the formula of equitable apportionment is rather generic and that its application in 1907 focused on agriculture simply because the dispute centered on agriculture.' (Personal communication 11 September 2015). Indeed, the doctrine has been utilized in other factual scenarios. See e.g., *North Carolina v. South Carolina* 558 U.S. 256 (2010) (drinking water). Nevertheless, the Supreme Court has yet to apply its allocation formula for multiple uses or solely for nature.

[37] The history of the Arkansas River follows the Europeans' colonialization and development of the United States. The river was discovered in 1541 by Francisco Vasquez de Coronado, who traversed it near what today is Dodge City, Kansas. See generally, The Arkansas River Historical Society Museum, An Outline History of the Arkansas River from Early European Explorations to the Development of the McClellan-Kerr Arkansas River Navigation System (1997), <http://www.tulsaweb.com/port/history.htm> accessed 17 January 2017; Francis Parkman, *La Salle and the Discovery of the Great West* (Random House USA 1999).

[38] United States Geological Survey, Geographic Names Information System, *Feature Detail Report for: Arkansas River* (last updated 2 April 2011), <http://geonames.usgs.gov/pls/gnispublic/f?p=gnispq:3:1932465391042904::NO::P3_FID:78956> accessed 17 January 2017.

[39] *Kansas v. Colorado*, 556 U.S. 98 (2009), 90 (Breyer, J.).

longest river in the United States and the second-longest tributary in the Mississippi-Missouri-River system.[40] The River's drainage basin envelops approximately 440 300 square kilometers (170 000 square miles).[41]

The area in dispute, in this case, is the upper 450 kilometers (280 miles) of the Arkansas River[42] and its tributaries in Colorado. These water bodies drain an area of 'approximately 22 000 square miles [56 979 km²] ... All [of which drains] east of the summit of the Rocky mountains and a large portion thereof in the mountains ...'[43] and the 'length of the river in Kansas [which] is about 310 miles [499 km]; its course is through a broad alluvial valley, and along its entire length in Kansas are alluvial deposits of great depth, amounting in the aggregate to about 2 500 000 acres [1 011 714 hectares], the greater part of acreage and the greater part of the course of the river lying in the western part of the state.'[44]

3.3 The Initial Salvo: The 1902 Action, *Kansas v. Colorado*, 185 U.S. 125 (1902)

The State of Kansas filed its complaint on 20 May 1901. Kansas' complaint sought an injunction enjoining and restraining Colorado from diverting, or permitting any corporation or other private person in Colorado from diverting, any of the Arkansas River's waters or any of its tributaries coursing through the state of Colorado, except for domestic use. Kansas also sought to enjoin Colorado from permitting any natural or corporeal person to lengthen or enlarge any of the canals or ditches that were then in existence, or to construct 'and operate any other canals, ditches, branches, laterals, or reservoirs in addition to those heretofore constructed and now in use in said state.'[45]

Kansas' complaint claimed that Colorado began to systematically appropriate and redirect the river's waters for the purpose of irrigating

---

[40] J.C. Kammerer, U.S. Geological Survey, *Largest Rivers in the United States* (rev. May 1990) <http://pubs.usgs.gov/of/1987/ofr87-242> accessed 17 January 2017.

[41] Ibid.

[42] 'The river has its origin in central Colorado, and is a mountain torrent for 130 miles to a point near Canon City where it enters a foothill region ending near Pueblo. Thence it traverses the high plains of eastern Colorado and western Kansas. In the areas mentioned the stream is non-navigable.' *Colorado v. Kansas*, 320 U.S. 383, 385 (1943).

[43] *Kansas v. Colorado*, 185 U.S. 125, 126 (1902).

[44] Ibid., 126–7.

[45] Ibid., 137.

non-riparian arid lands adjacent to the river, from Cañon City, Colorado to the Kansas border, a distance of approximately 360 km (190 miles). Moreover, Kansas alleged that, by 1891, Colorado appropriated the entire natural flow of the Arkansas and a considerable amount of its flood waters, so that the average flow (between the two) was reduced considerably and the natural flow was completely arrested.[46]

In response Colorado filed a demurrer, which contained ten points.[47] Two of these are relevant here. The first asserted that the Supreme Court had no jurisdiction over either the parties to or, the subject-matter of, the suit, because Kansas' claims on their face did not constitute a colorable controversy between the two states that is supported by the United States' Constitution.[48] The second asserted that Kansas' allegations do not arise between it and the state of Colorado, rather they arise, 'if at all, between the state of Kansas and certain private corporations and certain persons in the state of Colorado who are not made parties herein, and which matters so stated, if true, do not concern the state of Colorado as a corporate body or state.'[49]

The Court, by Chief Justice Fuller,[50] held that Colorado's filing of a demurrer cannot dispose of a case between two states, even though it would between two private parties. Ultimately, Justice Fuller observed that too many issues were unanswered at that juncture, including, (1) whether Colorado was apportioning too much water from the Arkansas River; and (2) whether certain Colorado corporations needed to be made defendants because they were indispensable parties. Accordingly, he held that there were too many 'intricate questions arising on the record [and therefore the Court was] constrained to forbear proceeding until all the facts are before us on the evidence.'[51] The demurrer was overruled, without prejudice, and the Court postponed its decision regarding any relief until a more complete factual record could be developed.[52]

---

[46] *Colorado v. Kansas*, 206 U.S. 46, 52 (1907).

[47] *Colorado v. Kansas*, 185 U.S. 137–9.

[48] Ibid., 138.

[49] Ibid.

[50] 'Mr. Justice Gray did not hear the argument, and took no part in the decision.' Ibid., 147.

[51] Ibid., 138.

[52] Ibid. Indeed, '[t]he court added that, before the developments in Colorado consequent upon irrigation were to be destroyed or materially affected, Kansas must show not merely some technical right but one which carried corresponding benefits.' *Colorado v. Kansas*, 320 U.S. 383, 385–6 (1943).

4 ROUND TWO: THE 1907 DECISION AND THE INTRODUCTION OF THE CONCEPT OF EQUITABLE APPORTIONMENT, *KANSAS V. COLORADO*, 206 U.S. 46, 49 (1907)

The case next came before the Court on 17 August 1903, when Kansas filed an amended complaint, naming 'as defendants Colorado and quite a number of corporations, who were charged to be engaged in depleting the flow of water in the Arkansas River.'[53] Here, Kansas again asserted that Colorado and the corporate defendants were illegally diverting water from the Arkansas River to the detriment of Kansas, and once more requested the Court to enjoin the defendants from diverting the water of the Arkansas River. However, prior to hearing the case, the Court required the parties to conduct discovery and 'take evidence'. This process concluded on 16 June 1905, and the Supreme Court's decision was issued on 13 May 1907.

In that decision, the Court noted the State of Colorado and several of the defendant corporations answered the complaint. Nevertheless, it did not consider the defenses of the corporations separately from those asserted by Colorado. Initially, Justice Brewer's opinion stated that the Court's jurisdiction, granted it under the Constitution's Article III Section 2, provides that 'the judicial power shall extend to all cases, in law and *equity*, arising under this Constitution'.[54] The Court next noted that under the Constitution only the states have the power to prescribe laws regarding the appropriation and distribution of waters from rivers in their territory. Furthermore, it is the states that establish their own water law regime, not the federal government.[55]

The Court also held that Kansas' attempt at enjoining Colorado's diversion of the waters of the Arkansas River were insufficient, although the diversions had a 'perceptible injury to portions of the Arkansas valley

---

53 *Kansas v. Colorado*, 206 U.S. 46, 49 (1907).

54 Ibid., 82.

55 Amy Kelly, 'Federal Preemption and State Water Law' (1996) 105 *J Contemp Water Res and Educ*, 4. ('One cannot fairly say that federal rules dominate in the field of water law … In this absence of any congressional interest in establishing a legal regime for water allocation during the decades of western expansion in the 1800s, the States took it upon themselves to choose which theory of water rights, riparianism or prior appropriation, best suited their needs; a practice that Congress did not choose to disrupt retroactively, and that the Supreme Court approved in *California Oregon Power Co. v. Beaver Portland Cement Co.*, 295 U.S. 142 (1935)').

in Kansas, particularly to those portions closest to the Colorado line.'[56] Moreover, as noted above, because Colorado water law was grounded in the western principle of prior appropriation, and Kansas adhered to the common-law principle of riparian rights, the Court, as was noted above, was confronted with a choice of law issue. Unable to shape a remedy that would encompass these two diverse water law norms, Justice Brewer sought an equitable remedy. Accordingly, he fashioned the new doctrine: 'equitable apportionment' to bridge the two norms, and observed that:

> [a]s Kansas thus recognizes the right of appropriating the waters of a stream for the purposes of irrigation, subject to the condition of an equitable division between the riparian proprietors, she cannot complain if the same rule is administered between herself and a sister state. And this is especially true when the waters are, except for domestic purposes, practically useful only for purposes of irrigation.[57]

In addition, the Court addressed the central tenets of flexibility, fairness and equity, in its usage of the doctrine. Justice Brewer reflected on 'the effect of what [had] been done upon the conditions in the respective states … [to] adjust the dispute upon the basis of equality of rights …' Similarly, in *Wyoming v. Colorado*, the Court was concerned about 'a basis of decision … consonant with the principles of right and equity …' Finally, the justices also sought to protect Colorado's prevailing agricultural economy. Accordingly, Kansas' complaint was dismissed without prejudice. Nevertheless, the Court held that Kansas had the right to institute a new suit at any point when she believed that her interests in the Arkansas River's waters would be substantially injured 'to the extent of destroying the equitable apportionment of benefits between the two states resulting from the flow of the river.'[58]

One of the key failings of the decision was that it did not address conservation. That issue was addressed in a subsequent dispute, *Colorado v. New Mexico*, where the Court decided that 'it is entirely appropriate to consider the extent to which reasonable conservation measures by New Mexico might offset the proposed Colorado diversion and thereby

---

[56] 206 U.S. 46, 117 (1907). Accordingly, Justice Brewer's opinion holds that whatever injury Kansas suffered is minimal when seen through the prism of the entire Arkansas river valley 'and regarding the interests of both states, and the right of each to receive benefit through irrigation and in any other manner from the waters of the Arkansas, Kansas has not made out a case entitling it to a decree …' (Ibid.).

[57] Ibid., 104.

[58] Ibid., 118.

minimize any injury to New Mexico users.'[59] It also determined that '[t]his approach comports with our emphasis on flexibility in equitable apportionment ...' The subject of conservation arose from a claimed inefficiency by the Vermejo Conservancy District in New Mexico.

Given that the judgment was entered over 70 years ago, prior to the age of environmentalism, the Court only saw two uses for the Arkansas River's waters: domestic purposes and irrigation. However, some years later, in *Nebraska v. Wyoming*[60] Justice William O. Douglas declared that:

> Apportionment calls for the exercise of an informed judgment on a consideration of many factors. Priority of appropriation is the guiding principle. But physical and climatic conditions, the consumptive use of water in the several sections of the river, the character and rate of return flows, the extent of established uses, the availability of storage water, the practical effect of wasteful uses on downstream areas, the damage to upstream areas as compared to the benefits to downstream areas if a limitation is imposed on the former—these are all relevant factors. They are merely an illustrative not an exhaustive catalogue. They indicate the nature of the problem of apportionment and the delicate adjustment of interests which must be made.

However, the Court has yet to utilize its informed judgment in the allocation of water.

In concert with Justice Douglas' priorities, we have learned a great deal more about riverine ecology. Thus, a modern decision should also consider water for fisheries, in stream-flow, bank and soil conservation, the planting of shrubs and trees to provide valuable habitat for birds, bees and other wildlife. For example, the physical and ecological processes that determine stream habitat for fish have been despoiled by human activity to the impairment of the fisheries. All of the world's streams and watersheds have been disturbed or polluted to some degree. Indeed, many have been significantly damaged to the point where fisheries and visually appealing characteristics have been severely degraded. On the other hand, streams that have the most constant flow processes and high levels of nutrients produce more fish and healthier ecosystems.

Nevertheless, although equitable apportionment is now part of customary international law and has been incorporated into numerous international instruments, e.g. the 1992 Helsinki Convention on the Protection and Use of Transboundary Watercourses and International Lakes,[61] it still does not include water for the benefit of streams, or river life or for

---

[59] 459 U.S. 176, 186 (1982).
[60] 325 U.S. 589 (1945).
[61] 1936 U.N.T.S. 269; 31 I.L.M. 1312 (1992).

nature. Rather, modern international law generally provides that the parties to water conventions shall take all appropriate action to '2. (c) ensure that transboundary waters are used in a reasonable and equitable way, taking into particular account their transboundary character, in the case of activities which cause or are likely to cause transboundary impact.'[62] Similarly, Article 5 of the 1997 Convention on the Law of the Non-Navigational Uses of International Watercourses[63] denominated 'Equitable and Reasonable Utilization and Participation', declares that '[w]atercourse States shall in their respective territories utilize an international watercourse in an equitable and reasonable manner.'[64]

However, the foregoing provisions do little to assist poor or downtrodden communities, who may be required to resolve contentious political issues involving water problems by large corporate interests in the transboundary context. These populations include aboriginals who seek water rights for fishing vs. large and industrial agricultural water rights in the United States' Klamath River Basin[65] or those living downstream on the Ganges River in South Asia, where India and Bangladesh are wrangling over the allocation of that river.[66] Additionally, they do not address the impacts of climate change in an age of drought and water shortage; nor do they deal with added pressures beyond the 'routine pressures' of attempting to balance water uses for human consumption, agriculture, cities, and the environment, in arid regions including the southwest US, the Middle East, sub-Saharan Africa and Central Asia.

---

62 Ibid. Art. 2(c).

63 36 I.L.M. 700 (1997); G.A. Res. 51/229, U.N. GAOR, 51st Sess., 99th mtg., UN Doc A/RES/51/229 (1997).

64 Ibid. Art. 5.

65 Tony Barboza, 'Water War between Klamath River Farmers, Tribes Poised to Erupt' *LA Times* (7 May 2013), <http://articles.latimes.com/2013/may/07/local/la-me-klamath-20130507> accessed 17 January 2017. For decades this rural basin has battled over the Klamath River's most precious resource: water that sustains fish, irrigates farms and powers the hydroelectric dams that block one of the largest salmon runs on the West Coast. Now, one of the nation's fiercest water wars is on the verge of erupting again. New water rights have given a group of Oregon Indian tribes an upper hand just as the region plunges into a severe drought.

66 IRIN, Analysis, Decades-Old Water Dispute Could Destroy Nation's Agriculture (20 September 2010), <http://www.irinnews.org/analysis/2010/09/20/decades-old-water-dispute-could-destroy-nation%E2%80%99s-agriculture> accessed 17 January 2017. ('Ongoing wrangling over vital waterways that pass through China and India – the two most populous countries in the world – could lead to agricultural devastation further downstream in Bangladesh, experts warn.')

5 CONCLUSION AND POTENTIAL ELEMENTS THAT INTERNATIONAL COURTS AND ARBITRAL TRIBUNALS SHOULD CONSIDER IN FUTURE ADJUDICATIONS

The doctrines of equitable allocation and equitable utilization are now firmly planted in the canons of transboundary water law. Nevertheless, nature and natural systems have yet to be considered in the formulaic distribution or calculus of water. Moreover, Justice Douglas' factors have not been considered in the allocation of water for over seventy years. However, the ever-present threat of climate change[67] ought to cause international judges and arbitrators to consider, or indeed, factor into their water allocation judgments and awards, the impacts of their apportionment on nature and natural systems.

Indeed, the very novel solution of equitable allocation, that Justice Brewer formulated in *Kansas v. Colorado*, should compel current adjudicators of transboundary water disputes to likewise find an innovative remedy that incorporates nature into their allocation schemes. Some factors that they might consider,[68] aside from those suggested by Justice Douglas, include:

1. Looking at the entire basin, rather than only assessing the water from a given river;
2. Assessing the geography of the river basin, particularly the coverage of the basin area in each state's territory;
3. A review of the hydrology of the basin at issue, as well as the contribution of the transboundary river's reach in each state;
4. Evaluating the basin's climate and micro-climates;
5. Assessing whether past or current water use ought to be considered;
6. Inquiring into whether there are other water resources that can be tapped, either through intra or inter-basin transfers;
7. Evaluating each state's economic and social conditions and needs;
8. Analysing the extent of each state's dependence on the transboundary river's or basin's water, *e.g.*, Iraq as the most downstream riparian may be more dependent on the waters of the Tigris and Euphrates Rivers than is Turkey;

---

[67] Elizabeth Kolbert, 'Climate Catastrophe, Coming Even Sooner' *The New Yorker* (31 March 2016) <http://www.newyorker.com/news/daily-comment/climate-catastrophe-coming-even-sooner> accessed 17 January 2017.

[68] A number of these suggestions are outlined in Murty (n 5).

9. Examining the conservation methods each state is taking and which would be most feasible under the circumstances; and
10. Assessing whether existing uses are consonant with the current facts on the ground.

In closing, equitable allocation or equitable utilization must be interpreted as equitable principles that include water for nature and natural systems and not simply for irrigation and other 'human needs'. To paraphrase John Lennon: 'All we are saying, is give nature a chance'.[69]

---

[69] John Lennon, 'Give Peace a Chance' 1969, included on the *Live Peace in Toronto 1969* album (Apple Records).

PART IV

Water allocation under the threat of environmental destruction

12. Climate change adaptation in water management – regulatory challenges and approaches

Moritz Reese

1 INTRODUCTION: CLIMATE CHANGE AND ADAPTATION AS A CHALLENGE FOR WATER MANAGEMENT

Climate science today leaves little room for doubt that the world's climate is changing as a result of greenhouse gas emissions and that these changes will affect living conditions in most parts of the world during the course of the twenty-first century.[1] Even if we manage to fulfil the ambitious commitments of the Paris Agreement to keep global warming below an average of 2°C, this will not be enough to avert significant impacts from climate change. In many regions of the world the following changes are already taking place in a measurable way:[2]

- an increase in average temperatures,
- a shift of precipitation patterns – e.g. from summer to winter,
- longer and more severe dry seasons and droughts, and
- an increase in extreme weather and flood events.

[1] Intergovernmental Panel on Climate Change (IPCC), Climate Change (2014): Synthesis Report, Contribution of Working Groups I, II and III to the Fifth Assessment, 40 <https://www.ipcc.ch/pdf/assessment-report/ar5/syr/AR5_SYR_FINAL_Front_matters.pdf> accessed 10 January 2017.

[2] Described in IPCC (n 1), 39ff.; see also IPCC, Working Group II, Impacts, Adaptation and Vulnerability (2014), 4ff.

It is expected that these developments will continue and will increasingly affect environmental conditions and, most notably, the water situation.[3] It is highly likely that climate change will exacerbate the global water crisis and intensify droughts and problems of scarcity, especially in arid regions. The primary challenges in these regions will therefore be to ensure a secure water supply and to prevent water conflicts. However, water-related impacts and challenges are sure to arise in other regions and in other ways (besides scarcity) as well. This means that water management will need to be adapted to climate change impacts in most regions in the world.

The present chapter looks at the implications of this adaptation challenge for water legislation and at how regulatory frameworks can facilitate and enforce greater adaptability of public water management. After a brief overview of the proven impacts and risks of climate change (Section 2), the chapter describes the corresponding governance requirements for adaptive water management in more detail, focusing on general rather than on specific sectoral requirements (Section 3). Taking this as a basis, it then analyses a number of key regulatory approaches to adaptive public water management both from a global perspective (i.e. the essential regulatory foundations of effective public water management – Section 4) and with particular reference to the European Water Framework Directive (WFD),[4] which provides an illustrative example of advanced water legislation (Sections 5–7). Using the example of the WFD and referring also to German water law the chapter will explain at some length how adaptation can be legally enforced in terms of risk/ impact assessment, the revision of management aims and measures, and the regulation and revision of water uses/permits (Section 8). The analysis focuses on the overarching challenges affecting integrated public management of water bodies, with its aim of reconciling human activities, aquatic ecology and flood control. Sector specific issues regarding, for example the management of flood risks, droughts and water infrastructures are not discussed in detail here.

---

[3] See IPCC (n 1), 66; IPCC WG II (n 2), 14ff.; with regard to Europe, see EEA, 'Impacts of Europe's changing climate – indicator based assessment', EEA Report No. 4/2008 (2008); and, EU Water Directors / EU Commission, Guidance Document No. 24 on the Common Strategy for the Implementation of the Water Framework Directive: River Basin Management in a Changing Climate (2009), 28ff.

[4] Directive 2000/60/EC of the European Parliament and the Council of 23 October 2000 establishing a framework for Community action in the field of water policy [2000] OJ L327/1.

2 POTENTIAL IMPACTS OF CLIMATE CHANGE ON WATER QUANTITY AND QUALITY

Climate change will affect natural water regimes in both quantitative and qualitative terms.[5] With regard to a range of key water uses – most notably drinking water, transport and agriculture – quantitative impacts (i.e., the declining availability of water) will often be of primary concern (2.1). While climate change is expected to increase the risk of water scarcity and drought, it will also bring a greater risk of heavy rainfall and flooding (2.2). Such changes in water quantity are – in addition to the heating of water bodies – a crucial factor in qualitative changes (2.3). In general terms, the effects on both quantity and quality will vary widely across geographic regions and depending on the specific natural conditions and human pressures present in a given location. The task of predicting the local effects of climate change on a water regime is therefore often subject to uncertainty. Thus, the main challenges for adaptive water management and water legislation are highly dynamic environmental conditions on the one hand and considerable uncertainty on the other hand.

2.1 Water Scarcity and Drought

Declining precipitation and growing risks of drought and water insecurity are widely regarded as likely consequences of global warming in many of the world's regions.[6] The latest assessment report of the Intergovernmental Panel on Climate Change (IPCC) confirms that global warming generally tends to exacerbate existing regional water conditions, meaning that dry regions will often become drier[7] and wet regions are likely to receive more precipitation.[8] Seasonal trends are also likely to be amplified, with dry summer seasons becoming longer and drier. This trend is also projected to affect Europe, where southern regions especially are likely to experience increasing water scarcity and risks of drought. The prospect of increasing water scarcity was specifically addressed by the European Union Commission in 2007 in its communication on 'the challenge of water scarcity and droughts in the European

---

[5] For a detailed account see: IPCC Technical Paper VI, Climate Change and Water (2008).

[6] IPCC (n 1), 65.

[7] IPCC WG II (n 2), 14.

[8] IPCC (n 1), 60.

Union'.[9] According to this communication, at least 11 per cent of the European population and 17 per cent of its territory have already been affected by water scarcity to date. The number of areas and people affected by droughts went up by almost 20 per cent between 1976 and 2006. One of the most widespread droughts occurred in 2003, when over 100 million people and a third of EU territory were affected. The cost of this to the European economy was at least €8.7 billion. The total cost of droughts over the past thirty years amounts to €100 billion, and the average annual cost quadrupled over the same period. In its communication, the EU Commission also presented a priority list of political and administrative measures that should be taken in order to mitigate the harm caused by droughts. These include water saving programmes and special drought risk assessments as well as maps and management plans in potentially affected regions. While such sector measures – as the Commission acknowledges – go beyond the scope of classical water management, it is clear that there is considerable overlap and that growing water scarcity also poses various challenges for the public management of surface water bodies and groundwater bodies.

2.2 Flood Risks

An increase in annual rainfall and a shift in the intensity of precipitation from the summer to the winter months has been a measurable feature in Central Europe for some time now, and this trend is predicted to continue.[10] Increasing amounts of rainfall during the winter period, glacial melting due to global warming, less storage of water as snow and earlier snowmelt are likely to lead to a further increase in the frequency and severity of flood events in many riverine areas, as already observed to a marked degree during the periods 1960–1980 and 1980–2000 when previous maximum levels doubled. This increase in flood risks is not due

---

[9] COM (2007) 414 final, 2ff.; see also Intergovernmental Panel on Climate Change (IPCC), 4th Assessment Report, WG II: Impacts, Adaptation and Vulnerability, 555 <http://www.ipcc.ch>; EEA (n 3) 99ff.; as to Germany, see Zebisch, M., Grothmann, T., Schröter, D., Hasse, C., Fritsch, U. and Cramer, W. *Klimawandel in Deutschland, Vulnerabilität und Anpassungsstrategien klimasensitiver Systeme*, UBA Berichte 08/05, 49ff.

[10] IPCC, Impacts, Adaptation and Vulnerability (n 2) – Chapter 8: Urban areas, 1276; EEA/JRC (European Environmental Agency/Joint Research Council) (2008), Impacts of Europe's Changing Climate – 2008 Indicator Based Assessment, EEA Report No. 4/2008, 94; DKKV, Hochwasservorsorge in Deutschland – Lernen aus der Katastrophe 2002 im Elbegebiet. Deutsches Komitee für Katastrophenvorsorge e.V., Bonn (2003).

in the first instance to climate change, however, but rather to residential developments and dykes encroaching on rivers, to the loss of retention areas, and to large-scale drainage of the landscape for agriculture. In this sense, climate change 'merely' reinforces an existing need for action, one that has been demonstrated dramatically by recent 100-year flood events.[11] This need for action consists specifically in effectively containing flood risks by a combination of technical mitigation measures on the one hand (raised dykes, polders, property protection) and prevention of causes (more space for rivers, increase of water retention on plains) on the other.

A potential increase in heavy rainfall events brings with it a greater risk of flooding in residential areas, partly due to overflowing drain systems. There have been many examples of such flooding after heavy rainfall in the past, so in this respect previous events cannot be attributed at all conclusively to climate change alone. The main cause is to be found instead in a type of urban and infrastructure development which has neglected the demands of effective urban drainage.[12] It is necessary here to design more efficient systems of drainage in residential areas on the one hand and to ease the burden on them by creating retention, infiltration and catchment areas as well as decentralized rainwater usage on the other. Well-planned drainage systems and a 'water-sensitive' form of urban planning that also combines land-related and infrastructural flood protection will be crucial in this.

2.3 Change and Deterioration of Water Quality and Ecology

With regard to water quality the following changes are described as a (possible) consequence of climate change in the latest assessment report of the IPCC:[13]

---

[11] See DKKV (n 10); DKKV, D.K.f.K., Das Hochwasser im Juni 2013: Bewährungsprobe für das Hochwasserrisikomanagement in Deutschland, DKKV-Schriftenreihe Nr. 53, Bonn (2015).

[12] See for example the analysis and suggestions by Bunster-Ossa, I.F. 'Sponge City', in S.T.A Pickett, M.L. Cadenasso and B. McGrath. (eds), *Resilience in Ecology and Urban Design: Linking Theory and Practice for Sustainable Cities* (Springer 2013), 301–306.

[13] IPCC, WG II (n 2), 234 (chapter on freshwater).

- Higher water temperatures lead to a reduction in the dissolved oxygen concentration, accelerate eutrophication and foster dissolution of pollutants from sediments in periods when water levels are low.
- Drier summers reduce the water volume and lead to higher concentrations of pollutants in water bodies.
- Increasingly dry summers and droughts may cause wetlands and upper river stretches to dry out.
- Lower water levels and higher temperatures make for reduced carrying capacity, with impacts on water extraction, pollutant discharges and uses of water for cooling purposes.
- Rising temperatures support the growth and proliferation of pathogens in water bodies.
- More frequent heavy storms and rainfall will increase soil erosion and the transport of pollutants from soils into surface water bodies and groundwaters.
- More frequent and more extreme flood events will likewise impact on water quality by flooding sewage systems, washing off contaminated surfaces and soils and shifting polluted sediments.
- In coastal areas, water bodies may suffer severely from salt water intrusion as a consequence of rising sea levels.
- All the above-mentioned effects may place greater pressures on aquatic animals and plants and significantly degrade or change aquatic ecosystems. Due to the expected increase in mean water temperatures alone, far-reaching changes may occur within the composition and abundance of local species.[14] As the lower-lying parts of rivers become warmer, it is likely that sensitive species will be forced to move upstream.

These primary impacts of climate change are very likely to be accompanied by 'secondary' effects arising from climate change mitigation or adaptation measures such as the following:[15]

---

[14] Merkel, W., Leuchs, W. and Odenkirchen, G. 'Herausforderungen des globalen Klimawandels für die Wasserwirtschaft in Deutschland: Praxisberichte, Handlungsfelder und Forschungsbedarf', in GWF Wasser Abwasser (2008), 334. On the effects of climate change on ecosystems and species and the implications for nature conservation (legislation), see Möckel, S. and Köck, W. (2009) *Natur und Recht* Vol 31, 318ff.

[15] The EU Water Directors assume that these secondary effects (especially from flood protection measures) will clearly outweigh the primary effects, at least in the medium term. See EU Water Directors (n 3), 49.

- construction projects aimed at maintaining the navigability of rivers in increasingly dry summer seasons,
- construction of additional dams and reservoirs as a means of regulating water levels, enacting flood prevention and securing water supply,
- construction of additional dykes as a flood-protection measure,
- increased use of fertilizers to cultivate biofuel crops,
- construction of additional hydropower plants as a means of climate neutral energy production.

Neither the primary nor the secondary impacts of climate change differ in type from those pressures that already exist independently of climate change. Climate change is not, then, causing entirely new kinds of pressures; rather, it is likely to exacerbate existing pressures and to lower the carrying capacity and resilience of the water bodies affected. One specific aspect of these worsening effects of climate change, however, lies in the fact that there is often considerable uncertainty surrounding their local extent and practical implications.

However, despite the uncertainties there is no doubt that water management geared towards nature conservation and a reduction of human pressures will increase resilience to the above-mentioned impacts of climate change.[16] In this context, the following pressures can be seen as the main factors in (reduced) resilience to climate change:

- abstraction and discharge of cooling water for power plants and industrial installations,
- modifications of water bodies and river banks by construction work, especially dams,
- pollution from diffuse sources and, in particular, pollution due to nutrients from fertilization,
- sewage spillovers.

Reduction of these pressures will not only improve water quality directly but also contribute significantly towards lessening the adverse effects of climate change. With regard to water management and water legislation, this means generally that a high level of environmental protection will often also contribute towards adaptation to climate change. However, such high levels of protection cannot be adequately ensured by global regulation alone. To a large extent it remains the responsibility of the

---

16 EU Water Directors (n 3), 47.

relevant local management authorities to establish adequate local solutions that are tailored to the specific situation and to the predicted local effects of climate change.

3 KEY REQUIREMENTS OF CLIMATE SENSITIVE WATER GOVERNANCE

As indicated above, climate adaptation in the water sector comprises a wider array of sectoral measures aimed, for example, at reducing nutrient emissions, renaturing watercourses, mitigating pollution and reducing flood risks. With regard to all these issues, each country and each unit of management needs to assess whether the standards and measures it has in place will be sufficient to boost resilience and to prevent the risks and adverse impacts of climate change. Depending on the regional situation concerned, this may involve questioning a whole array of existing standards and management objectives. The present chapter, however, does not examine specific standards and how they could be strengthened in view of potential impacts (in many respects, this can only be done in relation to specific local water situations). Instead, it highlights a number of more general, overarching governance challenges and relevant regulatory foundations of climate sensitive and adaptive water management. More precisely, it discusses which legal measures are needed to ensure that the knowledge base as well as the aims and measures of water management are regularly assessed and adapted in the light of potential impacts of climate change. As will be explained below, the 'adaptiveness' of water management and the legal framework will generally depend on the following system conditions:

(1) Water management will need to be based – more than it has been to date – on a thorough assessment and on projections of climatic developments and their local impacts on water quality and quantity.
(2) Management regimes need to contain elements of self-adaptation, including a regular review process in which new knowledge and projections are considered, so that the management programme can be adapted as necessary.
(3) Any regular review must also include management targets, including water status and quality targets. Assessments and decisions must be made about whether existing targets are to be 'defended' against adverse climatic developments or whether these developments should be regarded as changing reference conditions that require the management objectives to be adapted accordingly.

(4) New developments should be made resilient to the predicted impacts and risks of climate change by being designed in line with relevant management and safety objectives. Moreover, water legislation should enable appropriate interventions to be made in existing water uses whenever this appears necessary in order to maintain management objectives and safety standards in the face of risks and impacts from climate change.

The regulatory implications of adaptive water governance outlined above are analysed in greater detail below. This is done first with regard to regulatory foundations of adaptive water management established at a general global level (Section 4). These foundations are derived mainly from the UN Watercourses Convention (UNWCC)[17] as well as from the EU Water Framework Directive as a globally acknowledged regime of advanced integrated water management. Subsequently, each of the aforementioned key requirements of adaptive water governance is looked at separately (Sections 5–8).

4 REGULATORY FOUNDATIONS FOR RESPONSIVE AND ROBUST WATER MANAGEMENT AS AN INSTITUTIONAL BASIS FOR EFFECTIVE ADAPTATION ON CLIMATE CHANGE

In the light of the UN Watercourses Convention and in view of the advanced state of European and national water legislation, the following can be regarded as indispensable legal foundations for sufficiently effective and fair water management:

- *Principles of equitable and reasonable utilization and no significant harm*: These principles, contained in Articles 5 and 7 of the UNWCC, are internationally agreed as guiding principles for the management of transboundary waters. The notion of the equitable sharing of water resources – that scarce water resources should be

---

[17] Convention on the Law of the Non-navigational Uses of International Watercourses, adopted by the General Assembly of the United Nations on 21 May 1997, in force 17 August 2014. See General Assembly Resolution 51/229, annex, Official Records of the General Assembly, Fifty-first Session, Supplement No. 49 (A/51/49).

used reasonably and shared fairly between upstream and downstream users and that upstream activities should not cause significant harm to downstream users – can clearly apply equally within national boundaries and can thus be considered a standard objective of nationwide water management. At the same time, however, it is not entirely clear what 'reasonable and equitable utilization' means in practice. Consequently, the phrase is subject to a considerable margin of political discretion. For this reason, at the very least, it is crucial that this principle be safeguarded by procedural measures which ensure that conflicts over utilization are managed openly and in a coordinated manner. Most of the following features of modern water management can also be considered as prerequisites of equitable and reasonable utilization. It is clear that this basic principle will become increasingly important, given that water conflicts are likely to increase as a consequence of climate change.[18]

- *Permit requirements and procedures for water relevant uses*: It is clear that, with regard to projects, the above-mentioned procedural requirements include an ex-ante assessment of potentially significant effects on water quality and quantity. Environmental legislation needs to establish a formal obligation to thoroughly assess the effects of a given project, to inform and consult with stakeholders, and also to ensure that the project is realized only if it complies with existing standards and (equitable and reasonable) management measures.[19] Such permit and EIA procedures also form the basis for advanced climate proofing, as will be explained below (Section 9). Should the project prove to be incompatible with changing climatic conditions and impacts, legislation must provide for the possibility of revision and – where necessary – withdrawal in order to facilitate climate adaptation.
- *Integrated, coordinated river basin management planning/ programming*: Due to the complex interrelationships found in densely used catchment areas, it is impossible to implement the principles of equitable and reasonable utilization and to meet relevant coordination needs without an integrated, catchment-wide

[18] On the relevance of this basic principle of international freshwater law for adaptive management: Verschuuren, J. 'Climate change adaptation and water law' in Jonathan Verschuuren (ed.), *Research Handbook on Climate Change Adaptation Law* (Edward Elgar Publishing 2013), 257ff.

[19] See art 11 et seq. UNWCC.

planning process.[20] Adequate river basin management planning needs to be enforced by means of legal obligations and requirements regarding assessment, data exchange,[21] stakeholder involvement, coordination and transparent publication of the plan. This institution of formally regulated river basin management planning forms the basis of the management regime detailed in the EU WFD (Articles 13 and 11). The formal management plan is also an indispensable basis for meaningfully assessing the risks and impacts of climate change as well as developing appropriate adaptation strategies[22] (see below Sections 5 and 6).

- *Water quality objectives*: Article 21 of the UNWCC obliges the Parties to establish water quality standards, and in its Berlin Rules[23] (Article 28) the ILA has held that this obligation also flows from customary international law. Establishing water quality standards is an indispensable means of concretizing the principles of equitable sharing and no significant harm by determining the water quality that should be maintained by all upstream and downstream users. Ecological water quality objectives, in particular, are also a necessary means to effectively and efficiently implement an agreed level of ecological protection across the entire catchment area; they also provide some flexibility in the local choice of measures and in burden sharing. The EU WFD sets out an ambitious set of quality objectives regarding the ecological and chemical quality of surface water bodies as well as the chemical and quantitative status of groundwater bodies. The above-mentioned management plans and programmes of measures must set out appropriate measures in order to achieve these objectives within the prescribed deadlines (and by 2027 at the latest; for more details, see Section 7.1 below).

---

[20] See art 24, 2 (a) UNWCC and, in particular also No. 2.1 of the 2004 Berlin Rules of the International Law Association on the state of international water resources law: 'States shall, where appropriate … adopt efficient and adequate administrative measures, including management plans …' This provision is commented on by the ILA as follows: 'Foremost among the administrative measures will be to create and update management plans for the waters within their jurisdiction and control. Management planning normally will be based upon drainage basins, although there might be particular reasons for developing plans on some other basis, such as parts of a basin or combining all or parts of several basins.' <http://internationalwaterlaw.org/documents/intldocs/ILA_Berlin_Rules-2004.pdf> accessed 17 January 2017.

[21] See art 9 UNWCC.

[22] Verschuuren (n 18), 261ff.

[23] See above (n 18).

Water uses and relevant projects may only be permitted if they do not jeopardize the achievement of the quality objectives. The WFD also provides a number of exemptions for cases of *force majeur*, disproportionate costs, and overriding public interests (Article 4.4–7 WFD). This way of making water quality a mandatory guiding principle ensures that climate adaptation measures will be implemented insofar as the quality objectives are to be fulfilled even if additional efforts are required due to the adverse effects of climate change. If necessary, such additional (adaptation) measures must be included in the management plans and programmes of measures. However, in some cases climate change may substantially change natural conditions so that the original quality targets can no longer be considered appropriate management objectives. In such cases, adaptation of the management objectives needs to be considered (see below Section 7).

- *Waste water treatment and pollutant emission thresholds*: A regime of pollutant thresholds for waste water discharge is one of the basic features of modern water management and water legislation. As a rule, waste water standards are derived from the 'best available technique not entailing excessive costs' criterion (BATNEC) and, as such, are tailored to specific industrial sources and their respective pollution output and technical treatment potential. As a traditional instrument of public water law, waste water (treatment) standards have contributed greatly towards cleaning up rivers and lakes in the past decades. However, only a 'combined approach' of both emission and concentration standards – as implemented by the WFD, Articles 4 and 10 – can ensure that quality objectives and carrying capacities are not exceeded by accumulated emissions from multiple sources. At the same time, reducing emissions even below concentration thresholds according to BAT standards will strengthen the resilience of watercourses and water bodies, not least to the impacts of climate change.
- *Regarding flood risks: catchment-wide risk management plans, risk maps and land-use restrictions*: Flood risks, just like water quality and scarcity issues, need to be managed in a catchment-wide approach, taking account of the multiple interrelations and factors that cause and increase flood risks and hazards. Since flood risks are caused, to a great extent, by the reduction of flood plains in the upper parts of rivers, the principles of equitable management and no significant harm must apply accordingly and include the related procedural requirements outlined above (coordination, impact

assessment, data exchange and stakeholder hearings regarding relevant water construction projects etc.). Although it is obvious that effective and efficient flood protection and prevention requires close coordination across the catchment, adequate formal (planning) institutions have long been excluded from water legislation. With its Directive 2007/60/EC on the assessment and management of flood risks[24] (EU Floods Directive) the EU has obliged its Member States to introduce mandatory flood risk management plans which must be coordinated within the river basin and revised every six years in parallel to the management plans required under the WFD. The EU Floods Directive also provides for mandatory flood risk assessment and risk mapping as advanced instruments for managing the growing number of flood risks. Flood risk management is certainly a very important water-related challenge for adaptation to climate change and needs to be coordinated with general water management schemes. Due to the limited space available here, specific flood prevention issues will not be addressed in the following analysis.

5 LEGAL SAFEGUARDS FOR ADEQUATE CLIMATE IMPACT ASSESSMENT IN PUBLIC WATER MANAGEMENT

As noted above, it is necessary to gauge a number of factors as precisely as possible in order to adjust management processes to the potential impacts of climate change and to meet legally established management objectives even under the difficult conditions of climate change, should they arise. These factors are: how the regional climate is changing or will change; which primary impacts these changes (will) have on the resilience of water bodies in each instance; which secondary impacts can be expected to occur as a result of adaptation and climate protection measures (flood protection facilities, deepening of shipping channels, hydropower etc.); and which additional protection measures may need to be taken in order to achieve the quality targets nonetheless. In cases where it appears likely that the 'natural' reference conditions of the water body in question may change fundamentally, consideration should be

---

[24] Directive 2007/60/EC of the European Parliament and of the Council of 23 October 2007 on the assessment and management of flood risks [2007] OJ L288/27.

given to whether the corresponding management objectives and quality targets need likewise to be adjusted.

One way in which water legislation can contribute towards identifying the probable and possible impacts of climate change on water status is to require explicitly that this be done. This is not the case to date with the WFD. With its requirement of an initial review (Article 5) and, especially, a management plan (Article 13), the WFD requires a comprehensive assessment and depiction of the structure and status of water bodies, and of the effects of human activity on water status as well as an economic analysis of water uses. Changing climatic conditions, however, are mentioned neither among these requirements nor in those that define more precisely the key quality target of 'good status'. The same applies to Annex VII of the WFD, which determines more precisely what management plans are to cover. Climate change is not mentioned here as one of the 'significant pressures' or impacts on the status of surface waters pursuant to No. 2.[25]

The WFD does, however, set out a considerably broader assessment framework to form the basis of ongoing monitoring of water status required of Member States, and this framework explicitly includes 'long-term natural changes' (cf. Article 8 paragraph 2 in conjunction with 1.3.1 Annex V WFD). Annex V (No. 1.3), which stipulates more detailed requirements on this, states among other things that the monitoring programmes should provide information that enables an 'assessment of long-term changes in natural conditions'. These changes will necessarily include the permanent effects on watercourses entailed by climate change.

In 2009 the general principles of water management contained in Germany's Federal Water Act (*Wasserhaushaltsgesetz*, WHG) were augmented by a principle pertaining to precautionary action on the impacts of climate change (§ 6 Cl. 1 No. 5 WHG). This principle provides the rationale for imposing on the relevant German authorities a duty in principle to thoroughly ascertain and take account of the impacts of climate change, including the long-term ones. It should be emphasized here that, based on the German principle of taking a precautionary approach to water management in relation to potential impacts of climate

[25] Taking an extremely broad interpretative view, it is possible, of course, to consider the impacts of climate change as an anthropogenic effect – this indeed is the rather unconvincing approach taken by the EEA (n 3), 27, (2.2.2). It stands to reason, however, that this interpretation is neither what the legislative originally intended nor is it capable of becoming a sufficiently persuasive approach in practice.

change, attention should be paid not only to 'expected' or 'likely' effects; even just 'possible' effects of climate change are to be incorporated into management planning. German water legislation thus demands that all impacts of climate change that are relevant to water management and whose occurrence is deemed to be at least plausible are to be made the object of further assessments, considerations and, where appropriate, actual measures.

Admittedly, as experience shows, this kind of abstract obligation that exists merely 'in principle' is not sufficient to guarantee even and effective implementation. However, German water legislation also lacks more concrete requirements to gauge the effects of climate change that might guarantee more effective implementation. At the European level, though, the Commission and the Water Directors of the Member States published, in 2009, wide-ranging recommendations specifically on identifying the impacts of climate change and climate-related risks regarding water quality.[26] Prior to this, there had been a recommendation in a policy paper of June 2008 that the Member States should incorporate climate change as far as possible even in the first management cycle of the WFD (2009–2015):

> Climate change should at least be considered in the first RBMPs, both in national and international plans. This will serve the following purposes:
>
> - Facilitating stakeholder engagement and public consultation,
> - Improving general awareness of all actors for climate change trends and impacts,
> - Paving the way for more climate change related actions in 2nd and 3rd cycle,
> - Allowing for incorporating international, national and regional information on predictions,
>
> Particular emphasis should be placed on ensuring that the Programmes of Measures are sufficiently adaptive to future climate conditions. Therefore an attempt should be made to carry out a climate check of the Programme of Measures, based on readily available knowledge and data and common sense. Such a check should focus on which measures are enhancing or weakening the adaptive capacities, which measures can be considered as no regret or win-win-solutions, or which measures will be less robust in their effectiveness to achieve the WFD objectives because of climate change.'[27]

---

[26] In CIS Guidance Document No. 24 (n 3).

[27] EU Water Directors, WFD Common Implementation Strategy, Policy Paper 'Climate Change and Water', (2008), 7.

More far-reaching requirements of risk assessment are contained in the CIS Guidance Document No. 24. This contains a set of minimum demands for the second and third management period which state that management plans should express clearly and in a transparent manner

- how available climate predictions have been taken into account in the analysis of water status and pressures,
- how the monitoring programmes have been designed so that probable and possible effects of climate change on water quality can be recorded, and
- how the effectiveness of measures has been tested and taken into account with regard to future climate change impacts.[28]

Further, a series of 'guiding principles' and 'suggested actions' are formulated with regard to the need for action on risk assessments outlined above. These state the scope to be covered by the risk assessment and highlight the particular areas it should consider. They also urge inclusion of the best available information and of relevant actors and outline how, in basic terms, monitoring programmes are to be conducted. Specific 'best practice' examples are listed and sources of data and research results on regional climate impacts provided. The Water Directors consider it necessary, among other things, not only for qualitative information on specific climate impacts to be provided but also for quantitative analyses to be conducted of their predicted development over time and of impacts on water status; this should be done during the second and third management cycles at the latest.[29]

All these recommendations testify to the fact that the significance of climate change for water conservation and the importance of a thorough assessment of climate change impacts and adaptation needs have come to be taken almost as given. Precisely for this reason, then, it seems reasonable to question whether the best way of prompting analysis and consideration of the impacts of climate change is really solely by making non-binding references to forms of implementation. Experiences from European legislation as well as from national planning legislation show that, in order to implement policies – especially those that are part of longer term and more complex programmes of objectives – a crucial precondition is to lay down specific requirements regarding content, form and procedures for the process of gathering factual information and

---

28 EU Water Directors (n 3), 44.
29 EU Water Directors (n 3), 47.

planning measures.[30] This is why many different specific requirements regarding information gathering have been laid down by law in the WFD – albeit not in relation to climate adaptation. In particular, there is a clear instruction as to which influences on water quality are to be included in the review exercise and considered in the management planning (see Annex VI and VII of the WFD). Indeed, it is only on the basis of binding minimum requirements regarding the content of management plans that the Commission is able to enact effective and timely checks.

It is no surprise, then, that issues to do with climate change – in contrast to the above-mentioned CIS recommendations – have only been incorporated in one part of the draft management plans for the first planning cycle (2009–2015), and here largely only superficially.[31] There has been no closer consideration of the consequences for management plans and programmes of measures or indeed of possible long-term and perhaps even structural adaptation needs.

Overall, then, it seems worth considering the option of laying down specific duties of assessment and prediction in a binding way. The implementation recommendations contained in the CIS guidance document already contain helpful information about how the legal assessment programme might be augmented. One legal vehicle (beyond current legislation) might be a formal climate adaptation strategy that is added to the management plan module-by-module. Such a strategy might be one way of legally formalizing the requirements for assessing and transparently depicting the probable and possible impacts of climate change.

6 REGULAR REVISION OF WATER MANAGEMENT PLANS AND MEASURES

The management regime contained in the EU WFD offers a good example of how regular revision of local management strategies can be enforced by law. As explained above, the Directive obliges Member States to introduce, through national legislation, a cyclic management system with river basin management plans (RBMPs) and programmes of measures to be revised every six years (Article 11, 13 WFD). On the

---

[30] Reese, M. 'Qualitätsorientierung im Umweltrecht' in Martin Oldiges (ed.), *Umweltqualität durch Planung,* (2008), 40ff.

[31] WWF/EEB, 'What Future for EU's Water? – Indicator based Assessment of the draft River Basin Management plans under the Water Framework Directive' (2009), s 37 <http://assets.panda.org/downloads/what_future_for_eus_water___full_report.pdf> accessed 17 January 2017.

basis of mandatory ongoing monitoring programmes, each cycle is preceded by a systematic assessment of water status, sources of pollution and degradation, further action needed with regard to the management objectives and also the development of relevant natural reference conditions (Article 5 WFD). Within the revised RBMPs the relevant authorities must explain, where necessary, why management objectives have not so far been achieved and what will be done to achieve them in the future. Each revision process also comprises a mandatory round of public participation (public consultation exercise) where stakeholders can bring in their knowledge and concerns, not least regarding climate change resilience. It is therefore not surprising that the CIS assembly of national and EU water officials has judged the WFD regime to be well-equipped to promote climate adaptation: 'Climate change is not explicitly included in the Water Framework Directive (2000/60/EC). However, the step-wise and cyclical approach of the WFD makes it well suited to handle climate change.'[32]

7 ADAPTATION OF WATER MANAGEMENT/STATUS OBJECTIVES

7.1 Introduction

If the management objectives formulated to date no longer appear reconcilable with the changes occurring in natural reference conditions, then adaptation of these objectives themselves must be a genuine option. This applies especially with regard to the EU WFD approach, which derives its quality objectives from a point of reference that approximates to 'untouched nature'. Should this reference status change as a result of climate change, then in principle this must be followed by an adaptation of the quality objectives derived from it. The need for such an adaptation of objectives in the foreseeable future is, however, flatly denied in the CIS. It is understandable that, especially for those responsible for implementation who have worked over the years in laborious processes to get the quality objectives for all major water bodies established and specified, there is little interest in casually opening up these decisions to consider adaptations to supposed climatic changes. There is an understandable fear that climate change would be 'misused' as a welcome excuse to lower standards – the real reason for wishing to do so being

[32] EU Water Directors (n 3), 2.

economic contingencies. For this very reason, therefore, it seems necessary to establish the preconditions under which climate change may justify adaptations and deviations from existing objectives in a legally binding way. To illustrate this, we highlight a little more precisely the extent to which, based on current knowledge, decisions about adaptations of goals might become necessary even in the medium term (7.2). We then show to what extent such developments may (on the basis of existing management legislation) justify either adapting the reference conditions and specifics of objectives or deviating from the objectives (7.3).

7.2 Potential Need to Adapt Objectives – Even at an Early Stage

The overview of potential impacts of climate change offered in the introduction above has already made it clear that the quality of a watercourse or water body may, in certain circumstances, change entirely. In the long term, at any rate, it should be considered likely that quality objectives that have existed up to now will be called into question – either because the objectives would no longer be achieved even in a completely healthy (reference) water body or because it would require a disproportionate amount of effort under changed climatic conditions. Based on current projections, it can generally be considered indisputable that impacts such as these that are relevant to water conservation objectives are possible and that adaptations of the objectives may become necessary; certainly, the EU Commission and the Water Directors take this as given in their new CIS guidance document. With regard to the ultimate timescale for achieving the aims of the Water Framework Directive (2027), however, they strongly emphasized that such extreme impacts from climate change are not expected to be a regular occurrence within this period. Instead, they say, the effects of climate change are almost certain to be so minor initially that they cannot be distinguished significantly from the influences deriving from other anthropogenic pressures.[33] At the same time, they recognize that there may be exceptional cases in which, even in the medium term, considerable changes may occur even in the natural reference conditions: 'In only a few cases, climate change could impact status assessment in the relatively short term and influence the water body type and/or reference conditions … There are now a number of cases where tentative links to climate change are

---

[33] EU Water Directors (n 3), 44, 53.

claimed (see examples below).'[34] Among the impressive examples named in these exceptional cases are the decline in salt content of the Baltic Sea (entailing far-reaching consequences for the entire marine ecosystem) and the rising water level in the Ijsselmeer, which gives an idea of the considerable impacts of a rising sea level on many coastal waters. However, the most significant example of potential early impacts of climate change can be regarded as being the rampant spread of warmth-loving non-indigenous species, already long an observable feature in many European rivers. Admittedly, it remains unclear as yet to what extent these changes are due merely to the direct introduction of species or to a warmer climate. However, the fact that these non-indigenous species are predominantly those from warmer waters readily leads to the assumption that the heating up of the earth's waters is at least a factor in their spread. The significant role this factor plays in current water ecology is stated impressively in the following excerpt from the management plan for the Upper Rhine:

> The flora and fauna in Baden-Württemberg's water bodies have undergone a major transformation over the last few decades. This species change, involving the displacement of indigenous species, can in part be attributed directly to changed climatic conditions ... The effects are, in some cases, severe. In some parts of the Rhine, more than 90% of biomass derives from non-indigenous species. Mussels have frequently been observed to die in large quantities in connection with periods of heat and low water levels that are now occurring at shorter intervals.[35]

This quotation suggests that changes in the aquatic biocoenosis that are connected with a warmer climate, and in part have been promoted by the introduction of species, generally constitute more the rule than the exception even today; it gives us to assume that the dynamic of these changes is so strong that it may undermine traditional reference conditions more often than not in the medium term. This assumption is also supported by existing biological and nature conservation research dealing with the effects of climate change on flora and fauna in general. It is predicted, for example, that for every 1°C of increased temperature, areas of spread will shift by 200 to 300 km northwards and 200 vertical metres

---

[34] See n 3 above, especially 59ff. (examples 7 j–m).

[35] Extract from the 'Upper Rhine Management Plan', s 7.16 – 'Taking climate change into account' (Bewirtschaftungsplan Hochrhein, Berücksichtigung des Klimawandels – 2009), 220.

upwards.[36] A study by Germany's Federal Agency for Nature Conservation considers it likely that there will be a loss of between 5 and 30 per cent of all plant and animal species in Germany over the next few decades as a result of climate change; the most severe losses are expected to be among aquatic species.[37] These predictions are also – and especially – significant for water management, given that the latter has been geared crucially towards ecological quality objectives by the EU Water Framework Directive, and these in turn are based on the typical species composition of water bodies.[38]

7.3 Opportunities and Limitations of Adaptation of Objectives Based on Existing Legislation

The above comments point to the necessity for the management system also to enable appropriate adaptation of management objectives. To an extent, the Water Framework Directive already meets this prerequisite: It provides for a cyclical revision of management plans (as discussed above), and part of this revision – as will be outlined below (7.3.1) – is a review of the reference conditions and the water body-specific quality targets arising from them. In addition, the Directive provides for exemptions as a way of adapting objectives (7.3.2).

7.3.1 Adaptation of reference conditions

The WFD contains no specific regulations regarding the adaptation of quality targets to climatic changes in natural reference conditions. Upon closer inspection of the concept of defining quality targets, however, we can see that existing legislation makes it permissible – indeed obligatory – to adapt the relevant quality (target) components to changed reference conditions. The bases for this are the recurring reviews (Article 5 WFD) and the reference system of Annex II, which is therefore rehearsed here briefly:

---

[36] Heiland, S., Geiger, B., Rittel, K., Steinl, C. and Wieland, S. 'Der Klimawandel als Herausforderung für die Landschaftsplanung' (2008) *Naturschutz und Landschaftsplanung*, 37–8.

[37] Korn, H. and Epple, C. 'Biologische Vielfalt und Klimawandel' (*BfN-Skripten* 148, 2006), 11.

[38] See especially the biological quality criteria for good ecological water status acc. to Annex II 1.2.1. On 'ecomorphological orientation', Durner, W. 'Zehn Jahre Wasserrahmenrichtlinie – Bilanz und Perspektiven' (*NuR* 2010), 452, 457; with a special focus on potential conflicts: Reinhardt, M. 'Zum Verhältnis von Wasserrecht und Naturschutzrecht' (*NuR* 2009), 517ff.

Annex II stipulates, among other things, that 'type-specific reference conditions' are to be established for each water body in line with specifically defined criteria. In the case of surface waters, for example, type-specific reference conditions (pursuant to No. 1.3 i) of Annex II) are to be considered as the hydromorphological and physicochemical conditions 'representing the values of [those same] quality elements specified in point 1.1 in Annex V for that surface water body type at high ecological status as defined in the relevant table in point 1.2 in Annex V'. 'High ecological status' is defined almost throughout table 1.2 Annex V as the status that would exist if there were no (or very minor) anthropogenic influences ('undisturbed conditions').

In other words, water bodies are to be subdivided into types of water body, and for each type quality elements are to be established – based on the biological, hydromorphological and physical criteria detailed in Annex V – which this type of water body evidences or would evidence in the absence of anthropogenic pressures. Starting out from this reference status, the legally binding good ecological status to be achieved is defined – pursuant to Article 2 No. 22 in conjunction with Article 4 para. 1 ii) and Annex V WFD as a status that deviates only slightly from the 'water body type-specific' reference status. An analogous set of factors is used to describe the artificial and heavily modified water bodies with regard to the special goal of 'high ecological status' (cf. Annex II No. 1.3 ii)) that applies to these water bodies.

Since this entire typology of water bodies and the establishment of type-specific reference conditions are to be subject to review every six years, pursuant to Article 5 paragraph 2 WFD, it may prove necessary to adapt the way the general status objectives are described in relation to the water bodies concerned. If, for example, a repeat review were to show that the previously typical biocoenosis had changed considerably due to higher water temperatures and a lower water level, the ecological reference description for this type of water body would have to be modified accordingly, meaning that the corresponding ecological component would likewise no longer be the standard for very high and high ecological status and that from then onwards another type of water body would constitute the reference condition for high status.

The relevant CIS guidance documents also indicate that this would be the procedure for adjusting objectives: 'If there is strong evidence showing that the situation at unimpacted sites changes significantly because of changed climatic conditions (and not because of human induced pressures), reference conditions can be revised as part of the river basin characterization review.'[39] The Water Directors also assume that climate change might make it necessary to adjust the reference conditions over the longer term. CIS Guidance Document No. 24 considers above all the possibility of a change of type or even of category as a result of climate change. It states literally:

> Water bodies are 'typed' by a set of obligatory (e.g. topographic, geological, physical, hydrological) and optional (depth, mixing characteristic, nutrient status) descriptors. A number of these descriptors are climate sensitive. As any typology is a simplification compared to the natural continuum, some water bodies always remain on type boundaries and could, in theory, migrate from one type to another as a consequence of gradual climate change or sequence of extreme events. Over longer time horizons there is even the prospect that some sites could shift between different categories of water bodies such as a lake to a transitional water category.[40]

Additionally, both documents are clearly keen to set closely defined, tight conditions for any adaptation of reference conditions. This is accompanied by the general supposition – repeatedly emphasized – that it will not be possible to demonstrate significant climate-related changes in the reference conditions within the period of implementation to 2027. Thus, the CIS guidance document demands no less than 'overwhelming evidence' for such a demonstration. Minimum requirements regarding the database include, in particular, robust long-term monitoring of the reference water bodies:

> In fact, a sufficiently robust long-term monitoring network of reference sites, i.e. sites with missing or very minor anthropogenic impact, linked to meteorological data will be the only direct way to detecting responses of water bodies to climate change impacts. Only if such monitoring reveals long term coherent changes in the status of reference water bodies over large geographical areas, it will be a proof of changing reference conditions.[41]

There is a risk, then, in principle, that climate change might be recklessly 'misused' as an argument for lowering targets and be used as a cover for the real – economic – reasons. This is why the Water Directors rightly demand convincing evidence showing that the reference situation has changed significantly due to climate change. They refer correctly to the

---

[39] EU Water Directors (2008): Policy Paper 'Climate Change and Water' on the Common Implementation Strategy to the Water Framework Directive (2008), 6.

[40] EU Water Directors (n 3), 55.

[41] EU Water Directors (n 3), 54.

fact that such evidence requires a dedicated long-term monitoring programme. What is doubtful, however, is whether the non-binding recommendation of the CIS to avoid improper adjustments of the targets is sufficient. It would probably be more effective to develop a constructive legal regulation that explicitly permits such an adjustment only on the condition that convincing evidence for significant – climate-related – changes in the natural reference conditions of the water body type concerned has been presented.

The recurring review stipulated in Article 5 WFD will make only a reactive, corrective adaptation of the reference criteria and status objectives permissible. This is because – as stated explicitly in the CIS[42] – those natural conditions which have become established at the time of the most recent review will subsequently form the basis of the reference conditions. It is not permissible, by contrast, to establish a predictive reference condition derived on the basis of a temporal extension of management activities. Adjusting objectives in an anticipatory, preemptive manner is likely to entail considerable risks when it comes to maintaining a high level of protection because of the immanent danger of making errors and misusing the process.

As is noted correctly in the CIS guidance document, it is probably not sensible completely to ignore probable long-term changes in climatic conditions and natural reference conditions in the management strategy.[43] Such a management approach would inevitably run the risk of preserving even those stocks that, in all likelihood, cannot be preserved over the long term; it would risk prompting flawed developments in long-term infrastructure planning and investment as well as impeding ecological migratory movements. A constructive management regime should therefore be sure to include suitable regulations for enabling appropriate consideration to be given to prospective long-term 'natural' development.

An over-hasty adjustment of reference conditions and quality targets thus seems neither sensible nor necessary. Without putting protection standards at the mercy of predictions that are prone to error and misuse, the necessary long-term orientation could also be achieved by adding a climate adaptation plan to the management strategy in the form of a legally binding planning module geared towards the long term. Such a plan would form not only the formal framework for required projections and adaptation scenarios but could also formulate provisional long-term

42 EU Water Directors, CIS Guidance Document No. 10, River and Lakes – Typology, reference conditions and classification systems, EC (2003), 11.

43 EU Water Directors (n 3), 54ff.

perspectives for those quality components which will probably be subject to climate-related changes. This makes it possible, in the process of actually planning measures, to prioritize those measures geared towards the long term and to establish no regrets options.

7.3.2 Adaptation of management objectives by exemptions

The WFD contains a range of exemptions which make it possible in specific cases to deviate from the quality targets or from the deadlines for achieving them. The following may be the most relevant ones when it comes to climate change adaptation:

Deteriorating status due to extreme events: If implementation of the management objectives is prevented as a result of extreme events such as floods or droughts, there is a special exemption for exceptional circumstances of natural cause which could not have been foreseen (Article 4 No. 6 WFD). This exemption is clearly limited, however, to temporary deteriorations in status and is therefore linked with the (logically following) obligation to undertake all practically suitable measures to prevent any further deterioration in the status of water bodies and to re-establish the prior existing status (paragraph 2 No. 1 and 3). Given that extreme weather events are highly likely to increase as a result of climate change, this exemption provision could certainly become increasingly important. It will need to be established clearly how the conditions of 'exceptionality' and 'predictability' are to be interpreted, if – based on ever more detailed knowledge of regional climate impacts – such extreme events are likely to become increasingly severe and frequent.

Exemptions due to secondary impacts of climate change on the quality of water bodies: If, for purposes of climate change adaptation or in order to reduce greenhouse gas emissions, it should become necessary to take measures deleterious to achieving or maintaining high ecological status of water bodies or good ecological potential, an exemption can be made for new physicochemical changes pursuant to Article 4 paragraph 7 WFD and § 31 Cl. 2 No. 2 WHG (new version) under the conditions set out in more detail there. The exemption is likely to be relevant, for example, for urgent technical flood protection measures, but also for deepening shipping channels and regulating water levels to maintain shipping capability during dry periods. New hydropower stations may also be privileged by this exemption if they meet the other conditions of overwhelming public interest and a lack of alternative solutions. It is important to note that – even when this exemption is used – the obligation always applies to 'take all practically suitable measures to minimize adverse impacts on the status of water bodies' (§ 31 Cl. 2 No. 4 WHG, amended version).

8 FOSTERING CLIMATE COMPATIBILITY OF WATER USES AND RELEVANT LAND USE ACTIVITIES

8.1 Climate Compatibility of New Developments

Permit regimes are obviously the key legal instrument for adapting new projects or developments to the impacts and risks of climate change. In order to promote the climate proofing of new developments by legal means, the permit regime should stipulate that relevant projects must

- comply with applicable environmental and safety standards over their entire life span, including those relating to expected or possible impacts of climate change (substantial climate compatibility requirement), and
- be duly assessed as to their climate compatibility (climate assessment requirement).

To a certain degree, substantial climate compatibility is, of course, included implicitly in the applicable environmental permit requirements. As a rule, it is required that applicable standards be observed over the entire life span of a development, and a permit cannot be granted if there is evidence that the project will necessarily come into conflict with such standards due to expected changes in its environment. However, what is often unclear is which philosophy of risk to apply with regard to uncertain developments and to what extent adaptation can or must be left to later amendments of the project and the permit respectively. In this regard, legal clarifications can be helpful.

In European water law compatibility with the WFD's management objectives is a crucial requirement for the admissibility of projects. As has been underlined recently by the ECJ, Member States must not approve developments which jeopardize fulfilment of the quality targets and, in particular, the prohibition of deterioration in water status.[44] As these status targets are not limited in time, they also require long-term compatibility of any new development. Hence, approval must be denied if there is clear evidence that water conditions will probably change in ways that will render the respective use incompatible with WFD management objectives. This could be the case, for example, with a planned power plant that relies on cooling water extracted from a river that is

[44] Case C-461/13 *Bund für Umwelt und Naturschutz Deutschland eV v Bundesrepublik Deutschland* [2015] ECR I-433.

predicted to see declining water volume as a consequence of climate change. In EU water law too, however, it is not clear how to deal with uncertainty. There is also no jurisprudence to date that deals with such long-term risks resulting from a dynamic environment. What seems clear, though, is that uncertain environmental developments to which the project can reasonably adapt at later stages should not in principle hinder approval. In this regard, the permit-based philosophy of risk is also strongly dependent on the legal means of adapting or withdrawing the permit after it has been issued (see Section 9 below).

A climate assessment requirement can be inferred from the substantial obligation to ensure 'climate compatibility'. However, a merely implicit assessment requirement is often not effective, especially if it comes to uncertain developments and risks which require more effort to examine and analyse. This is why – as is well known – the EU and its Member States (just like many other countries around the globe) have long established mandatory Environmental Impact Assessment (EIA) schemes which include a formal procedure, stakeholder participation and a transparent environmental report as a sophisticated basis for the project decision. The EIA Directive 2011/92[45] requires such formal assessments for virtually all projects with potentially negative effects on the environment, including the aquatic environment. Annex I and II of the Directive list nearly all important water related uses and construction projects and thus make them subject to formal assessment requirements under the Directive. This includes industrial installations with significant waste water output. Hence, in the EU a formal assessment of the environmental effects of a project is guaranteed by law.

In a recent amendment to the EIA Directive[46] the Union has clarified that the impacts of climate change shall also constitute an important element in the assessment.[47] Pursuant to Annex IV No. 5 f) of the Directive, the environmental report must in future also describe the 'vulnerability of the project to climate change'. This means that not only the impacts of the project on the (changing) environment must be assessed but also the potential impacts of the changing environment on the project – quite a novelty in EIA law. However, the EIA Directive

---

[45] Directive 2011/92/EU of the European Parliament and of the Council of 13 December 2011 on the assessment of the effects of certain public and private projects on the environment [2012] OJ L26/1.

[46] Directive 2014/52/EU of the European Parliament and of the Council of 16 April 2014 amending Directive 2011/92/EU on the assessment of the effects of certain public and private projects on the environment [2014] OJ L124/1.

[47] Ibid., see the recitals.

gives no further indication as to how thoroughly these future climate risks must be investigated/projected. It is questionable, though, whether a thorough assessment of climate risks will be ensured without more concrete requirements on the depth, quality and transparency of projections and scenarios. In practice, developers and administrations regularly try to limit the assessment effort, and in Germany, for example, the Federal Administrative Court has also sought to restrict the scope of the assessment and has limited it to providing information on the basis of available knowledge (methods, empirical evidence and predictive knowledge) about effects for humans and the environment.[48] The Court decided with regard to the environmental impact assessment that it is 'not conceived as a search procedure that serves the purpose of revealing environmental impacts which elude conventional means of knowledge acquisition'.[49] Above all, it states, the environmental assessment is not an instrument for further developing the state-of-the-art of science.[50] Accordingly, then, no demand can be made in the context of the EIA to commission, for example, regional climate models or scientific studies on the local impacts of climate change.

Developing regional climate models and scenarios and conducting research into local impacts of expected climatic changes will, however, regularly overwhelm the operators of individual projects, and it makes more sense that such knowledge is generated and publicly provided by the relevant public authorities as a common information basis for future developments. This again supports the idea of a formal climate proofing module as a special part of river basin management plans, where a more general assessment of the local impacts of climate change could be provided.

8.2 Adaptation of Existing Water Uses

Declining water levels, rising water temperatures, higher flood risks and other water-related effects of climate change may also make it necessary to adapt, limit and even abandon existing water and land uses. In legal

---

[48] For more detail, see Reese, M., Möckel, S. and Köck, W. in Moritz Reese, Stefan Möckel, Jana Bovet and Wolfgang Köck (eds), *Rechtlicher Handlungsbedarf zur Anpassung an die Folgen des Klimawandels* (2nd edn, UBA-Berichte 7/2016), 348ff.

[49] Official Journal of the German Administrative Court (BVerwGE) Volume 100, 238, 248.

[50] Official Journal of the German Administrative Court (BVerwGE) Volume 100, 238, 248 and 370, 377.

terms this assumes that the authorities are invested with sufficient powers to amend or withdraw permits and to impose necessary adaptation measures. Conflicts and critical obstacles to effective adaptation may arise, however, where the law provides strong protection for approved land or water uses. If a limitation or withdrawal of the using right is subject to mandatory compensation, this can also hinder adaptation considerably.

In this regard, the German permit and control system can be presented here as an example of a rather flexible framework that gives clear priority to public management interests and provides a range of options for regulating and adapting existing water uses to changing conditions without compensation.[51] The German water law permit system is based on the – constitutionally approved – principle that water is basically to be considered a public good that needs to be publicly managed in the interests of common welfare. As a consequence, all types of water uses – including discharge into waters – are subject to a permit requirement and, more importantly, a normal water use permit (*Erlaubnis*) enjoys no special legal protection under German water law. The normal permit is generally subject to the condition that the water use remains compatible with management objectives and implementation strategies (§ 12 section 1 Federal Water Act). Otherwise it may be limited or withdrawn without compensation (§§ 13, 18 section 1 Federal Water Act). Moreover, it is largely left to the discretion of the relevant authorities how to distribute increasingly scarce water resources and decreasing carrying capacities among competing water uses. This broad discretionary leeway also provides the authorities with strong 'adaptive powers', even with regard to existing water uses. Yet the authorities are not at liberty to restrict existing uses arbitrarily, and they are still obliged – irrespective of the discretion they may exercise in terms of risk evaluation and prioritization – to legitimize any intervention in private water use using cogent arguments and justifiable management – i.e. climate adaptation – strategies. In this respect, again, it would be extremely helpful for a distinct climate impact assessment and adaptation concept to be included in formal water management planning.

[51] See Reese, M. and Möckel, S. in Reese *et al.* (n 48), 153 for more details.

9 CONCLUSION

Increasing aridity and flood risks are doubtless the most worrying water-related impacts of climate change, especially in regions that are already experiencing a water crisis or suffering from frequent flood events. Securing water supply and strengthening flood protection in the most vulnerable regions is thus the first adaptation priority, and this can certainly be done by means of legal requirements regarding the relevant infrastructures and precautionary schemes. However, climate change will have a significant impact on water quality and quantity beyond these major threats and the most obviously affected regions; in the long run, at least, this may call for diverse countermeasures and changes in the relevant water and land-use arrangements. This accelerated dynamic of the water situation makes it necessary to establish a more dynamic, adaptive and also risk-oriented mode of water management. As shown in detail above, this can be strongly supported by a regulatory framework which provides, above all,

(1) adequate minimum standards regarding the assessment/projection of relevant climate impacts and adaptation needs and options,
(2) a target oriented management planning scheme with a mandatory review mechanism,
(3) distinct rules on when to adapt management objectives and water quality targets to changing natural conditions, and
(4) a permit and control regime which includes a climate compatibility check with regard to new developments and also enables the adaptation of existing permits.

To some degree, these regulatory drivers of adaptive water management are already present in modern water legislation, as represented here by the EU WFD. However, as is the case with the WFD, there will frequently be room for improvement and refinement as well. In this regard, much will depend on clear and explicit standards regarding the assessment, monitoring and projection of regional climate impacts as well as the screening of adaptation options and scenarios. With regard to all the above-mentioned key requirements of adaptive water governance, it appears that mandatory climate proofing – as part of cyclic water management planning – could be a suitable legal means of enforcing climate sensitive water management. In particular, this could help to create a robust information basis not only for the relevant authorities and public planners but also for private users, as well as in the context of permit procedures and interventions in existing uses.

13. Regulatory framework of China's water quality management

Liping Dai

1 INTRODUCTION

Despite China being a water-rich country in terms of the total volume of water resources, the per capita availability – estimated in 2007 – is only a quarter of the world average and among the lowest for a major country.[1] Unfortunately, this limited water resource is further compromised by high levels of pollution.[2] Water pollution further exacerbates water scarcity. The overall cost of water scarcity associated with water pollution has amounted to 147 billion RMB ($21.3 billion) per year, or about 1 per cent of GDP.[3] Pollution also poses serious risks to human health. Research indicates that approximately 190 million people fall ill and 60,000 people die from a range of diseases and illnesses associated with water pollution in China each year.[4,5] After more than three decades

---

1 See Jian Xie, Andres Liebenthal, Jeremy Warford, John Dixon, Manchuan Wang, Shiji Gao, Shuilin Wang, Yong Jiang and Zhong Ma, *Addressing China's Water Scarcity: Recommendations for Selected Water Resource Management Issues* (The Word Bank 2009), 1; see also Liping Dai, 'Identifying and understanding the main challenges for sustainable water resource management in China' (2015) *The Journal of Water Law* 24:5/6, 249–64.

2 Ibid.

3 See the World Bank, 'Cost of Pollution in China: Economic Estimates of Physical Damages' (2007), 1–151.

4 See Tao Tao and Kunlun Xin, 'Public health: a sustainable plan for China's drinking water' (2014) *Nature* 511, 527–8.

5 See Qing Wang and Zhiming Yang, 'Industrial water pollution, water environment treatment, and health risks in China' (2016) *Environmental Pollution* 218, 358–65.

of rapid economic growth, China's water pollution has become too serious to ignore.[6]

According to the 2015 China Environmental Situation Bulletin, among the evaluated 423 rivers and 62 major lakes (including reservoirs), only 2.8 per cent of surface water reached Class I. Class II, III, IV and V accounted for 31.47 per cent, 30.3 per cent, 21.1 per cent and 5.6 per cent respectively. Inferior V had reached 8.8 per cent,[7] which is unfit for potable use or human contact. The inter-provincial waters with water quality classified as better than Class III and inferior Class V represented 66.0 per cent and 17.5 per cent, respectively, based on 530 important inter-provincial sectors evaluated. The main parameters exceeding the standards in evaluated rivers and lakes were chemical oxygen demand (COD), biological oxygen demand (BOD5) and total phosphorous (TP), and in inter-provincial waters they were COD, TP, and ammonia nitrogen.[8] The contributors of COD discharges from the greatest to the least were agriculture wastewater, domestic sewage and industrial wastewater; the contributors of Ammonia Nitrogen discharges in turn were domestic sewage, agriculture wastewater and industrial wastewater.[9]

The central government has taken numerous steps with regard to preventing and controlling water pollution during the last decade, such as setting reduction targets for the above-mentioned parameters. Some positive changes have been achieved. This chapter analyses the regulatory framework applied in China's water quality management and aims to provide an overview of the Chinese approaches in this regard.

Before turning to the analysis, it is perhaps useful to say a few words about the Chinese regulatory system. As a one-party state, politics and law in China deeply influence and penetrate each other. The Communist

---

[6] Xinhua News, 'China Headlines: Water pollution plan to bring long-term gain' (English.news.cn, 17 April 2015) <http://news.xinhuanet.com/english/2015-04/17/c_134160737.htm> accessed 26 December 2016.

[7] Ministry of Environmental Protection, '2015 China Environmental Situation Bulletin' (2016). The classification of surface water quality is based on the Environmental Quality Standard for Surface Water, GB2002-3838: Class I is for pristine water sources (such as river headwaters and protected natural catchment areas); Class II is for water source protection areas for centralized drinking supply; Class III is for water source protection areas for drinking supply and recreation; Class IV is for industrial water supply and recreational water with no direct human contact; Class V is for limited agricultural water supply.

[8] Ibid; see also Dajun Shen, 'Water Quality Management in China' (2012) *International Journal of Water Resources Development* 28:2, 281–97.

[9] Ministry of Environmental Protection, '2014 Statistical Bulletin of the National Environment' (2015).

Party has a profound influence on social development, as well as on water quality management. This makes it necessary to discuss both Party policy and legal documents when analysing the regulatory framework.

The following section (Section 2) starts with the policy framework consisting of two Five-Year Plans (FYP) and their corresponding sub-plans. Then the discussion turns to the legal framework including concrete laws and regulations at the central level (Section 3). Implementation will be analysed in Section 4. A conclusion and an outlook are then provided at the end.

2 POLICY FRAMEWORK FOR WATER QUALITY MANAGEMENT IN CHINA

This section reviews the policy development of water quality management in China during the last decade. Key policies in two periods, the 11th FYP period (2006–2010) and the 12th FYP period (2011–2015), are discussed. Those plans and policies set out the general strategy for water development and reflect the Party's ruling ideology in this regard.

2.1 The Period Covered by the 11th Five-Year Plan

The 11th FYP is considered to be a milestone in the history of environmental protection in China. For the first time it set binding goals for environmental management, i.e. applying total pollution load control to two major pollutants, sulphur dioxide (SO_2) (a main source of air pollution) and COD, respectively. Concrete reduction targets for those two pollutants were prescribed by its sub-plan, the 11th National Environment Protection Plan. The main approaches for achieving the COD reduction included accelerating the facility construction of urban sewage treatment and recycling, and strengthening industrial wastewater treatment. According to the Plan, by the end of 2010, all cities were to be equipped with sewage treatment facilities, and at least 70 per cent of urban sewage was to be treated. For controlling COD discharges from industry, the Plan focused on the major industries, which account for 65 per cent of China's total industrial-COD wastewater discharges, but also paid attention to the small heavily-polluted industries which have a low equipped level and poor environmental protection facilities.

Besides controlling COD, addressing drinking water security and improving water pollution management in major river basins are the other two tasks in the 11th FYP. In response to those two challenges, the Ministry of Environmental Protection (MEP), jointly with other related ministries, published a National Plan of Urban Drinking Water Source Protection (2008–2020) and six prevention and control plans in major river basins (2006–2010) (Table 13.1). The former is the first national plan for urban drinking water source protection and applied to 4002 drinking water source areas nationwide. It identifies the main pollution sources for different levels of the drinking water protection zones and provides corresponding management measures for pollution prevention and control. The overall goals and the indicators for evaluation are also prescribed by the Plan.

There are six pollution prevention and control plans which apply in six corresponding major river basins (Table 13.1). Each prescribes specific water quality targets and COD reduction targets. The COD targets in each plan are disaggregated and the constituent parts further assigned to provinces located within the basin in order to allocate clear responsibility (Section 4).

Table 13.1 Policy Framework for Water Quality Management in China

FYP	Document name	Water pollution – related targets or tasks
11th FYP Period	11th National Environment Protection Plan	• Achieving the reduction targets of COD and SO_2, respectively; • Safeguarding the safety of surface drinking water sources; • Enhancing water pollution prevention and control in major river basins.
	National Plan for Urban Drinking Water Source Protection (2008–2020)	By 2020, • Fully controlling environmental pollution in urban drinking water source areas; • Enhancing the capacity of emergency monitoring and water supply; • Comprehensively improving the water quality in urban drinking water source areas.
	Water Pollution Prevention and Control Plan in Huai River, Hai River, Liao River, Chao Lake, Dianchi and Upper and Middle Reaches of the Yellow River (2006–2010)	• Setting water quality targets in corresponding river basins; • Setting pollution control targets.

FYP	Document name	Water pollution – related targets or tasks
12th FYP Period	12th National Environment Protection Plan	• Achieving the reduction targets for COD, SO_2, Ammonia Nitrogen and Nitrogen Oxides, respectively; • Strictly protecting drinking water sources; • Enhancing water pollution prevention and control in major river basins; • Paying attention to water pollution prevention and control in other river basins; • Promoting groundwater pollution prevention and control; • Enhancing rural environmental protection.
	National Plan for Groundwater Pollution Prevention and Control (2011–2020)	• Investigating groundwater pollution status; • Safeguarding the safety of groundwater drinking water sources; • Strictly controlling urban pollution which may contaminate groundwater quality; • Strengthening pollution prevention control of major industries which may contaminate groundwater quality; • Classifying and controlling agricultural diffuse pollution which may contaminate groundwater quality; • Strengthening soil pollution prevention and control in order to protect groundwater quality; • Carrying on groundwater restoration step by step; • Developing a monitoring system for groundwater quality.
	Water Pollution Prevention and Control in Major River Basins (2011–2015)	• Setting water quality targets in corresponding river basins; • Setting total pollution load control targets for COD and Ammonia Nitrogen for industrial and domestic wastewater in corresponding river basins.
	12th National Pollution Prevention and Control Plan for Livestock and Poultry Farming (2011–2015)	• Improving existing regulations, policies and standards of livestock and poultry farming; • Promoting pollution abatement in livestock and poultry farming; • Strengthening zoning management and pollution source control; • Demonstrating and promoting techniques of water pollution prevention and control in livestock and poultry farming; • Developing monitoring capacity in livestock and poultry farming.
	No. 1 Document of 2011	• Three Red Lines

Table 13.1 (continued)

FYP	Document name	Water pollution – related targets or tasks
	Action Plan for Prevention and Control of Water Pollution	By 2020, • Periodically improving national water quality; • Substantially reducing seriously polluted water bodies; • Continuing to enhance the security level of drinking water; • Strictly control the over exploitation of groundwater; • Preliminary containment groundwater pollution; • Improving the water ecological environment in important regions. By 2030, • Striving to improve the overall quality of national water quality; • Restoring, on a preliminary basis, the system function of water ecology. By 2050, • Comprehensively improving the national ecological environment.

2.2 The Period Covered by the 12th Five-Year Plan

The 12th FYP went a step further towards environmental protection by adding two new binding pollution reduction targets for Ammonia Nitrogen and Nitrogen Oxides (a main source of air pollution), respectively. To achieve the reduction targets for COD and Ammonia Nitrogen, besides continually improving urban sewage treatment facilities and strengthening water pollution management in major river basins, the main approaches in the 12th FYP period included applying total pollution load control in the main polluted areas and polluting industries, and promoting intensive livestock and poultry farms.

The MEP published a comprehensive Pollution Prevention and Control Plan in Major River Basins (2011–2015) (Table 13.1), in which four additional basins (including major lakes and reservoirs) were added in comparison to the 11th FYP period. The Plan divides the ten river basins into 36 Controlling Zones and 318 Controlling Units based on the natural characteristics of water catchment and the needs of administrative management. The reduction targets for COD and Ammonia Nitrogen are broken down into layers via the three-tier system of River Basin-Controlling Zone-Controlling Unit. The Controlling Units usually overlap with the lowest administrative level or are covered thereby. The Controlling Units are further divided into Priority Controlling Units and General Controlling Units; the water bodies in Priority Units are again divided into different types, i.e. water bodies for preservation, pollution risk prevention or pollution control. Different measures are applied in corresponding different water bodies. The purpose of this system is to achieve meticulous management in river basins.

The discharge of COD and Ammonia Nitrogen contributed by livestock and poultry farms accounted for 45 per cent and 25 per cent respectively of the total discharge amount, and 95 per cent and 79 per cent of the total agricultural discharge amount in 2010.[10] To cope with this, the MEP, jointly with the Ministry of Agriculture, published the 12th National Pollution Prevention and Control Plan for Livestock and Poultry Farms (2011–2015) (Table 13.1), in which the reduction targets for COD and Ammonia Nitrogen and the targets for waste treatment facilities in the livestock and poultry farms are explicitly formulated. Those targets are broken down and assigned to each province as well.

Groundwater attracts more attention in the 12th FYP period. It was included in the category of water environment quality in the Plan, giving it equal status with the category of drinking water security, instead of ranking under that as in the 11th FYP. In response to this new situation, the MEP issued a National Plan of Groundwater Pollution Prevention and Control (2011–2020) (Table 13.1), which identifies the main challenges of groundwater pollution management and formulates both short-term and long-term targets, i.e. basically understanding the pollution situation and initiating restoration pilot schemes by 2015, and comprehensively monitoring typical pollution sources and expending the restoration scope by 2020. The main approaches include conducting investigations into pollution sources, strictly implementing environmental laws, and developing risk prevention mechanisms and so on.

A very influential policy in the 12th FYP period is the so-called 'No. 1 Document of 2011', jointly issued by the Party's Central Committee and the State Council.[11] It highlights the sustainable use of water resources over a ten-year period and sets the famous 'Most Stringent Standards' which are usually called the 'Three Red Lines', i.e. establishing a control

---

[10] Ministry of Environment Protection, Ministry of Agriculture, '12th National Pollution Prevention and Control Plan for Livestock and Poultry Farms (2011–2015)' (2013).

[11] The Central Committee of the Communist Party of China, the State Council, 'Decision on Promoting Water Conservancy Reform and Development' (2011).

limit for water resource exploitation and utilization, a control limit for water use efficiency, and a pollutant-carrying limit for water function zones. A year later the State Council published Guiding Opinions on the Implementation of the Most Stringent Water Management, which set up corresponding targets for the Three Red Lines. In terms of the pollutant-carrying limit, the discharged pollutants in water function zones must be controlled within the water pollutant-carrying capacity, and 95 per cent of the water bodies must achieve the required water quality standards by 2030. Targets are broken down and assigned to provinces (Section 4).[12]

The most comprehensive plan, the Action Plan for Prevention and Control of Water Pollution, was issued by the State Council at the end of the 12th FYP. The new plan, which includes a list of measures to tackle water pollution, forms part of China's efforts to combat all types of pollution. It follows a plan released in 2013 to tackle smog, which has gained initial success.[13] This water action plan aims to improve the water quality of surface water and groundwater in the whole country by 2030. It is also called the '10-Point Water Plan' since it contains ten strategic measures, which are the following:

- Taking overall control of pollutant discharges;
- Promoting the transformation of the economic structure (to lower pollution intensity);
- Focusing on the protection of water resources through saving water;
- Strengthening support for science and technology;
- Allowing market mechanisms to impact water and pollution levies;
- Strengthening environmental law enforcement and supervision;
- Strengthening the management of the overall hydrological cycle;
- Ensuring the security of water for ecological and environmental purposes;
- Confirming and implementing responsibilities for all parties involved;
- Strengthening public participation and social supervision.[14]

---

[12] General Office of the State Council, 'Notice of Measures for the Examination of the Most Stringent Water Resources Management System' (2013).

[13] Xinhua News, 'China Headlines: Water pollution plan to bring long-term gain' (English.news.cn, 17 April 2015) <http://news.xinhuanet.com/english/2015-04/17/c_134160737.htm> accessed 26 December 2016.

[14] Dongmei Han, Matthew J. Currell and Guoliang Cao, 'Deep challenges for China's war on water pollution' (2016) *Environmental Pollution* 218, 1222–33.

Each measure prescribes corresponding approaches, indicators, scope and timeframes. For example, in order to achieve the purpose of taking overall control of pollutant discharges, small factories in sectors such as paper, insecticides and tanning which do not meet national industrial policy requirements must be shut down by the end of 2016. Additionally, special pollution prevention and control programmes should be established for ten major industries including fertilizers and pesticides; industrial agglomeration areas should be equipped with centralized sewage treatment facilities and fitted with automatic online monitoring devices by the end of 2017. Moreover, existing urban sewage treatment facilities should be altered according to local conditions and reach corresponding discharge standards or recycling requirements before the end of 2020, to name just a few.

This Action Plan is regarded as the blueprint for water quality management in the coming decade. It differs from previous approaches by stressing the philosophy of total water cycle health including groundwater, surface water and marine water, and acknowledging their connections. Furthermore, it aims to give rise to the development of an extensive environmental protection industry and an environmental protection services sector, in order to create a new area to boost sustainable economic growth.[15]

3 LEGAL FRAMEWORK OF WATER QUALITY MANAGEMENT IN CHINA

The legal framework of water quality management, under the umbrella of the Constitution and the Environmental Protection Law, consists of four categories, which are divided according to their functions (Table 13.2). They are natural resource laws, laws and regulations on pollution prevention and control, laws and regulations on environmental management, and normative documents such as environmental quality standards, emission standards and technical guidelines. This section discusses these categories. The discussion is limited to the national level.

---

[15] Ibid.

Table 13.2 Legal Framework for Water Quality Management in China

Categories	Laws, Regulations, Measures and Standards	
Natural resource laws	Water Law, published in 1989, revised in 2002, amended in 2009, 2016; Water and Soil Conservation Law, published in 1991, revised in 2010; Fisheries Law, published in 1986, amended in 2000, 2004, 2009, 2013; Mineral Resources Law, published in 1986, amended in 1996, 2009; Grasslands Law, published in 1985, revised in 2002, amended in 2009, 2013; ...	
Laws and regulations on pollution prevention and control	Water Pollution Prevention and Control Law, published in 1984, amended in 1996, revised in 2008, 2016; Atmospheric Pollution Prevention and Control Law, published in 1987, amended in 1995, revised in 2000, 2015; Law on the Prevention and Control of Radioactive Pollution, published in 2003; Law on the Prevention and Control of Environmental Pollution Caused by Solid Wastes, published in 1995, revised in 2004, amended in 2013, 2015, 2016; Regulation on the Prevention and Control of Pollution from Large-scale Breeding of Livestock and Poultry, published in 2013; Regulation on Environmental Pollution from the Recycling of Ships, published in 1988, revised in 2016; ...	
Laws and regulations on environmental management	Law on an Environmental Impact Assessment, published in 2002, amended in 2016; Regulation on the Environmental Impact Assessment of Planning, published in 2009; Regulations on the Administration of Pesticides, published in 1997, revised in 2001; ...	
Normative Documents	Water-related Environmental Quality Standards	Environmental Quality Standards for Surface Water GB 3838-2002; Quality standards for groundwater GB/T14848-93; Standards for Drinking Water Quality GB 5749-2006; Water Quality Standards for Fisheries GB 11607-89; Standards for Irrigation Water Quality GB 5084-92;
	Water-related Discharge and Emission Standards	Discharge Standards for Pollutants for Municipal Wastewater Treatment Plants GB 18918-2002; Discharge Standards for Pollutants from Livestock and Poultry Breeding GB18596-2001; Emission Standards for Pollutants from the Petroleum Refining Industry GB 31570-2015; Emission Standards for Pollutants from the Inorganic Chemical Industry GB 31573-2015; ...
	Technical Guidelines	Technical Guidelines for an Environmental Impact Assessment General Provisions HJ 2.1-2011; Technical Guidelines for the Environmental Impact Assessment of Surface water quality HJ/T 2.3-93; Technical Guidelines for the Environmental Impact Assessment of Groundwater quality HJ 610-2016; ...

3.1 The Constitution and Environmental Protection Law

The Chinese legal system is primarily based on the Civil Law model, and the primary source of law is statutes. The Constitution is at the top of legal system. It links water pollution prevention and control as a responsibility for the state. 'The state protects and improves the environment in which people live and the ecological environment. It prevents and controls pollution and other public hazards'.[16]

The Environment Protection Law is the fundamental basic law in the field of environmental activities. The new Environmental Protection Law, which entered into force in 2015, sets the principles for environmental protection, for example giving priority to protection, focusing on prevention, comprehensive treatment, engaging the general public, and enforcing accountability for damage, and so on. Specific to environmental pollution prevention and control, approaches prescribed in the Law include promoting clean production and the recycling of resources, implementing the 'Three Synchronizations' (the facilities for pollution prevention and control must be conducted simultaneously with the main projects with regard to their design, construction and operation),[17] establishing a liability regime, establishing a pollution levy system, and implementing a total pollution load control system. This law focuses more on principles and the allocation of responsibility; more concrete regulations with regard to environmental protection are spread among the following different categories.

3.2 Natural Resource Laws

Natural resource laws are formulated for the purpose of the development and utilization of natural resources, although they also touch upon pollution prevention and control. For example, the Water Law lays down its purpose in Article 1 by stating that 'this law is formulated for the rational development, utilization, preservation, and protection of water, for the prevention and control of water disasters, and for the sustainable utilization of water resources in order to meet the needs of national economic and social development'. Its scope of application lies 'in the development, utilization, preservation, protection, and management of water resources and in the prevention and control of water disasters within the territory of China' (Article 2). Although it later deals with water pollution prevention, for example in Article 34 which states that the

---

[16] Article 26 of the Constitution 1982.

[17] See Liping Dai (n 1).

'construction of any outlet for sewage discharge in the protection zones of drinking water sources is prohibited', it explicitly states that 'water pollution prevention and control shall be implemented in accordance with Water Pollution Prevention and Control Law' (Article 81).

Other laws in this category share the same character in that they mainly focus on the development and utilization of corresponding natural resources. For example, the Water and Soil Conservation Law intends 'to prevent and control water and soil loss, protect and reasonably utilize water and soil resources, reduce flood disasters, drought and sandstorms' while the Fisheries Law has been formulated 'for the purpose of enhancing the protection, increase, development and reasonable utilization of fishery resources' and the Grasslands Law intends 'to protect, develop and make rational use of grasslands'.

Although conservation and sustainable utilization could indirectly contribute to the improvement of environmental quality, this is not really the focus of this category.

3.3 Laws and Regulations on Pollution Prevention and Control

The main purpose or responsibility of the second category is to prevent and control environmental pollution. The most relevant law for water quality management is the Water Pollution Prevention and Control Law, which has been formulated for the purpose of 'preventing and controlling water pollution, protecting and improving the environment, safeguarding drinking water safety ...' It applies to 'rivers, lakes, canals, irrigation channels, reservoirs and other surface waters and groundwater'. The law regulates important aspects of water quality management, including the policymaking bodies for water quality standards and discharge standards, the supervision bodies for water pollution prevention and control activities, and both general and specific measures for preventing and controlling industrial pollution, urban domestic pollution, agricultural diffuse pollution, pollution from ships and the pollution of drinking water sources and other special water areas such as scenic areas and important fishery water bodies.

The Water Pollution Prevention and Control Law is currently being revised. The proposed draft, in comparison with the previous version, has added 51 new articles including a new chapter on information disclosure and public participation. It suggests enhancing the supervision of trans-provincial river basin pollution, increasing penalties for water pollution which exceeds emission standards, extending the scope of the application of pollutant discharge permits, and strengthening pollution source control and so on, which are all existing barriers to effectively manage water quality in practice. Once adopted, the new Water Pollution Prevention and Control Law will facilitate water quality management to a large extent.

Other laws and regulations in the second category, although they do not directly manage water pollution, nevertheless contribute to water quality management since water is highly intertwined with many other environmental activities. Combating one type of pollution often produces gains as regards another form of pollution. For example, the implementation of the Law on the Prevention and Control of Environmental Pollution Caused by Solid Wastes could directly contribute to the prevention of water pollution in related water bodies since, according to the law 'no entity or individual may dump solid wastes into or pile them up in rivers, lakes, ditches, reservoirs, bottomlands, banks or slopes under the highest waterline ...' Similarly, the implementation of the Regulation on the Prevention and Control of Pollution from Large-scale Breeding of Livestock and Poultry could also prevent water pollution as the Regulation explicitly requires that 'livestock husbandry development plans shall take the environmental carrying capacity into consideration' and 'setting up livestock and poultry farms and breeding areas in protection zones of drinking water sources is forbidden'.

3.4 Laws and Regulations Concerning Environmental Management

The third category consists of laws and regulations that relate to environmental management. The Law on an Environmental Impact Assessment (EIA) is one of the most important laws in this regard. It has been formulated in order to 'prevent the unfavourable impacts of programs and construction projects upon the environment after they are carried out'. Together with its technical guidelines for an environmental impact assessment in surface water and groundwater (the fourth category), the law provides for principles, working procedures, working methods and requirements for a water quality assessment of possible impacts on surface water and groundwater caused by construction projects.

Among the notable regulations in this category are the Regulations on the Administration of Pesticides. China is the largest consumer of pesticides although it has only 7 per cent of the world's arable land. The annual usage is 2½ times higher than the global average. The misuse of pesticides, as well as fertilizers, has caused serious water pollution.[18] The

[18] Liping Dai, 'Regulating water pollution in China and the European Union in terms of agricultural pollution' (2015) *Journal of Water Law* 24:3/4, 150–55.

Regulations on the Administration of Pesticides and their Implementation Measures regulate the supervision and administration of the production, marketing and application of pesticides. To a large extent they standardize the use of pesticides at the national level, and if they are effectively implemented they will be of assistance in reducing diffuse water pollution to a certain extent. The overuse of fertilizers is another trigger for water pollution and the formulation of a regulation for fertilizer usage has been proposed for many years; however, such a regulation has not yet been given its final shape.

3.5 Normative Documents on Water Quality Management

This category contains a number of normative documents and three sub-categories are included therein. They are the Water-related Environmental Quality Standards, the water-related emission standards and related technical guidelines (Table 13.2). These normative standards and guidelines provide a scientific basis for environment-related decision-making, such as the formulation of environmental plans and discharge reduction targets.

The Water-related Environmental Quality Standards stipulate indicators and indices for water quality in different water bodies. The Environmental Quality Standards for Surface Water GB3838-2002 define five classes for water bodies according to their use. Some 109 parameters, including 24 basic parameters, 5 supplemented parameters, and 80 specific parameters are prescribed by the Standards. According to these Standards, those basic indicators are implemented for rivers, lakes, canals, channels, and reservoirs, while supplementary and specific indicators are applied to drinking water resources, and specific indicators could be determined and selected by local governments according to local environmental characteristics and protection requirements.[19] These parameters provide clear references for establishing water targets and water quality monitoring in daily life.

The water-related emission standards define quantitative limits for wastewater discharges from specific sources over specific timeframes. A total of 46 national discharge and emission standards have been published for a wide range of industries since the 12th FYP, ranging from chemicals to coal to pharmaceuticals and to sugar. These standards play a very important role in water quality monitoring and assessment. Together

[19] Jing Su, Danfeng Ji, Mao Lin, Yanqing Chen, Yuanyuan Sun, Shouliang Huo, Jianchao Zhu and Beidou Xi, 'Developing surface water quality standards in China' (2016) *Resources Conservation and Recycling* (2017) 117, 294–303.

with the above environmental quality standards, they work as yardsticks for judging whether the existing regulations have been adhered to. Those who violate these standards must bear corresponding legal responsibility.

The technical guidelines provide a standard code for implementing the related regulations. For example, the Technical Guidelines for the Environmental Impact Assessment of Surface Water Quality (HJ/T 2.3-93) provide for terminology, standard symbols, working procedures, and evaluation methods for the environmental impact assessment in different surface water bodies. They standardize operational norms in this regard and have practical significance for water environment management.

4 IMPLEMENTATION

The central government has striven to improve the country's water quality during the past decade, and some positive results have been achieved. By the end of the 11th FYP period, according to the 12th FYP, the reduction targets for SO_2 and COD had been exceeded. The treatment rate of urban sewage had increased to 72 per cent in 2011, as compared to 52 per cent in 2005. The 12th FYP targets for the reduction of SO_2, COD, Ammonia Nitrogen and Nitrogen Oxides had also been exceeded. In 2015 the average concentration of COD in surface water had fallen by 32 per cent when compared to 2010, which has directly contributed to the improvement of water quality. Evaluated nationwide in 2015, the quality of surface water in water bodies classified as better than Class III was 14.6 per cent higher than in 2010, and that classified as inferior Class V was 6.8 per cent lower.[20]

However, according to central government the decline in pollutants without binding reduction targets, such as TP, has been much slower in comparison with the above four pollutants.[21] Whether or not the targets are binding plays a key role in success or failure in achieving the necessary reductions. In fact, around 80 per cent of the COD reduction target had not been accomplished during the 10th FYP period as that was only a recommended target at that time. The binding targets are implemented by the Government Objective Responsibility System, which is a core operating tool in environmental management in China.

The Government Objective Responsibility System primarily finds its legal place in the Environmental Protection Law, and is then further

---

[20] State Council, 'Report on the Environmental Situation and the Implementation of Environmental Targets in 2015' (2016).

[21] Ibid.

elaborated by a number of special implementation plans in different fields. The Law explicitly states that

> The State implements the objective-oriented responsibility system and the evaluation system for environmental protection. The people's governments at the county level or above shall incorporate the accomplishment of environmental protection objectives, which shall be an important basis for evaluation of the personnel and departments which are responsible for the supervision and management of environmental protection at the same level, as well as the personnel and departments at lower-level (Article 26).[22]

One of the key strengths of the Chinese approach to social development is its capacity for planning at the national level. The central government establishes binding targets for key areas of development priority in the national plans, and implements them through the Government Objective Responsibility System. Through this System, the overall established targets are assigned to provincial governments, which then reassign their quotas to lower levels (municipal and county levels). From there the quotas are further allocated to local major polluted enterprises or wastewater treatment plants and so on. Party leaders and government officials are required to sign responsibility contracts with the upper level and their performance will be assessed by the latter.[23]

For example, following the establishment of the COD reduction targets in the 11th FYP environmental plan, during the year thereafter the National Environmental Protection Bureau (which subsequently became the Ministry of Environmental Protection in 2008) signed responsibility contracts with nine major COD emission provinces to specify their emission targets, and gradually with other provinces. To apply the Three Red Lines mechanism (Section 2.2), the Ministry of Water Resources, jointly with ten other Ministries, issued an Implementation Plan Concerning the Most Stringent Water Management Assessment. The process, contents, and scoring methods of the assessment are specified in the Plan.[24] In the Action Plan for the Prevention and Control of Water

---

[22] Environmental Protection Law, issued by the Standing Committee of the National People's Congress in 1989, revised in 2014.

[23] Liping Dai, 'A New Perspective on Water Governance in China – Captain of the River' (2015) *Water International* 40:1, 87–99.

[24] Ibid.

Pollution (Section 2.2), 238 concrete measures have been assigned to departments and offices.[25]

Officials who fail to meet the binding targets will be held accountable. For example, according to the above-mentioned Action Plan, the result of the assessment will form an important basis for evaluating the leadership and leading cadres. In the case of a failure to pass this annual assessment, the officials responsible will be interviewed by the upper government, and limited approval will be given for the EIA of construction projects in related areas. The responsible officials might be disciplined or demoted, depending on the severity of the failure. Action may still be taken against retired officials due to their dereliction of duty while in office. The further allocation of funds from central government with respect to water pollution management will also be affected.

Conversely, outstanding performances in the assessment in general can be rewarded through promotion (in rank or position), additional wages or bonus payments, or other material benefits, including administrative benefits (such as free transport, entertainment, training, and travel), and other allowances for cadres (such as subsidized housing, health care, and opportunities for further education).[26]

Such career pressures motivate governmental officials to devote attention to tackling water pollution. Water quality management in China therefore largely relies on governmental administration rather than on legal regulation.[27] Although the legal framework has been given more attention than before during the last decade – and it can be observed in Table 13.2 that a number of laws have been amended or revised during this period – it is still weak in practice. The reasons for this have been well documented in previous research, such as the fact that the nature of the law is vague and ambiguous in China, which makes implementation difficult;[28] the judicial system is weak since it is dependent on provincial and local governments for its funding, and these governments often interfere with judicial decisions in order to protect local industries or

---

[25] Xinhua News, 'China Headlines: Water pollution plan to bring long-term gain' (English.news.cn, 17 April 2015) <http://news.xinhuanet.com/english/2015-04/17/c_134160737.htm> accessed 26 December 2016.

[26] K. Genia, 'Barriers to the Implementation of Environmental Policies at the Local Level in China' (2014) World Bank Policy Research Working Paper. See also Liping Dai (n 23).

[27] Jason J. Czarnezki and Wenxuan Yu, 'Challenges to China's Natural Resources Conservation and Biodiversity Legislation' (2013) *Envtl. L.* 43, 125, <http://digitalcommons.pace.edu/lawfaculty/916/>.

[28] Liping Dai (n 23).

litigants;[29] and the general public, when they encounter environmental disputes, usually choose to report these issues to government departments rather than bring the case before the courts.[30]

5 CONCLUSION AND OUTLOOK

China has been witnessing continuous high economic growth during the last decade. The rapid economic growth and urbanization are ushering in a new phase of continuous pressure on the environment. Water pollution is developing from traditional pollution with conventional pollutants to a compound type of pollution with new and old pollutants interacting with each other, and from point-dominant pollution to the coexistence of point pollution and diffuse pollution.[31] Traditional pollution management schemes are not able to meet these challenges.

In order to adapt to the new situation, many efforts have been made by the central government. A number of new instruments have been developed to tackle different water pollutants and pollution sources. However, there are still some concerns with respect to the implementation of those instruments.

For example, there is too much focus on binding targets for particular pollutants and this leads to the prevention and control of other important pollutants being neglected. This 'one falls, another rises' effect could ultimately result in a failure to improve water quality in general. Furthermore, the improvement in water quality during the last decade has actually largely relied on taking engineering measures such as improving the facilities in wastewater treatment plants or livestock and poultry farms. Soft measures such as cooperation and coordination between different sectors have not been extensively developed, which is still one of the main obstacles in general water management in practice. Implementation strongly relies on governmental administration; the judicial system has lagged behind to a large extent.

The new Action Plan for the Prevention and Control of Water Pollution offers a turning point with respect to water pollution management. It sets

---

29 Marco Rubio, *Judicial Independence in the PRC* <http://www.cecc.gov/judicial-independence-in-the-prc>.

30 Liping Dai (n 23).

31 Yi Wang, *China's water issues: transition, governance and innovation* (Earthscan Publications 2012).

out comprehensive strategies and ambitious targets. It is however more like a blueprint at the current stage, although more concrete implementation measures are expected to be developed in the coming years.

14. Does it take three to tango? The practitioner's viewpoint to three EU governance instruments addressing the agricultural runoff dilemma

Tiina Paloniitty

1 INTRODUCTION

The EU has had a multi-faceted approach in addressing agricultural runoff, specifically when relating to the eutrophication of the Baltic Sea: the Common Agricultural Policy[1] ('CAP'), the oldest of the endeavours,

[1] The EU legislation on CAP has been issued in four regulations (colloquially referred to as pillars) on rural development, horizontal issues, direct payments, and market measures. See respectively Regulation (EU) No 1305/2013 of the European Parliament and of the Council of 17 December 2013 on support for rural development by the European Agricultural Fund for Rural Development (EAFRD) and repealing Council Regulation (EC) No 1698/2005 [2013] OJ L347/487; Regulation (EU) No 1306/2013 of the European Parliament and of the Council of 17 December 2013 on the financing, management and monitoring of the common agricultural policy and repealing Council Regulations (EEC) No 352/78, (EC) No 165/94, (EC) No 2799/98, (EC) No 814/2000, (EC) No 1290/2005 and (EC) 485/2008 [2013] OJ L347/549; Regulation (EU) No 1307/2013 of the European Parliament and of the Council of 17 December 2013 establishing rules for direct payments to farmers under support schemes within the framework of the common agricultural policy and repealing Council Regulation (EC) No 637/2008 and Council Regulation (EC) No 73/2009 [2013] OJ L347/608; and Regulation (EU) No 1308/2013 of the European Parliament and of the Council of 17 December 2013 establishing a common organization of the markets in agricultural products and repealing Council Regulations (EEC) No 922/72, (EEC) No 234/79, (EC) No 1037/2001, and (EC) No 1234/2007 [2013] OJ L347/671.

was followed in 2000 by the Water Framework Directive[2] ('WFD') and its combined approach to pollution sources. Most recently the EU Strategy for the Baltic Sea Region[3] ('Strategy', 'SBSR') was inaugurated as a representative of macroregional governance having agriculture as its *ratio legis* – or rather, when noting the limited legal stringency of the apparatus, *ratio moderatio*. In what follows, the policies and instruments are examined in the legal system of Finland – a country that is one of the main culprits behind the poor state of the sea – with the aim of scrutinizing the practical, legal and normative ramifications of these governance tools. How does their implementation restrict the individual practitioner's right to exploit and cultivate his or her land? Does the latest amendment or transformation of EU water law – namely the so-called Weser ruling (CJEU C-461/13)[4] of the European Court of Justice ('Court', 'CJEU') in the summer of 2015 – have an impact also on the WFD's combined approach to water pollution and thus also the regulation of agricultural runoff? What are the practical effects of the financial or environmental protection instrument, the CAP, for the farmer – or for the water quality?

Examination of these questions begins with taking a look at the Strategy and its consequences for practical agricultural undertakings. Second, the CAP is studied with the objective of finding out whether its reforms have 'greened' agricultural undertakings and, also, the efficacy of the instrument when compared to monetary input. Third, the WFD is put in the spotlight: its current tendencies towards legal normativity are explained and whether this new normativity can be applied to the diffuse pollution source called agricultural runoff is then evaluated. In the final section, the influences of the three governance instruments are summed up, especially from the viewpoint of the practitioner. The question asked is: which governance tools, if any, have an impact on farming practices or the farmers' daily undertakings?

---

[2] Directive 2000/60/EC of the European Parliament and of the Council of 23 October 2000 establishing a framework for Community action in the field of water policy [2000] OJ L327/1 (hereafter WFD).

[3] Presidency Conclusions of the Brussels European Council (29/30 October 2009), 15265/1/09 REV 1; Communication concerning the European Union Strategy for the Baltic Sea Region, COM(2009) 248 final (hereafter Strategy).

[4] Case C-461/13 *Bund für Umwelt und Naturschutz Deutschland eV v Bundesrepublik Deutschland* [2015] ECR I-433 (The Weser Case).

2 THE STRATEGY AND AGRICULTURE AS *RATIO MODERATIO*

The Council approved a strategic road map, the EU Strategy for the Baltic Sea Region, in 2009, followed by the Commission's communication on the matter.[5] The EU developed the approach focusing on regions as an attempt to balance the enlargement and with external relations in mind.[6] The Commission presented the Strategy in 2009 after a joint drafting process with the stakeholders and the Member States. Interestingly, the results of the collaborative process were virtually congruent, all participants favouring an integrated and multi-sectoral approach.[7] What integration in this context refers to is a contested question but can be summarized as understanding that regional drivers of growth are distinct but interdependent.[8]

The Baltic region was a natural choice for a territorial cohesion project: its nature as a geographical space between the international and national levels is so well established that analytical concepts such as 'meso-' or 'macroregion' or a definition of 'peripheral subregion' describe the region well. Macroregion, the most established of these concepts, refers to 'a meso-level bringing together a group of units that are at the same time part of (or related to) a more comprehensive political

---

[5] Strategy (n 3).

[6] Kristine Kern, 'Governance for Sustainable Development in the Baltic Sea Region' (2011) 42(1) *Journal of Baltic Studies* 21, 22. The history of cooperation in the region can be dated back to the era of the Hanseatic League, ibid., 29.

[7] Strategy (n 3), 1, 4. Similar strategy processes are also ongoing in the Danube region, the Adriatic-Ionian region and the Alpine region, see respectively the Communication concerning the European Union Strategy for the Danube Region – 08.12.2010 – COM(2010) 715 final, the Communication concerning the European Union Strategy for the Adriatic and Ionian Region – 17.06.2014 – COM(2014) 357 final, and the Communication concerning the European Union Strategy for the Alpine Region – 28.7.2015 – COM(2015) 366 final. There have been deliberations whether collaboration in e.g. the North Sea or Carpathian regions might evolve into formal macroregional strategies. The latest evaluations are, however, rather hesitant and careful. See DG Internal Policies: Policy Department B, 'New role of Macro-Regions in European Territorial Cooperation', IP/B/REGI/FWC/2010_002/LOT02-C01/SC01 (2015), 73–74, and Annex, 100–101.

[8] For a summary of the questions which emerged, see Tamás Kaiser, 'The added value of the integrated approach: the case of Hungary' (Conference on the EU Strategy for the Danube Region: Challenges and Chances 2014–2020) (2015), 100, 102.

entity'.[9] Irrespective of the label used, the Baltic Sea is undisputedly the most established area for co-operation in Europe and has the longest tradition in the field.[10] That is not to say that the Baltic Sea region is a homogenous area, rather the contrary: regarding socio-economic factors the region is a heterogeneous one. The key common feature is the highly polluted sea shared by the littoral states.[11]

These features add to the Baltic Sea region's suitability for a macro-regional governance approach, macroregional governance being a step forward from territorial-based governance to functional governance. The functional aspect is present in the Strategy especially with regard to agricultural runoffs and the sea, and one of the justifications underpinning the Strategy is addressing the issue in an integrated manner throughout the whole region.[12] Since macroregional governance must have at least one shared problem that justifies the governance apparatus, this key problem could be called *ratio moderatio* (ex analogia to *ratio legis*). That would make the state of the Baltic Sea the *ratio moderatio* of the Strategy.

Irrespective of the disparities during the Strategy's drafting process, the environmental aspects remained a key feature, establishing the first of four pillars in the Strategy. The Strategy is presented in four pillars for structural purposes only; their order or arrangement is not supposed to convey any independent meaning – on the contrary, the Commission encourages seeing the four pillars as entwined, as an integrated approach suggests.[13] Within these environmental concerns, the question of nutrient runoffs is of primary importance, while others deal with biodiversity protection, hazardous substances, clean shipping and climate change mitigation and adaption.[14] The Strategy is accompanied by a list of

---

9 Stefan Gänzle and Kristine Kern, 'Macro-region, "Macro-regionalization" and Macro-regional Strategies in the European Union: Towards a New Form of European Governance?' in Stefan Gänzle and Kristine Kern (eds), *A 'Macro-regional' Europe in the Making: Theoretical Approaches and Empirical Evidence* (Palgrave Macmillan 2016), 3, 5.

10 Stefan Gänzle, 'Introduction: Transnational Governance and Policy-Making in the Baltic Sea Region' (2011) 42(1) *Journal of Baltic Studies* 1, 1, 4.

11 Alexandre Dubois, Sigrid Hedin, Peter Schmitt and José Sterling, *EU macro-regions and macro-regional strategies – A scoping study* (Nordregio 2009) 43, 25–9, <http://www.nordregio.se/Publications/Publications-2009/EU-macro-regions-and-macro-regional-strategies/> accessed 5 December 2016.

12 Gänzle and Kern (n 9), 3.

13 Strategy (n 3), 6.

14 The other three being economic prosperity; accessibility and attractiveness; and safety and security, Jonathan Metzger and Peter Schmitt, 'When Soft

priority areas. Out of a total of 17 groups two are on agriculture: the first on sustainable agriculture in general and the eleventh on nutrient runoffs to the sea.[15] Even though the order of the pillars is not decisive in any regard and the priority list is flexible and reviewed on a regular basis, the structure supports claims that the position of agri-environmental causes in the Strategy is strong.[16] As planned, the priority list was updated in 2015.

The relation between the Strategy and the work of HELCOM is of interest. The Helsinki Convention and HELCOM as its secretariat have aimed at improving the ecosystem of the Baltic Sea since the 1970s. It is an aptly named intergovernmental steering body when it comes to cooperation concerning environmental issues.[17] HELCOM's main objective is an environmental one: to protect, restore and enhance the marine ecosystem that is notoriously fragile and polluted.[18] Even though the efficacy of its work has been discussed, its significance as a regional network has not.[19] HELCOM represents intergovernmental tradition, partly due to its age – HELCOM was initiated during the strong East-West divide. Even though frames of mind change more slowly than societal systems emerge and collapse, transnational governance and collaborative actions were better equipped after the end of the East-West divide, especially so in the environmental and economic sectors. HELCOM's leading role in intergovernmental activities was later enhanced with other environmental initiatives which also aim to include private and non-governmental actors in the governance processes, a development which has earned the Baltic region the title of the laboratory for environmental governance.[20]

Spaces Harden: The EU Strategy for the Baltic Sea Region' (2012) 44(2) *Environment and Planning* A 263, 272.

[15] Strategy (n 3), 3, 6–7 and The SBSR priority areas, 1, <http://ec.europa.eu/regional_policy/sources/cooperate/baltic/doc/priority_areas.pdf> accessed 5 December 2016.

[16] Rikard Bengtsson, 'An EU Strategy for the Baltic Sea Region: Good Intentions Meet Complex Challenges' (2009) 9 *European Policy Analysis* 1, 3–4.

[17] Convention on the Protection of the Marine Environment of the Baltic Sea Area (Helsinki Convention); Gänzle (n 10), 2. Even though recommendations are non-binding, they are HELCOM's main policy tool; Stacy D. VanDeveer, 'Networked Baltic Environmental Cooperation' 42(1) *Journal of Baltic Studies* 37, 40, also serves as a short summary of HELCOM, its working practice and place in the network of operators.

[18] Kern (n 6), 26–7.

[19] Gänzle (n 10), 4–5.

[20] Coalition Clean Baltic, Union of the Baltic Cities, and Baltic Sea Action Group being some of them, Gänzle (n 10), 4–5; Kern (n 6), 23–4. An extensive,

The environmental collaboration appears in networks defined by the theme, resulting in overlapping participation – one organization can be and often is a member of multiple networks each having its own, distinct political influence. It is notable concerning the agricultural runoff problem that organizations aiming to tackle the dilemma do not form one single (regional or environmental) network but many. The aim of these organizations might be unanimous but the way in which they interact in the developments of normative or scientific data varies. This flux of influence and dense networks makes measuring their impact difficult: even if the development of a concrete matter occurs, pinpointing the development to the actions of a certain organization or network is nearly impossible. This challenge has not gone unnoticed and the role of implementation and efficacy has gained weight among the goals of the actors.[21] This also applies to the improvements of environmental quality.

However, the multitude of actors and initiatives, coupled with the difficulty of establishing cause and effect patterns, has prompted questions about the Strategy's rationality, especially due to its emphasis on agri-environmentality.[22] Being built on 'three noes' has raised questions about whether the Strategy can bring any added value to the region – the 'three noes' sum up the reality that the European Commission accepted no new institutions, legislation, or instruments while initiating or implementing the Strategy.[23] From the Union's perspective, due to the 'three noes' all value from the strategies is added value, since no significant input is required. Regarding the relation between HELCOM and the Strategy, the Commission found that HELCOM has not been successful enough in finding support from all relevant sectors. This is due to HELCOM's notoriously environmentalist approach, which has caused, for example, the agricultural sector to shy away from HELCOM's work. In the Commission's view the Strategy has succeeded better in this regard; the Strategy's cross-sectoral approach has proved worthwhile.[24]

---

even though still partial, list of environmental initiatives in the area is given in Table 1 of VanDeveer (n 17), 37.

21 VanDeveer (n 17), 38–9.

22 Bengtsson (n 16), 6.

23 Metzger and Schmitt (n 14), 272–3.

24 Other strategic benefits were the ability to tackle issues shared in a certain region but not in the whole Union, Á. Kelemen, 'Assessing the added value of macro-regional strategies – Environment Discussion paper', 46, a stance directly quoted in the Commission Staff Working Document – 27.6.2013 – COM(2013) 468 final, European Commission 2013, 6.

Even when the Strategy does not allow for new institutions or legislation, it nonetheless is a form of new governance, and as such questions of legitimacy are justified. Multi-level and multi-functional governance apparatuses are at risk of being vague or equivocal, a logical consequence of the fact that the results of the strategies are hard to evaluate. Accountability can be established as a 'master value' securing the legitimacy of the governance actions taken – in the case of the Strategy, it would be ruled as ineffectual without the accountability of those responsible for the agricultural runoffs.[25] However, the macro-regional sphere necessitates a redefinition of accountability from its legal counterpart: due to the large number of parties involved, accountability is seen to be more of a 'work-in-progress', a continuous process between different actors, existing in their discussions and exchanges and being also itself subject to constant change. This understanding of accountability is comprehended as the only sensible one when it comes to macro-regional governance: since the terrain itself is flexible and in constant evolution, amendments to the understanding of accountability must also be made. That means that accountability is seen as a part of this unending process of various actors and acts, not as a yardstick used to evaluate the process.[26]

The Strategy can be taken to represent an *interactionist* perspective by admitting that multiple interests are at stake in the agricultural runoffs issue, contradictory even to the extent of colliding, and that the Strategy is the balancing tool to settle the predicament. This description is deemed the most suitable since in the Strategy eutrophication is not a problem to be tackled but a dilemma that needs to be balanced. Also, other solutions taken into the Strategy enhance this interpretation: the Strategy is an operation between different actors and their viewpoints, not an endeavour to enhance scientific knowledge itself or justify decisions merely based on scientific information.[27]

The different analyses and viewpoints examined above share an understanding of the Strategy: in all its flexibility, predicting its efficacy

---

25 Rasmus Kløcker Larsen and Neil Powell, 'Making Sense of Accountability in Baltic Agro-Environmental Governance: The Case of Denmark's Green Growth Strategy' (2013) 33(2) *Social and Environmental Accountability Journal* 71, 72, 75, <http://dx.doi.org/10.1080/0969160X.2012.743276> accessed 5 December 2016.

26 The authors readily admit that the promulgated understanding of accountability is not in congruence with normative responsibility or the strictly legal understanding of the concept, ibid., 75–6.

27 Ibid., 77–9.

is, if not directly impossible, at least a challenge but even so the Strategy seems to deserve its place in the laboratory of governance that is the Baltic Sea region. However, due to the 'three noes' approach it can hardly have an influence in the practitioner's daily life, if participating as a stakeholder is not included. Time will tell how the Strategy influenced the Baltic Sea's environmental status – at present such evaluations would be premature given the slow cycles of environmental and socio-political change.

3 'GREENING THE CAP' OR BLUE-GREENING THE SEA?

The Common Agricultural Policy has been the EU's prime policy field for over half a century, especially budget-wise: over the years the CAP has eaten up as much as 70 per cent of the Community budget.[28] During its lengthy history in EU policy-making the CAP has undergone a number of repairs: in order to contextualize the newest reform and the aspirations attached to it a short examination of these repairs follows.

Remodelling done in the late 1980s is known as the MacSharry reform; later ones are the Agenda 2000 and Fischler reforms. Methodologically speaking, analysing the three has proved to be a challenge as one standard tool for policy research, discourse analysis, has not been considered adequate: in earlier years, policy makers were not expected to justify their actions to the same extent as is the case later in history.[29] The greatest explanatory force has come from combining multi-level, multi-functional and multi-lateral approaches in order to grasp an overall picture of the policy change.[30] However, turning to multi-functional analysis constitutes a shift in agricultural regulation research, since its

---

[28] The title of the section refers to the newest CAP reform, known as 'Greening the CAP' and concurrently also to the result of agricultural runoffs, eutrophication and blue-green algae, one of its outcomes. For an excellent historical and overall review of the CAP, see David Harvey, 'What does the history of the Common Agricultural Policy tell us?' in Joseph A. McMahon and Michael N. Cardwell (eds), *Research Handbook on EU Agricultural Law* (Edward Elgar Publishing 2015).

[29] Isabelle Garzon, *Reforming the Common Agricultural Policy: History of a Paradigm Change* (Palgrave Macmillan 2006), 7.

[30] Ibid., 8–9. The multi-level players in the field range from the Member States via EU organizations to the international level; multi-functional alludes to the various issues at stake; and the multi-lateral aspect deals with the bargaining involved.

aim is to incorporate agriculture as a field of industry into environmental law. 'The general objective of the approach is to regulate the environmental impacts of agriculture in a similar manner to the way in which the environmental impacts of other industries are regulated.'[31] However, in the absence of more suitable solutions, in what follows multi-functional analysis is also considered to be an appropriate one.

In the MacSharry reform the agri-environmental issue was an unconventional approach – all but market imbalance or price support factors were conventional.[32] Perhaps due to these broadened horizons the MacSharry reform has been described as 'the most important CAP reform up to that time'.[33] The international trade negotiations of the era, viz. the Uruguay Round of the GATT, strongly influenced the reform: it has been suggested that the MacSharry reform accommodated the negotiation's twists and turns.[34] Thus, in terms of multi-lateral analysis, the GATT negotiations not only affected matters at their own level but also as a horizontally-connected issue to the reform.[35] Interestingly, the producers' interest groups were the only part of civil society driving the discussions.[36] It must not hence come as a surprise that in the MacSharry reform agri-environmentality was not a decisive matter but a novelty.

The weight of civil society was augmented in and during the negotiations of the next reform, known as Agenda 2000.[37] Not only did the

---

31 Massimo Monteduro, 'Environmental Law and Agroecology: Transdisciplinary Approach to Public Ecosystem Services as a New Challenge for Environmental Legal Doctrine' (2013) 22(1) *European Energy and Environmental Law Review* 2, 5. Naturally, multi-functional analysis also has other objectives, the aims of the analysis being multiple, but the diversity does not account for the underpinning methodological issue. Monteduro defines the aim of incorporation to be at least 30 years old.

32 Garzon (n 29), 51. Finding environmental causes a novelty in the 1990s might be slightly surprising, since environmental concerns had already been in the public sphere for a good 20 years. The explanation goes that environmentality influenced the Union agriculture policy before the MacSharry reform but was only incorporated into the CAP then. Garzon (n 29), 53–4.

33 Arlindo Cunha and Alan Swinbank, *An Inside View of the CAP Reform Process: Explaining the MacSharry, Agenda 2000 and Fischler Reforms* (OUP 2011) 68.

34 Ibid., 72, 100.

35 Garzon (n 29), 74.

36 Garzon (n 29), 75.

37 Agenda 2000 was seen as a package of measures preparing the Union for enlargement and other novelties the new century might bring, Cunha and Swinbank (n 33), 115.

producers' organizations diversify but environmental groups also mobilized themselves, partly due to the new practice according to which environmental concerns were to be assimilated into all Union policies.[38] Budgetary stringency was rudimentary as were considerations on cohesion, that is to say issues emerged along with the prospective enlargement of the Union.[39] In this round, rural development was taken into the CAP by incorporating the agri-environmental measures into its second pillar.[40] In general, Agenda 2000 has been recalled either as a disappointment or as a reform of modest outcome. Regarding agri-environmentality, there might be more to the story than mere dissatisfaction: Agenda 2000 enlarged the space given to environmental NGOs in the negotiation process and as such found its place in the sequence of reforms.

The negotiations for the third of these reforms, the Fischler reform, commenced shortly after Agenda 2000 was completed.[41] Some hitherto elementary issues, such as budgetary rigour, were not advanced in the Fischler round whereas others, such as agri-environmentality, continued to improve their position. The latter was due to the increased significance of civil society after severe food crises and, also, to an acknowledgement of the impact Union agriculture had on developing countries – as if the Union food policy started to develop a conscience.[42] In spite of that, the diversification of the value base in the CAP negotiations was concretized in the Fischler reform: in this round, not only food security but also environmental security were fully considered in the process. However, the broad choice of implementation options given to the Member States might have diluted the results – the Fischler reform has been nicknamed *the cafeteria CAP*[43] – but at least there was something to be diluted!

Studies of the CAP over the past 20 years have noted that the earlier reforms consist of three different discourses: productivist, neo-liberal and

---

[38] Garzon (n 29), 88–90.

[39] Garzon (n 29), 93, 95.

[40] Harvey (n 28), 19; Garzon (n 29), 54.

[41] Berlin European Council 24–25 March 1999 Presidency Conclusions, paras 19 and 20.

[42] E.g. ethical issues started to have a say, Cunha and Swinbank (n 33), 126; Garzon (n 29), 99–100. The international dimension played a significant role in this reform: the Doha Agreement in 2001 caused external pressures, and this time the future prospects of agreements in agricultural policies between e.g. the US and the EU were at stake, Garzon (n 29), 100–101.

[43] Alan Greer, *Agricultural Policy in Europe* (Manchester University Press 2005), 208; Cunha and Swinbank (n 33), 133.

multi-functional.[44] In short, the first-mentioned emphasizes the food security and farm income function of agriculture, the second is willing to admit more diverse functions of the industry, and the last focuses on the internal and external pressures this highly capitalized field faces, rising retrospectively from the Union budget negotiations and international trade negotiations.[45] The three discourses, in summary, represent the 'holy trinity' of the CAP. Interestingly, the environmental causes are deemed to be a diversificator of (the allegedly) previously stable and (apparently also) unbiased situation, posing as a telltale sign of the assumptions in which the funding of food production can be detached from the environment in which the food is produced.

The Commission started the negotiations for the most recent reform with a document entitled 'Greening the CAP'.[46] Even though this reform, under the name of Commissioner Cialoş, is quite fresh and newly-minted, whether it deserves its nickname has already been examined by a study of whether and how the discourses used have been transformed into policy processes and instruments.[47] Using multi-functional discourse in which the environmental aspects were highlighted, it was found that the new direct payment scheme was justified. The same combination of discourses was also present in the later negotiations on the CAP reform. Discourses on the basic payment scheme were apparently 'blank', including mainly bureaucratic formulations or numbers. Nevertheless, the researchers were able to track down traces of productivist discourse – an

---

44 Even though discourse analysis of the earliest negotiations has been challenging to undertake, credible results have been sourced from other reforms, see text at n 29.

45 Karmen Erjavec and Emil Erjavec, '"Greening the CAP" – Just a fashionable justification? A discourse analysis of the 2014–2020 CAP reform documents' (2015) 51 *Food Policy* 53, 55.

46 Communication from the European Commission to the European Parliament, the Council, the European Economic and Social Committee and the Committee of the Regions: The CAP towards 2020: Meeting the food, natural resources and territorial challenges of the future (COM(2010) 672 final); Cordula Rutz, Janet Dwyer and Jörg Schramek. 'More New Wine in the Same Old Bottles? The Evolving Nature of the CAP Reform Debate in Europe, and Prospects for the Future' (2014) 54(3) *Sociologia Ruralis* 266.

47 Harvey (n 28), 24; Erjavec and Erjavec (n 45), 54, 56. On the procedural side, enhanced monitoring and evaluation have been found to be needed in the 'greening' process, see David Baldock, 'Twisted together: European agriculture, environment and the Common Agricultural Policy' in Joseph A. McMahon and Michael N. Cardwell (eds), *Research Handbook on EU Agricultural Law* (Edward Elgar Publishing 2015), 130; simultaneously, however, they increased the costs in adding to compliance checks, Harvey (n 28), 36.

interpretation seconded by commentators who found the main purpose of the reform was a justification of the excessive spending on agriculture.[48] Thus the 'greening' seems to come into question only after the basic level of production and income are secured: significant amounts of productivist discourse were utilized to justify support measures for young farmers, coupled payments and flexibility options for direct payment systems.[49]

It might, however, be unwise to hasten to a conclusion that the greening of the CAP can be equated with white- or green-washing it. It took until as late as the 1990s for environmental concerns to arrive as a novel new aspect starting to make its way into the decision-making process of the financial distribution tool. The CAP consists of trade, food supply and security, and environmental causes, all of which have their own objectives in the decision-making process. In retrospect, the progress and position of environmental concerns is evident. The same conclusion was reached when the distribution of funds was taken into account in the policy analysis. After assigning measures to certain discourses and comparing their share of the total CAP budget it was found that productivist discourse accounted for 60 per cent of it. Multi-functional discourses earned 30 per cent of the total, while neo-liberalist discourses were almost non-existent. While the 'greening' of the CAP was the key element in multi-functional discourse, it might not be an exaggeration to say that its share in the end result is greater than in the earlier versions of the CAP. The danger of oversimplification lingers, though: earlier CAP reform coupled 'greening' elements with payments, and in the latest reform, greening is arbitrary.[50] Also, when it comes to the CAP reform(s), the language in the policy documents cannot necessarily be condensed into equivalent measures and budget distributions.[51] It is thus relatively safe to declare that over the course of the decades the CAP has showed signs of greening: ecological considerations have emerged from the fringes and arrived at the centre of policy-making.

Another question is whether the money is effectively spent at the Member State level whilst implementing the CAP. The efficacy of spending in relation to the environmental quality of the Baltic Sea can be examined using economic tools and such studies were also conducted in

---

48 Erjavec and Erjavec (n 45), 57; Harvey (n 28), 24.

49 Erjavec and Erjavec (n 45), 58–9. The outcome has also been explained by the order of events: after securing the budget and Pillar I, support for the environmental issues in Pillar II underwent a long haul, Baldock (n 47), 148.

50 In the reform of 2003, cross compliance rules were introduced to direct payments, and were an element of 'greening'.

51 Erjavec and Erjavec (n 45), 61.

Finland before the newest reform. In counterfactual analysis, the focal point is the link between the money spent on agri-environmental programmes and the environmental benefits gained with the funded measures. The results of the 'social cost-benefit analysis' were rather discouraging: the environmental gains of reduced nutrient runoffs were considerably less than the money spent.[52] The result is less surprising when one notes that over the years that Finland has had its agri-environmental programme the amounts of nitrate runoffs have increased, not decreased, against the programme's aims.[53] Even when there might be justified reasons to oppose cost-benefit analyses in policy research, for analysing benefit distribution a cost-benefit analysis nonetheless suits well.[54]

The complexity of the CAP makes clear answers concerning its effectiveness difficult. Ecological realities steer the evaluator in the same direction: since nutrients in water ecosystems compound, eutrophication is caused not only by current or recent emissions but also by emissions from previous years, even decades back, that have become sedimented in the seabed. The complexity of the issue is thus multifaceted and makes evaluating the efficacy of the current – or earlier – measures, in economic analysis or otherwise, especially troublesome. The CAP, however, strongly influences farming practices in the Member States, Finland included, by also being the main source of administrative work for practitioners. Even when the results of the CAP or the efficacy of it might be controversial, its impact on practitioners' lives and practices is robust: part of the compensation from their work comes in the form of financial subsidies distributed within the CAP regime. Hence the claim that CAP, unlike the previously studied Strategy, has had a tangible influence on the practitioner's life.

---

[52] Jussi Lankoski and Markku Ollikainen, 'Counterfactual approach for assessing agri-environmental policy: Theory with an application to Finnish water protection policy' (56) 2011 University of Helsinki, Department of Economics and Management Discussion Papers. The latest Union-level implementation report suggests improvements, even though uncertain ones, Report from the Commission to the Council and the European Parliament on the implementation of Council Directive 91/676/EEC concerning the protection of waters against pollution caused by nitrates from agricultural sources based on Member State reports for the period 2008–2011, (COM(2013) 0683 final).

[53] The amounts of soil phosphorus have increased in some parts of the country, also proving the inefficacy of the programme, Lankoski and Ollikainen (n 52), 5.

[54] S.A. Shapiro and C.H. Schroeder, 'Beyond cost-benefit analysis: A pragmatic reorientation' (2008) 32(2) *Harvard Environmental Law Review* 433.

4 COMBINING THE WFD'S NEWLY-FOUND NORMATIVITY WITH ITS COMBINED APPROACH

The examination moves now from financial tools and governance mechanisms towards the realms of 'the legal': the WFD in its contemporary mien. Could the WFD cover the needs for normativity in the field? Could – or should – its newly-found normativity have an impact on the individual farmer's undertakings? After all, WFD Article 10 requires employment of a combined approach to all pollution sources, diffuse and point-source alike, and in Article 10(2)(c) mentions the best environmental practices as one mechanism through which the approach should be instigated.[55] How binding are these articles, legally speaking?

Land-based water pollution from agriculture is a prime example of the ecosystems' holistic reality, namely the fertilizers used in cultivation runoff to the nearby waters and eventually to the sea. From the 1970s on these holistic ecosystems have been researched and managed using an adaptive management paradigm, a nowadays dominant nature resources management theory which also underpins the WFD.[56] Adaptive management systems are complex and resilient and as such they have proved to be a trial for regulators – lawyers have recoiled from adaptive management as it fundamentally contradicts predictability and certainty.[57] Regulating adaptive management is a challenge, especially in civil law

[55] WFD Art 10(3) reaffirms that the WFD does not allow for any deterioration from the previous water quality requirements or emission controls set out in earlier EU legislation. It is also noted at the political level that lack of compatibility between the WFD and the CAP may explain the poor achievement of the former's objectives. European Commission *A Blueprint to Safeguard Europe's Water Resources* Brussels – 14.11.2012 – COM(2012) 673 final, 4 and William Howarth, 'Integrated water resources management and the European Union's Common Agricultural Policy' in Joseph A. McMahon and Michael N. Cardwell (eds), *Research Handbook on EU Agricultural Law* (Edward Elgar Publishing 2015), 246–247, 250, who couples the failure to the fact that the WFD as a whole regulates water quality rather than quantity, ibid. Howarth, 260.

[56] The oft-cited C.S. Holling and others have been credited for initiating the paradigm shift from previous natural resources management, which in its practicality was less attached to the complexity of ecosystems, C.S. Holling (ed.), *Adaptive Environmental Assessment and Management* (Wiley-Interscience, 1978), 377.

[57] For an excellent analysis of the law's relation to adaptive management, see e.g. Bradley C. Karkkainen, 'Adaptive Ecosystem Management and Regulatory Penalty Defaults: Toward a Bounded Pragmatism' (2002) 87 *Minnesota Law Review*, 943; Bradley C. Karkkainen, 'Collaborative Ecosystem Governance:

systems that revolve around rules (at most discussing their interpretation or characteristics) which never relinquish their elementary position in the system. Much-debated principle or rule distinctions or dialogue over flexible norms might have enlivened the discourse, but one thing remains adamant: predictability is part of the law's epistemology.[58]

The incomparability of adaptive management and the normative is obvious when it comes to the regulation of large-scale adaptive water management, well illustrated by the Water Framework Directive, which has received much criticism from scientists, managers, and legal scholars, including claims that the WFD does not make the best use of ecological knowledge. In the WFD the ecological status of a water body is defined using biological 'quality elements', classified by scientists in the Member States and unified at the EU level. However, when gauging the 'good ecological status' of waters, these elements are actually secondary features – quality is more about structures and processes of aquatic ecosystems than the presence or absence of single elements.[59] In the WFD the aim is that the quality of all water bodies should not deteriorate and that they should attain a good ecological status, objectives which are meticulously defined in the Directive (Article 4(1) and Annex V). Law scholarshave expressed interest in this 'good status' objective, deliberating over its normative impetus.[60] The so-called 'non-deterioration' principle has shied away from the limelight: it has been commonly understood to refer to the quality of EU waters at the time the Directive was issued.[61]

Scale, Complexity, and Dynamism' (2002) 21 *Virginia Environmental Law Journal* 189.

58 This fact is shared by both common and civil law systems. What differs is the mechanism for balancing predictability and flexibility and the struggles involved in the task.

59 A thorough analysis of these questions is provided by Henrik Josefsson and Lasse Baaner, 'The Water Framework Directive – A Directive for the Twenty-First Century?' (2011) 23(3) *JEL* 463.

60 Andrea M. Keessen, Jasper van Kempen, H. Van Rijswick and Chris Backes, 'European River Basin Districts: Are They Swimming in the Same Implementation Pool?' (2010) 22(2) *JEL* 197; J.J.H. van Kempen, 'Countering the Obscurity of Obligations in European Environmental Law: An Analysis of Article 4 of the European Water Framework Directive' (2012) 24(3) *JEL* 499.

61 William Howarth, 'Aspirations and Realities under the Water Framework Directive: Proceduralisation, Participation and Practicalities' (2009) 21 *JEL* 391, 411.

4.1 The WFD in the Court's Reading

In 2015, in the Weser ruling by the European Court of Justice, this interpretation of the 'non-deterioration' principle was shown to be mistaken. The nature of adaptive management did not hinder the European Court of Justice from turning to legal formalism in its assessment of the Directive. In the Weser Case the Court had its chance to have a say on the normativity of the established management system. Concurrently, the Court resolved the most suitable mode of regulation for adaptive water management, as far as its scope of review and limits of discretion allowed.[62] But as always with EU law, enforcement lies in the hands of the Member States, which are thereby also responsible for enforcing the legal state the Weser ruling established.

Regarding the Directive, the original enforcement was variegated: the Member States opted for a variety of implementation strategies. The Member States' understanding of the normativity of the 'good status' objective might have been either the *objective of best effort* or the *objective of result*, or an amalgamation of both.[63] Implementation of the WFD in Finland is an invigorating example of these variations: in Finland, what is taken into a river basin management plan ('RBMP') 'shall be taken into account' in legal and administrative procedures relating to environmental matters.[64] Due to the country's geographical realities as a 'land of a thousand lakes', the demands of EU water law are found to be especially onerous, tentatively also explaining the original reluctance to give normative clout to the WFD.

The preliminary reference that prompted the Weser ruling was skilfully established – the questions set focused solely on the normativity of individual undertakings. The preliminary reference gave the Court latitude in deliberating how the 'good status' objective and the non-deterioration principle ought to be interpreted and how they relate to individual undertakings and their authorization.[65] The referring court

---

[62] The Weser Case (n 4).

[63] Keessen *et al.* (n 60), 206.

[64] Finnish Environmental Protection Act (527/2014) 51 § and Finnish Water Act (587/2011) Chapter 3, 6 §, both sections are part of the regulation on permit consideration. The Environmental Protection Act still lacks its formal English translation, but the unofficial English version of the Water Act is available online: <http://www.finlex.fi/en/laki/kaannokset/2011/en20110587.pdf> accessed 5 December 2016.

[65] The request for a preliminary ruling, The Weser Case (n 4).

insisted especially on the meaning and significance of the status classifications: the referring court wished to know which aspect of this process was decisive when defining the acquisition of the 'good status' objective.[66] Unlike the preceeding scholarly literature, the Court did not employ a distinction between the objectives of best effort and the objectives of result in its reasoning, but referred to two distinct but intrinsically linked obligations taken into the Directive: *the obligation to prevent deterioration* and *the obligation to enhance*.[67] The Court continued with the understanding that the obligations came into being because of deliberate choices of the Union legislator, and hence protecting their normativity best served the legislator's intention. Thus the Member States must refuse to authorize projects that might undermine the attainment of the said objectives, providing no exemption is granted according to the Article 4(7).[68]

The second aspect in which the Court parted from the scholar's ponderings was its focus on the non-deterioration principle.[69] It established the principle as an independent key objective of the Directive. To achieve such an interpretation the Court dismissed the popular 'status class' theory as one with merely instrumental value and opted for the 'status quo' version instead.[70] Here the Court departed from the previous understanding for the third time: instead of understanding the 'status quo' as one that prevailed at the time the Directive was issued, the Court took 'status quo' to refer to the highest water quality achieved at any point of time.[71] Hence the quality of a waterbody does not meet the desired objective whenever deterioration occurs and, consequently, deterioration of the waterbody becomes accessible only when derogation is granted.

---

66 WFD, Annex V, ss 1.3. and 1.4.1. Much of the criticism the Directive has gained is rooted in aspects of this mechanism, see e.g. Josefsson and Baaner (n 59).

67 The Weser Case (n 4) para 39.

68 The Weser Case (n 4) paras 22, 50–51. The obligation prevails irrespective of the Member States' possible negligence in compiling the programmes of measures, which include the definitions of 'good status' objectives, i.e. data according to which the deterioration is assessed.

69 The non-deterioration principle has been noted but often not further elaborated since it has usually been interpreted as applying only between statuses or to the most severe cases, such as agricultural runoffs. See, for example, van Kempen (n 60), 527–8; Keessen et al. (n 60), 210–12.

70 The Weser Case (n 4) para 52 and, in more detail, Case C-461/13 *Bund v Germany* [2015] ECR I-433, Opinion of AG Jääskinen, para 90.

71 The Weser Case (n 4) para 55.

When giving full normative clout to the environmental objectives and the non-deterioration principle the Court in other words imposed traditional, formalistic legal logic on the adaptive water management regime: in the Court's reading the environmental objectives are legal norms in the strictest sense and should be interpreted as such.[72] The EU water quality law now includes clear rules and, as with any rule, there is also the possibility of exemptions. Due to the WFD's integrated and holistic approach as well as its desire for a combined approach, agricultural runoffs are equally part of the newly-established demands of increasingly stringent water quality demands. In particular, the new normativity of the non-deterioration principle may be a challenge to implement, and not only regarding agricultural runoffs. The reasons behind the Court's stance related to the long traditions and established notions of water law. As AG Jääskinen stated, 'the concept of deterioration is a well-established concept of water law, having a general, rather than technical or detailed scope'. In other words, the Court attempted to distance itself from the much-criticized technicalities in the WFD and examine the issue from a broader perspective.

4.2 Individual Agricultural Undertakings, the Scope of Derogation and Environmental Permitting

As always, when legal formalism is in question, alongside clear rules comes the question of exceptions. The conditions on which derogation can be granted are listed in WFD Article 4(7).[73] It is noteworthy that the exception clause only applies to new endeavours causing a decline in water quality. Thus, if an existing activity is replaced with a new one and the detrimental effects remain equal or diminish, no exemption is needed.[74] This rudimentary aspect of the derogation regime sets the stage for a mechanism that could be called 'replacement measures'. These are

---

72 The Weser Case (n 4) ruling paras 1–2; Tiina Paloniitty, 'The Weser Case: Case C-461/13 BUND V GERMANY' (2016) 28(1) *JEL* 151, 159.

73 Art 4(7) is not the only derogations regime established in the Directive. Art 4(3) on artificial or highly moderated waterbodies can be understood as one mechanism for ensuring the environmental quality objectives. Also Articles 4(4)–(5), which regulate the extensions granted for deadlines and the possibility of altering the environmental objectives, can be understood as leeway mechanisms.

74 Common Implementation Strategy for the Water Framework Directive (2000/60/EC), Guidance Document No. 20, Guidance Document on Exemptions to the Environmental Objectives. Office for Official Publications of the European Communities 2009, 24.

the measures applied when a novel undertaking in a certain river basin district buys out existing establishments with detrimental effect, in order to ascertain that the water quality either improves or remains in status quo. Another variation is that the project that causes detrimental effect is coupled with activities that improve the water quality in the said river basin district, causing the overall level of the assessed quality elements not to deteriorate.

Both these possibilities predict an interesting future for the agricultural industry. As a field of industry it is often among the biggest pollutants but in terms of individual farms it is a small player. Moreover, the rudimentary issue of agricultural runoffs not falling under the scope of any environmental permit again becomes relevant: the pollution from agricultural activities is not evaluated in environmental permits.[75] Since the river basin management plans include all activities impacting the waters, agriculture is however included in the assessments and the fulfilment of the environmental objectives. Coupling this reality with the WFD's normativity, and especially its derogation regime, becomes increasingly important now when the total pollution load has become ever more important. In a standard situation when deciding upon derogations, the non-deterioration principle and the good status objective are treated differently: a derogation can be granted if the proposed undertaking causes detrimental effects due to 'new modifications to the physical characteristics of a surface water body' or when the status of a surface water body is estimated to deteriorate from 'high' to 'good' through an endeavour that is deemed sustainable.[76]

Thus a novel undertaking may be granted an exemption to cause effects resulting in the plunge from 'good status' only if the deterioration is not caused by emission but by direct physical alterations of the water body. Detrimental effects caused by the emission-pollution mechanism can be exempted only when statuses above 'good' are in question, providing that the undertaking is pronounced sustainable.[77] This reality might cause increasing tensions in the watershed between different

---

[75] Parts of agricultural production, e.g. animal shelters exceeding a certain headage, are regulated as point-source polluters in the IE Directive (Directive 2010/75/EU of the European Parliament and of the Council of 24 November 2010 on industrial emissions (integrated pollution prevention and control) [2010] OJ L334/17, 17–119) but the distribution of manure to fields as fertilizer is not taken into account when considering the said permits. In Finland, the regulations are taken into the Environmental Protection Degree (713/2014) 2 § 11a).

[76] WFD Art 4(7).

[77] WFD Art 4(7).

actors, all of whom cause water pollution while only some fall into the scope of environmental permits.

Could agriculture as a field of industry be eligible for exemption? Apart from the two general notions mentioned above, there are four more qualifications to be fulfilled if derogation is desired. Exemption from the rules is to be treated as a last resort, the reasoning behind the planned modifications must be detailed in the RBMPs, and conditions such as technical feasibility or disproportionate cost must not allow for a significantly better environmental option.[78] The fourth condition is the most complex one. According to Article 4(7)(c), reasons such as human health, safety or sustainable development might outweigh the attainment of the non-deterioration principle or the good status objective: the article establishes a weighing and balancing mechanism to the exception regime. It is worth noting that weighing and balancing here is not a general 'pro and con' adjudication with no strings attached but a considerably more restricted activity between the attainment of the set objectives in Article 4(1) and benefits for human health, safety, or sustainable development. Only 'an overriding public interest' makes this weighing and balancing unnecessary. That being the case, the condition of Article 4(7)(c) concerns the public interest only and all other requirements can be omitted.

Article 4(7) employs broad concepts, all familiar elsewhere in environmental law: health and safety, public interest, and sustainability, but the article offers no further advice on interpretation. Since the case law on the WFD is yet to develop its own guidelines, some analogical aid can be sought from the case law of nature directives, especially the Habitats Directive,[79] a pivotal piece of EU nature conservation legislation utilizing similar concepts in its endeavour. The Habitats Directive established a Union-wide network of Natura 2000 areas, aiming to safeguard ecosystems from endangering developments.[80] Article 6(4) of the Habitats Directive regulates the exemptions from the protection, allowing administrative permissions to undertakings with 'imperative reasons of overriding public interest' – social and economic ones included – when alternative solutions are not available. The derogation regime in the Habitats Directive thus loosely resembles the one created in the WFD. The significance of the planned project to economic development and employment has been taken into account in a harbour case in Hamburg,

---

78 WFD Art 4(7)(a)–(b), (d).

79 Council Directive 92/43/EEC of 21 May 1992 on the conservation of natural habitats and of wild fauna and flora [1992] OJ L206/7.

80 Ibid. Art 6(3).

Germany;[81] also, elsewhere, the severe employment situation and harsh economic conditions have been decisive when ruling in favour of an undertaking.[82] In Rotterdam, the Netherlands, a major port was again granted an exemption due to its importance to the EU transport network.[83] When deciding upon an artificial lake development in notoriously dry southern Europe, securing a water supply for local and regional water consumption, agriculture and industry have been determining factors, providing again that other feasible solutions have been absent.[84] Thus the planned projects must have *significant* benefits before derogation can be granted – it is impossible to be more specific in summarizing the scope of derogation at the EU level.

The agricultural industry's standing in the 'post-Weserian' EU water law thus seems to be rather multifaceted. To begin with, the environmental objectives of the WFD are made legally binding in individual undertakings – but agricultural water pollution lacks the administrative-legal procedure in which this norm could be imposed upon it. The strict legal norms now in the WFD come with exceptions for significant undertakings in which extreme public interest is involved. Domestic food production might be effortlessly counted as one significant undertaking in many Member States but, unfortunately, only as a field of industry, not if individual farms are evaluated separately. If the Member States had

---

[81] Commission Opinion of 6 December 2011 delivered upon request of Germany pursuant to Art 6(4) sub par. 2 of Council Directive 92/43/EEC of 21 May 1992 on the conservation of natural habitats and of wild fauna and flora, concerning the deepening and widening of the ship fairway Unter- and Außenelbe (river Elbe) to the port of Hamburg (Germany), C(2011) 9090 final.

[82] Opinion of the Commission of 24 April 2003 delivered upon request of Germany according to Art 6(4) sub par. 2 of Council Directive 92/43/EEC of 21 May 1992 on the conservation of the natural habitats as well as the wild animals and plants, concerning the approval of an operational master plan ('Rahmenbetriebsplan') of the Prosper Haniel Colliery operated by Deutsche Steinkohle AG (DSK), for the period 2001–2019, 24 April 2003.

[83] Opinion of the Commission delivered pursuant to Art 6.4 § 2 of Council Directive 92/43/EEC of 21 May 1992 on the conservation of the natural habitats and of wild fauna and flora (Habitats Directive), concerning the 'Request by the Netherlands for advice and exchange of information with the European Commission within the framework of the Birds and Habitats Directives', in relation to the 'Project Mainport Rotterdam' Development Plan, 24 April 2003.

[84] Opinion of the Commission pursuant to Art 6.4 § 2 of Council Directive 92/43/EEC of 21 May 1992 on the conservation of natural habitats and of wild fauna and flora, concerning the 'Request by the Kingdom of Spain in relation to the La Breña II reservoir project' C(2004)1797, 7 May 2004 (only the proposed opinion is made available).

permission granting systems for land-based pollution the farms would be evaluated individually, and none of them would fall under the scope of derogation according to WFD Article 4(7). As a result, the regulation of agricultural runoffs might be against the newly-established norm of non-deterioration but mechanisms to enforce the norm are absent.

5 AGRICULTURAL WATER POLLUTION IN THE TRIANGLE OF INFLUENCE, FINANCES AND NORMATIVITY

From the agricultural practitioner's point of view the EU Strategy for the Baltic Sea Region has proven to be relevant mainly as a channel for stakeholder participation, and has no tangible influence on the daily life of a practitioner. Its role as a balancing act, however, is significant, giving a voice to previously silenced stakeholder groups and bringing a new, presumably more impartial, terrain for discourses about the development of the region. The importance of these features should not be dismissed. The CAP, then again, is as practical in its aims as possible: more or less the whole of a farmer's income consists of the financial aid distributed according to its varied schemes, no matter to what extent the aims of the programme may acknowledge something else or how much this reality is called 'unequivocally both de facto and de jure misguided'.[85] However, the efficacy of the CAP has been criticized, its proven results in enhancing environmental quality being somewhat narrow. The criticism must nonetheless be read with a full understanding of the ecological underpinnings, namely the reality of nutrients becoming concentrated in the seabed.

Then again the WFD, especially in its 'post-Weserian' form, might serve the needs of efficacy and bring results for anyone desiring that water quality be enhanced and polluters be held accountable. However, the diffuse nature of pollution would seem to escape the above-described new normativity of the environmental quality objectives: as long as

---

[85] Harvey (n 28), 34, who fiercely disputes that the CAP would, could, or should be supporting farm incomes. This has never been the openly expressed objective of the scheme, nor do the economic theories underpinning the programme allow for that interpretation. However, in actual reality the subsidies granted in the CAP schemes result in financial means being transferred from the Union to the farmers, a reality which prevails irrespective of how many logical steps from theory to practice one needs to take to acknowledge the state of affairs.

agricultural runoffs do not fall under the scope of any administrative procedure aiming to control environmental pollution, the farming practices and their consequences circumvent the normativity of EU water quality regulation. Since agricultural runoffs and other diffuse pollution sources are often the most significant reasons behind the failure to attain the desired environmental quality the 'post-Weserian' EU water law leaves agricultural pollution in a tight spot: it *de facto* acts as a hindrance or even obstacle for permitting other undertakings, resulting in a lack of mechanisms to counteract the situation. Given that the norm the Court established in the Weser ruling is rigorously implemented in the Member States, those struggling with the agricultural runoffs issue are likely to come up with some counterbalancing mechanisms in order to enable a variety of societal undertakings with water impacts.

15. Shale gas development and water in South Africa: regulatory aspects

Hennie Coetzee and Louis J. Kotzé

1 INTRODUCTION

The development of unconventional gas (like most other fossil and some renewable energy sources) is heavily dependent on water access and use.[1] This is due to the extraction method known as hydraulic fracturing (fracking) whereby large quantities of water are mixed with chemical additives and pumped down a horizontal well under high pressure to enable the extraction of the trapped gas.[2] Of particular recent interest is

---

[1] Kalanithy Vairavamoorthy, Jochen Eckart, George Philippidis and Seneshaw Tsegaye, 'Water and Energy in the Urban Setting' in Anders Jägerskog, T.J. Clausen, T. Holmgren and K. Lexén (eds), *Energy and Water: The Vital Link for a Sustainable Future* (Report Nr 33 SIWI Stockholm 2014). Studies indicate that the use of horizontal wells during a typical hydraulic fracturing process can require up to five times the water used by vertical wells. See Stephan Lutter, Christine Polzin, Stephan Giljium, Tamás Pálfy, Thomas Patz, Monika Dittrich, Lisa Kernegger and Adriadna Rodrigo, 'Under Pressure – How our material consumption threatens the planet's water resources' Global 2000 (Vienna 2011), 11.

[2] Catriona Ross, 'Unconventional Gas: Coal Seam Gas, Shale Gas and Tight Gas', Research Paper for Parliament of Victoria (December 2013). Environmental Protection Agency June 2015 report: 'Assessment of the Potential Impacts of Hydraulic Fracturing for Oil and Gas on Drinking Water Resources' External Review Draft EPA/600/R-15/047a ES1. The Chartered Institute of Water and Environmental Management (CIWEM) estimates that a single well (drill and fracture) will use 11–22 M.l (1Ml = 1000m3). This does not account for the fact that a well will typically be fractured around 4 times; hence water usage will be around 44–88ML (equivalent of 35 Olympic swimming pools of water). See Jenna Brown, Chad Staddon and Enda Hayes 'Water Security and shale gas exploration in the UK' (2014) *Environmental Scientist*, 51. Others estimate water use to be 10ML to 30ML – see Laura Grant and Alastair Chisholm 'Shale gas and water: An independent review of shale gas exploration and exploitation in

the geographical area of the Karoo in South Africa,[3] where it is believed shale gas reserves exist.[4] A number of companies have already applied for exploration rights in the concession area of the Karoo.[5] South Africa is the thirtieth driest country in the world and its Karoo region, which is one of the driest in the country, faces a continuous critical challenge of water scarcity.[6] Still gripped in its worst drought to date,[7] it is estimated

---

the UK with a particular focus on the implications for the water environment', The Chartered Institution of Water and Environmental Management (CIWEM) 2014.

[3] The Karoo is best defined by its vegetation, which consists of assorted succulents and low scrub bushes spaced from one foot to several feet apart. The area is devoid of surface water, and its name is derived from the Khoisan word meaning 'land of thirst'. Its sub regions include the intermontane vales of Little Karoo and Great Karoo in the lesser-elevated areas of Western Cape and Eastern Cape provinces and the main Karoo (or Upper Karoo) spanning the vast highland plateau of Northern Cape province to the north and east of the Great Escarpment. Altogether the Karoo occupies about 153,000 square miles (395,000 square km), about one-third of the total area of South Africa.

[4] It is estimated that potential shale gas reserves of between 19 and 23 trillion cubic feet (tcf) exist over an area of roughly 100 000 km$_2$. See the Academy of Science of South Africa (ASSAf) Report, 'South Africa's Technical Readiness to Support the Shale Gas Industry' (2016) at xv <http://research.assaf.org.za/bitstream/handle/20.500.11911/14/fracking_report_2016.pdf?sequence=8&isAllowed=y> accessed 1 November 2016. Also see 'Petroleum Potential of the Karoo Basins' as per the Petroleum Agency of South Africa (PASA) <http://www.petroleumagencysa.com> accessed 18 October 2016; Surina Esterhuyse, Marthie Kemp and Nola Redelinghuys, 'Assessing the existing knowledge base and opinions of decision makers on the regulation and monitoring of unconventional gas mining in South Africa' International Water Resources Association (2013) <http://www.tandfonline.com/doi/pdf/10.1080/02508060.2013.818478?needAccess=true> accessed 15 July 2016.

[5] Shell International, Bundu Gas, and Falcon Oil and Gas (in partnership with Chevron). Exploration and production applications are made to the Petroleum Agency of South Africa (PASA) in terms of the Mineral and Petroleum Resources Development Act 28 of 2002 (MPRDA) – an application for an exploration right in terms of sections 79–82 of the MPRDA and Regulations 28–33 to the Act. See PASA <http://www.petroleumagencysa.com> accessed 4 September 2016.

[6] Only 8 per cent of land area provides 50 per cent of South Africa's surface water; South Africa's rainfall (490 millimetre per annum) is half the global average and most of available reliable water (98 per cent) has already been allocated. South Africa faces high to extremely high water stress where its shale gas is located (Karoo). See Paul Reig, Tianyi Luo and Jonathan N. Proctor 'Global Shale Gas Development – Water Availability and Business Risks' *World Resources Institute* <http://www.wri.org/sites/default/files/wri14_report_shalegas.

that South Africa could face significant water shortages by 2020 due to an expected increase of water needs/usage and the impacts of climate change.[8] This paints a dire picture of particular water scarcity challenges, notably involving the possible trade-offs with fracking's water use and water resource degradation due to pollution and the multiplier effect of climate change, and unsurprisingly, several calls have been made for 'an appropriate and comprehensive' regulatory regime that will guarantee the protection of the Karoo's scarce water resources.[9]

While the current South African regulatory framework applicable to shale gas development (SGD), and more specifically the water related aspects of SGD, have undergone some changes – most notably the draft regulations on SGD being replaced by new regulations in June 2015, other more recent law and governance developments have taken place which could indicate further enhancements to the regulatory framework.[10] These include the commissioning of two reports on aspects of SGD, namely, a Strategic Environmental Assessment (SEA) and a report on South Africa's Technical Readiness for SGD.[11] The scientific assessment phase of the SEA was published on 15 November 2016 and the

---

pdf> 34–37; and 2013 WWF Report 'An Introduction to South Africa's Water Source Areas', 23 <http://awsassets.wwf.org.za/downloads/wwf_sa_watersource_area10_lo.pdf> accessed 22 September 2016.

[7] Since rainfall records began in 1904. See Jan-Jan Joubert, 'SA drought not broken after driest year in history' Times Live <http://www.timeslive.co.za/local/2016/09/08/SA-drought-not-broken-after-driest-year-in-history> accessed 9 September 2016.

[8] The impact of climate change is particular severe in sub-Saharan Africa and drier areas such as the Karoo. See Albert I.J.M. Van Dijk, 'Water resources, climate change and energy' in Jamie Pittock Karen Hussey and Stephen Dovers (eds), *Climate, Energy and Water* (Cambridge University Press 2015), 11.

[9] Centre for Environmental Rights (CER) 'Minimum Regulation for the Regulation of Environmental Impacts of Hydraulic Fracturing in South Africa' 2 <http://cer.org.za/wp-content/uploads/2014/04/CER-Minimum-Requirements-for-the-Regulation-of-the-Environmental-Impacts-of-Fracking-Web.pdf> accessed 5 September 2015.

[10] See 'Proposed Technical Regulations for Petroleum Exploration and Exploitation' in Government Notice No 36938 of 15 October 2013 (as per the Mineral and Petroleum Resources Development Act, 28 of 2002. Replaced by The Regulations for Petroleum Exploration and Production GN R466 in Government Gazette 38855 of 3 June 2015.

[11] The SEA Report, Bob Scholes, Paul Lochner, Greg Schreiner, Luanita Snyman-Van der Walt and Megan de Jager (eds) 'Shale Gas Development in the

second report was completed and published during October 2016.[12] Both the SEA and the Technical Report seem to agree on four main findings and/or suggestions. First, that base line studies and continuous monitoring is urgently needed for all water related aspects of SGD,[13] including air quality and all surface and groundwater aspects related to the water cycle.[14] Steps should be taken to ensure that all regulations relating to SGD keep track of ongoing rapid technological developments and hence also the need to include requirements for the endorsement and implementation of 'internationally recognised state-of-art best practices'.[15] They also highlight the need for alignment and cooperative governance between different government departments and alignment between different pieces of legislation.[16] Related to the latter, there is a critical need for establishing either a new or existing government agency to enable and facilitate alignment and cooperative governance.[17] The discussion below will equally highlight some of these concerns as they permeate the water-SGD governance paradigm in South Africa.

Another development is the declaration of exploration and production of natural oil and gas resources as controlled activities by the Department of Water and Sanitation (DWS), which means that these activities will in future require an environmental impact assessment (EIA).[18] In bringing the exploration and production processes of SGD within the explicit

---

Central Karoo: A Scientific Assessment of the Opportunities and Risks' November 2016 <http://seasgd.csir.co.za/individual-chapter-pdfs/> accessed 15 November 2016.

[12] For example: Scholes *et al.*, *Shale Gas Development in the Central Karoo: A Scientific Assessment of the Opportunities and Risks* CSIR, Pretoria and *South Africa's Technical Readiness to Support the Shale* Gas *Industry*. See <http://seasgd.csir.co.za/individual-chapter-pdfs/> accessed 15 November 2016. Also see The South African Academy of Engineering (SAAE) consensus report *South Africa's Technical Readiness to Support the Shale* Gas *Industry* <http://dx.doi.org/10.17159/assaf.2016/0003> accessed 25 October 2016.

[13] Scholes *et al.* (n 11), 1–69. Also see ASSAf (n 4) xiii–xxii.

[14] Scholes *et al.* (n 11), 23–24, 30–31, 53. Also see ASSAf (n 4) xiii-xxii, 43, 49, 58, 64–6, 68–9, 70–75, 88, 99, 101, 104–105, 107–109; 111–13.

[15] ASSAf (n 4), 171. Also see Scholes *et al.* (n 11), 12–19.

[16] Winkler *et al.*, 'Air Quality and Greenhouse Gas Emissions' in Scholes *et al.*, *Shale Gas Development in the Central Karoo: A Scientific Assessment of the Opportunities and Risks*, CSIR, Pretoria. Also see ASSAf (n 4), xiv, 12, 91.

[17] ASSAf (n 4), xvi, 9, 105. Also see Winkler *et al.* (n 16), 63–4.

[18] Under the National Water Act, 36 of 1998 – see notice 999 of 2015 by DWS <http://cer.org.za/wp-content/uploads/2010/05/gg39299_nn999.pdf> accessed 8 August 2016. Also see K. Pietersen, T. Kanyerere, A. Levine, A. Matshini and H.E. Beekman, 'An analysis of the challenges for groundwater

ambit of the National Water Act[19] (NWA), the DWS is in the process of developing regulations to more pertinently control the water related impacts from SGD.[20]

Importantly, government has been intimately involved in conducting the above studies and drafting these reports; no fewer than five national government departments formed part of the Technical Team to the Ministerial Task Team that commissioned the SEA.[21] These include the Department of Environmental Affairs (DEA), the Department of Science and Technology (DST), the Department of Mineral Resources (DMR), the Department of Energy (DE) and the DWS.[22] The commissioning of these assessment reports seems to indicate that the government will be taking the findings seriously – particularly since there is no legislative requirement in South Africa to conduct an SEA.[23] This commitment by the government to gather data and gain knowledge at a strategic level arguably indicates the likelihood of further amendments or changes to the regulatory framework. More importantly for present purposes, these developments also provide an opportunity to appraise South Africa's regulatory framework related to water issues in the SGD paradigm, which is the main objective of this chapter.

As a contribution to such an appraisal, in this chapter we consider a number of regulatory issues related to water aspects of SGD from a legal perspective, and we assess the extent to which South African law provides regulation of these issues. We do so by focusing on the water life cycle approach for SGD which consists of the following: (i) water acquisition, (ii) chemical mixing, (iii) well injection, (iv) flowback and produced water, and (v) waste water treatment and disposal. Before we

---

governance during shale gas development in South Africa' (2016) *Water SA* 42(3), 426 <http://www.scielo.org.za/pdf/wsa/v42n3/07.pdf> accessed 23 August 2016.

19 The National Water Act, 36 of 1998.

20 Pietersen *et al.* (n 18), 426.

21 Joint statement on launch of strategic environmental assessment of shale gas development, 12 May 2015 <http://www.gov.za/speeches/statement-ministers-environmental-affairs-mrs-edna-molewa-science-and-technology-naledi> accessed 10 February 2016; also see preface to SEA final report by Scholes *et al.* (eds) <http://seasgd.csir.co.za/wp-content/uploads/2016/11/1_Foreword-Contents_14Nov2016_LR.pdf> accessed 15 November 2016.

22 Ibid.

23 Jan Glazewski, 'Sustainable development and proposed shale gas extraction in South Africa: prospects and challenges' in John C. Dernbach and James R. May (eds), *Shale Gas and the Future of Energy Law and Policy for Sustainability* (Edward Elgar Publishing 2016), 223.

commence with the analysis, we first provide a brief background to the existing regulatory framework applicable to the water aspects of SGD, while we also offer some background on the commissioning of the two assessment reports.

2 BACKGROUND TO THE REGULATORY FRAMEWORK

While the Mineral and Petroleum Resources Development Act,[24] (MPRDA) contains the specific regulations applicable to SGD in the form of the Regulations for Petroleum Exploration and Production,[25] (MPRDA Regulations) it must be read with all other relevant legislation, namely the National Water Act[26] (NWA), the Water Services Act[27] (WSA), the National Environmental Management Act[28] (NEMA), the National Environmental Management: Waste Act[29] (NEMWA), and the National Environmental Management: Air Quality Act[30] (NEMAQA).[31] All of these laws deal with one or other aspect of water in the context of SGD. The different laws are administered by different government departments including: the DMR, the DEA, and the DWS.[32] While this sectoral approach is supposed to be mitigated by the co-operative governance mechanism provided for in the Constitution,[33] particularly the requirement that all organs of state must inform and consult one another

24 Act 28 of 2002.

25 Government Notice No 466, 3 June 2015.

26 Act 36 of 1998.

27 Act 108 of 1997.

28 Act 107 of 1998.

29 Act 59 of 2008.

30 Act 39 of 2004.

31 Sections 85(2) and 86(1) of the Regulations.

32 DMR administers the mining and petroleum sector activities in terms of the MPRDA (including the Regulations for Petroleum Exploration and Production). DWS administers the NWA and WSA to include the issuing of water use licenses and water/sanitation services. The DEA administers the NEMA and a number of related laws dealing with, among other matters, waste management (NEMWA) and air quality (NEMAQA).

33 Chapter 3 of the Constitution of South Africa, 1996 section 41(1)(iii)–(iv). The main legislation which gives effect to the constitutional co-operative governance imperative is the Intergovernmental Relations Framework Act, 13 of 2005.

on matters of common interest (including the co-ordination of legislation), there is currently no inter-governmental committee or agency explicitly dealing with SGD in South Africa.[34] This could be problematic since a fragmented approach could increase governance inefficiencies, encourage problem-shifting, and lead to a situation where crucial aspects of water governance are not tackled in an integrated and holistic fashion.[35] Apart from the above noted regulations which deal specifically with SGD, none of the other laws have been amended to specifically incorporate aspects of SGD.

A number of stakeholders have pointed to other potential shortcomings of the regulatory framework.[36] One argument turned on the notion that science should inform the setting of a regulatory framework on SGD, particularly since the biophysical, geological and biodiversity aspects of the Karoo are unique and therefore present unique considerations.[37] This eventually led to the commissioning of the above-mentioned assessment reports which focused on the country being in a unique position in that it can learn from the experience of other shale gas jurisdictions and determine and collect base line data before it commences with SGD.[38] Most jurisdictions that are extracting shale gas have not recorded base line data on critical aspects such as water quantity and quality.[39] This is mainly due to the many changes in land use over the years.[40] Not having any baseline data on aspects such as water use and possible water pollution makes it extremely difficult to determine resource status and it

---

[34] Glazewski (n 23), 227; ASSAf (n 4), 105.

[35] For a general discussion, see Louis Kotzé, *A Legal Framework for Integrated Environmental Governance in South Africa and the North-West Province* (Wolf Legal Publishers, 2007).

[36] For example the Centre for Environmental Rights (CER) <http://cer.org.za/> and Treasure the Karoo Action Group (TKAG) <http://www.treasurethekaroo.co.za/> accessed 2 August 2016.

[37] Glazewski (n 23), 210, 219–20. Also see CER and TKAG (n 36).

[38] ASSAf (n 4), 10–11. Also see Scholes *et al.* (n 11), 1–69.

[39] P. Hobbs *et al.*, 'Water Resources' in Scholes *et al.* 2016 *Shale Gas Development in the Central Karoo: A Scientific Assessment of the Opportunities and Risks* 5–57 <http://seasgd.csir.co.za/wp-content/uploads/2016/11/Ch-5_Water_13Nov2016_LR.pdf> accessed 16 November 2016.

[40] Seth B. Shonkoff, Jake Hays and Madelon L. Finkel 'Environmental Public Health Dimension of Shale and Tight Gas Development' (2014) *Environmental Health Perspectives* 122(8), 787–95 <http://ehp.niehs.nih.gov/wp-content/uploads/122/8/ehp.1307866.pdf> accessed 3 August 2016.

militates against the idea that science should inform the design of a regulatory framework on these aspects.[41] The Karoo area is relatively underdeveloped due to its land use not having changed much – agriculture remains the main economic activity, hence the possibility to record base line data *ex ante* SGD.[42] The need to provide base line data was therefore arguably one of the main drivers behind the commission of the reports and assessments described above.

3 THE WATER CYCLE OF SGD

A number of potential instances exists whereby SGD could affect water resources.[43] These could be distinguished between those taking place either above or below ground.[44] Known hydraulic fracturing methods that could affect surface and ground water include water withdrawals at times or in locations of water scarcity, spills of hydraulic fracturing fluid and chemicals, and inadequate treatment and discharge of hydraulic fracturing waste water.[45] Methods impacting groundwater include movement of liquid and gas via the production well into underground water resources,

---

[41] See the Strategic Environment Assessment for Shale Gas Development in South Africa: <http://seasgd.csir.co.za/> accessed 3 June 2016. Also see Pietersen *et al.* (n 18), 428–9.

[42] See the SEA for Shale Gas Development in South Africa which was commissioned by the DEA with a project team consisting of the Council for Scientific and Industrial Research (CSIR); the South African Biodiversity Institute (SANBI); and the Council for Geoscience (CGS) and a Report by the Academy of Science of South Africa (ASSAf) in collaboration with the South African Academy of Engineering (SAAE) commissioned by the Department of Science and Technology (DST), *South Africa's Technical Readiness to Support the Shale Gas Industry*. For the former see <http://seasgd.csir.co.za/> (http://seasgd.csir.co.za/scientific-assessment-chapters/) and for the latter <http://research.assaf.org.za/handle/20.500.11911/14> accessed 17 November 2016 and 20 October 2016.

[43] Hobbs *et al.* (n 39), 63, 70 <http://seasgd.csir.co.za/wp-content/uploads/2016/11/Ch-5_Water_13Nov2016_LR.pdf accessed> 16 November 2016.

[44] In other words, not limited to the act of fracturing i.e. the pumping of fluids under high pressure.

[45] Technical report by WWF 2015, *Shale Gas: Introduction to water impacts – Overview of current state of the knowledge on water sourcing, wastewater management and possible water contamination risks related to shale gas operations and their implications for shale gas development in South Africa's Karoo Basin*, 33–49. Also see EPA, 'Hydraulic Fracturing Drinking Water Assessment' (June 2015) EPA/600/R-15/047a, ES-23 <https://

and movement of liquid and gas from the fracture zone to these resources via pathways in subsurface rock formations.[46] Potential impacts of SGD on water resources can be mapped by way of a water project lifecycle approach. A typical water project lifecycle for SGD consists of water acquisition, chemical mixing, well injection, flow back and produced water, and waste water treatment and waste disposal.[47]

3.1 Water Acquisition

Water acquisition from surface and groundwater resources obviously depends on water availability (quantity), while withdrawal also impacts water quality.[48] Studies indicate that water use can vary from 8000 to 100 000 m^3 per shale gas well.[49] This variability in water use is due to a number of factors such as differences in well borehole configuration and volume, the type of hydrocarbon, the specific target gas reservoir, and the drill year of the well.[50] In instances where groundwater withdrawals exceed natural recharge rates, water storage in aquifers decreases, thereby activating contaminants or allowing the infiltration of lower quality water from the land surface or nearby formations.[51] In addition, withdrawals could decrease groundwater discharge to streams, thereby possibly affecting surface water quality.[52] Excessive water use may lead to a fall in the availability of public water supply and ecosystem degradation, and it may have adverse effects on aquatic habitats, erosion, and changes in water temperature.[53]

---

www.epa.gov/sites/production/files/2015-06/documents/hf_es_erd_jun2015.pdf> accessed 2 February 2016. Also see Pietersen *et al.* (n 18), 421–2.

46 Pietersen *et al.* (n 18), 421–422; EPA (n 45) ES-23; WWF 2015 (n 45), 33–49.

47 See Corrie E. Clark, Robert M. Horner and Christopher B. Harto 'Life Cycle Water Consumption for Shale Gas and Conventional Natural Gas' (2013) *Environmental, Science and Technology* 47(20), 11829–36.

48 Typically water for hydraulic fracturing comes from surface water, groundwater, or reused hydraulic fracturing waste water. See EPA (n 45), ES-7.

49 Pietersen *et al.* (n 18), 422; and ASSAf (n 4), 44–5.

50 Pietersen *et al.* (n 18), 422.

51 Monica Freyman, 'Ceres Report Hydraulic Fracturing & Water Stress: Water Demand by the Numbers' February 2014; EPA (n 45), ES-10.

52 EPA (n 45), ES-10. Withdrawals of surface water could also affect water quality because such withdrawals may lower water levels and change stream flow, thereby decreasing a stream's ability to dilute contaminants.

53 Pietersen *et al.* (n 18), 422–9.

High water use for fracking or consumption alone does not necessarily result in impacts to water resources.[54] It is rather the combination of water use or consumption and water availability at local scale that result in impacts.[55] Areas where significant groundwater pumping occurs are most likely to experience impacts, especially drought prone regions with limited groundwater recharge such as the Karoo area.[56] A study by Du Toit[57] found that 31 of the 34 sizeable towns within the Karoo shale gas exploration concessions rely wholly or partially on underground aquifers. Therefore, the possible impact that hydraulic fracturing could have on groundwater in the Karoo area is significant.[58]

The MPRDA Regulations have provisions on general water management such as water balances, the protection of water resources and water use, which will apply in the context of SGD.[59] Recently the DWS declared the water use aspect of SGD as a controlled activity:[60]

> The exploration and or production of onshore naturally occurring hydrocarbons that requires stimulation, including but not limited to fracking and or underground gasification, to extract, and any activity incidental thereto that may impact detrimentally on the water resource.

This declaration of SGD as a controlled activity by DWS, confirms that water use licences will be required and that conditions could be attached to any water use licence.[61] The latter turns on the term 'use of water' which includes a number of activities, and more importantly, the potential for pollution. The significance of this is that in addition to the standard water use licence conditions, the responsible authority can add conditions '... specifying the waste treatment, pollution control, and monitoring equipment to be installed, maintained and operated; and specifying the

---

54 EPA (n 45), ES-9.

55 Ibid.

56 EPA (n 45), ES-10.

57 Julienne Du Toit, 'Fracking: Groundwater threats to Karoo towns' <http://karoospace.co.za/fracking-groundwater-threats-to-karoo-towns/> accessed 1 February 2016.

58 See Mark Broomfield, 'Support to the identification of potential risks for the environment and human health arising from hydrocarbons operations involving hydraulic fracturing in Europe' (Report No. ED57281-17c) 2012 European Commission DG Environment.

59 See the Regulations for Petroleum Exploration and Production 121–3. Government Notice No. 466, 3 June 2015.

60 By invoking section 38 of the NWA. General Notice No. 999 as section 37(e) of the NWA.

61 Hobbs *et al.* (n 39), 52.

management practices to be followed to prevent the pollution of any water resource.'[62] In other words the DWS is empowered to formulate its own regulations on SGD.[63] Other applicable sections of the NWA include: the determination of a reserve before the issuing of a water use licence; the prevention of pollution incidents and process to be followed in case of emergency incidents; and the monitoring, assessment and distribution of information.[64] The NWA states that a national monitoring and information system must be established – to include a national register of water authorizations, and an information system on quality and quantity of water resources, which could provide valuable data on SGD and water related aspects.[65] The NWA also calls for consultation with water management institutions and existing and potential water users.[66]

3.2 Chemical Mixing

To create the hydraulic fracturing fluid, water, chemicals and proppant are mixed.[67] Due to the large volume of fluids required for the hydraulic fracturing operation, thousands of litres of chemicals can potentially be stored on site.[68] The on-site storage, mixing and pumping of chemicals could lead to accidental releases in the form of spills or leaks.[69] Estimates of the overall median per chemical per well are in the range of 2500 litres; on the further assumption of 14 unique chemicals used per well, an estimated 34 000 litres of chemicals may be injected per well.[70]

---

62 ASSAf (n 4), 15–16.

63 Hobbs *et al.* (n 39), 52.

64 NWA see sections 19–20 and 137–45.

65 Ibid.

66 Ibid.

67 The purpose of the hydraulic fracturing fluids is to create and extend fractures, to transport proppant and to place proppant in the fractures. See EPA (n 45), ES-10.

68 EPA (n 45), ES-10. Large variabilities in the composition of hydraulic fracturing fluids result from factors such as the geology of the specific formation, availability and cost of different chemicals and service operator preference.

69 EPA (n 45), ES-10.

70 EPA (n 45), ES-12. These estimates have been developed from information on FracFocus (chemical registry for hydraulic fracturing – it is managed by the ground Water Protection Council and Interstate Oil and Gas Compact Commission in the USA. For FracFocus see <http://fracfocus.org> accessed 7 October 2015.

A report by the American Environmental Protection Agency (EPA)[71] indicates that causes of spills and leakage of fracking fluid and chemicals tend to include equipment failure, failure of container integrity, human error and to a lesser extent weather and vandalism. Corrosion and failed valves on blowout preventers seem to be the most common cause of equipment failure.[72] Mobility, solubility and volatility are all properties affecting the likelihood of a spilled chemical reaching and impacting a water resource.[73] There are a number of ways in which a spill can potentially contaminate water resources. They include overland flow to nearby surface water, soil contamination, and infiltration and contamination of underlying groundwater.[74] Impacts of such spills could occur quickly, be delayed for short or long periods of time, or have a continual effect over time.[75]

Regulatory oversight of hydraulic fracturing injection fluids differs from one jurisdiction to another and also within the same jurisdiction.[76] A comparison of a number of jurisdictions confirms that the requirements for disclosure of chemicals used vary highly. Factors that play a role in the effectiveness of disclosure rules include the timing of disclosure; the type of information provided (i.e. what is disclosed); and whether the information is accessible to both the regulator and the public.[77] The United States (US) does not have national requirements on the disclosure of chemical formulations and tends to leave it to the different states to implement their own regulations.[78] Although a chemical disclosure registry exists in the US in the form of FracFocus, it is a voluntary disclosure site for a particular area.[79] While generic information is

---

[71] EPA (n 45), ES-11.

[72] Ibid.

[73] Some chemicals, such as choline chloride and tetrakis phosphonium, tend to move quickly through the environment with the flow of water, while others have the potential to stay in the environment as long-term contaminants. Not many of the chemicals volatilize and hence a large portion of most hydraulic fracturing fluids tends to stay in water. See EPA (n 45), ES-12.

[74] EPA (n 45), ES-13.

[75] Ibid.

[76] Pietersen *et al.* (n 18), 423.

[77] Ibid.

[78] Pietersen *et al.* (n 18), 424.

[79] 28 states require disclosure of hydraulic fracturing fluids (chemicals) while 23 states direct its reporting to FracFocus. See Kate Konschnik and Archana Dayalu, 'Hydraulic fracturing chemicals reporting: Analysis of available data and recommendations for policymakers' (2016) *Energy Policy*, January 88, 504.

provided on the commercial names of the identified chemicals, details tend to be lacking on the actual quantities of chemicals used.[80] The absence of agreed national or state-by-state approaches to the disclosure of chemicals used has led to a number of challenges such as nondisclosure of proprietary or 'trade secret' mixtures, insufficient penalties for reporting inaccurate or incomplete information, and timelines that allow for after-the-fact reporting.[81]

Canada also employs a version of FracFocus whereby chemical fluids must be identified and disclosed within 30 days of the completion of a fracturing operation.[82] This leaves little opportunity for stakeholders to raise concerns before actual injection of chemical fluids occurs.[83] In contrast, both Western Australia and the United Kingdom (UK) require complete disclosure of chemical fluid additives before the start of fracturing operation(s).[84]

Although a number of fracturing fluid substances are prohibited from being used in South Africa,[85] the backdoor is left open to do exactly that, namely in instances where it is not technically feasible to use existing fluids, an application can be made to the designated agency to add or use additional additives.[86] In addition, while the use of hydraulic fracturing fluids must be reported to the designated agency, no obligation exists to make this information public. The only foreseeable way for the public to obtain this information will be to make use of the Promotion of Access to Information Act 2 of 2000 (PAIA). However, a likely ground for refusal to such a request could be that of 'confidential commercial information' and hence there is the possibility of protracted court proceedings to obtain the information.[87] Disclosure of the composition of the fracking fluid additives is critical in ensuring that correct treatment

---

80 Pietersen *et al.* (n 18), 424.

81 A. Maule, C. Makey, E. Benson, I. Burrows and M. Scammell, 'Disclosure of Hydraulic Fracturing Fluid Chemical Additives: Analysis of Regulations' (2013) *New Solutions* 23(1), 167.

82 Ana Komnenic, 'National Energy Board Tells Companies to Publicly Disclose Hydraulic Fracturing Fluid Recipes' *Mining and Energy*, November 2013 <https://www.biv.com/article/2013/11/national-energy-board-tells-companies-to-publicly-/> accessed 15 July 2016.

83 Pietersen *et al.* (n 18), 424.

84 Ibid. Western Australia requires this disclosure in the form of companies' environmental plans.

85 See schedule 1 to MPRDA regulation 113(1) GN R 466 of 3 June 2015.

86 Regulation 109(2).

87 W. Du Plessis, 'Regulation of Hydraulic Fracturing in South Africa: A Project Life-Cycle Approach?' (2018) *PELJ* 18(5), 1466–7.

options are identified and followed in cases of pollution,[88] and South African law must make clearer and more comprehensive provision for such disclosure.

3.3 Well Injection

Foremost ways of possible contamination of water resources by the injection of hydraulic fluids include the unintended movement of liquids or gases from the well or along the outside of the well via deficiencies in the well's casing or cement into a water resource, and movement of fluids from the production zone through subsurface geological formations into a water resource.[89] A particularly important well construction feature is cemented surface casing to keep water resources isolated from liquids and gases that may move through the subsurface.[90] Inadequate design or construction issues of casing and/or cement could lead to possible contamination of water resources and are of critical importance.

Construction issues, sustained pressure on casings and the presence of natural faults and fractures could work together to create pathways for fluids and gases to move into water resources.[91] The possibility of the likelihood of subsurface migration of liquids and gases to water resources could also depend on the vertical distance between the base of the water resources and the top of the shale formation. Less vertical distance or separation between the two creates a proportionately greater risk of liquids or gas migrating to water resources. Shale gas resources and water resources could also in some instances co-exist in the same formation. In these instances the injection of hydraulic fracturing fluids will affect the water resource, because some of the fluid will likely stay in the formation after completion of the hydraulic fracturing.[92] Wells could also intersect (generally known as 'frack hits' or 'well communications'), possibly resulting in the flow of liquids and gases into nearby wells (or on the same well pad) or their fracture networks.[93] Older or inactive wells may pose significant risks as they are more likely to be

---

[88] S. Oelofse, J. Schoonraad, and D. Baldwin, (2016) 'Impacts on Waste Planning and Management' in Scholes *et al.*, *Shale Gas Development in the Central Karoo: A Scientific Assessment of the Opportunities and Risks* CSIR, Pretoria.

[89] A combination of the two is also possible. See EPA (n 45), ES-13.

[90] EPA (n 45), ES-14.

[91] EPA (n 45), ES-15.

[92] Ibid.

[93] EPA (n 45), ES-16.

negatively affected by stresses caused by nearby hydraulic fracturing operations that have not been plugged properly (i.e. without cement).[94]

The MPRDA Regulations indicate that the American Petroleum Institute (API) standards must be used for certain aspects, such as well construction, cementing, compressive tests, blowout prevention, hydraulic fracturing equipment etc.[95] These standards do not necessarily represent international practice in setting out best international regulatory requirements, particularly as standards (and studies) have been produced by other authorities, such as the International Energy Agency, the UK's Department of Energy and Climate Change and the European Commission.[96] The typical reasoning is that because hydraulic fracturing has taken place in the US for some time, other countries should look to and learn from the US's experience. While this is true, it does not mean that the standards set by the API are the only standards and they cannot unequivocally apply universally to other jurisdictions and circumstances. Any standard subscribed to on well construction and cementing should take into account the unique geological character of the Karoo area (as a whole), including the different geological conditions of the different regions within the Karoo.[97]

---

94 Ibid.

95 Regulation 96 on well construction standards, regulation 102 on cement requirements and compressive tests, regulation 105 on blowout prevention, regulation 111 on hydraulic fracturing equipment. For general API standards see <http://www.api.org/Publications-Standards-and-Statistics/Standards/WhatsNew/Publication-Updates/New-Exploration-And-Production-Publications> accessed 2 February 2016.

96 Commission Recommendation of 22 January 2014 on minimum principles for the exploration and production of hydrocarbons (such as shale gas) using high-volume hydraulic fracturing 2014/70/EU <http://eur-lex.europa.eu/legal-content/EN/TXT/PDF/?uri=CELEX:32014H0070&from=EN> accessed 23 November 2015, <https://www.gov.uk/government/uploads/system/uploads/attachment_data/file/283834/Regulation_v3.pdf> accessed 5 December 2015, <https://www.gov.uk/government/publications/about-shale-gas-and-hydraulic-fracturing-fracking/developing-shale-oil-and-gas-in-the-uk> accessed 5 December 2015.

97 ASSAf (n 4), 27.

3.4 Flowback and Produced Water[98]

After the injection pressure is released following a hydraulic fracturing operation, water flows back from the well. This water is initially similar to the fracturing fluid, but with time the composition is affected by the characteristics of the specific formation and possible reactions between the formulation and the fracturing fluid.[99] Depending on the amount of time since the hydraulic fracturing was completed and on the specific well, the amount of produced water varies, averaging from 10 per cent to 25 per cent of injected volumes.[100] However, in some instances the total volume of produced water can be equal to or exceed the injected volume of the fracturing fluid.[101] If produced water is spilled it has the potential to impact water resources – spills could also occur away from the well, such as during truck and pipeline transport. The EPA[102] estimates that the median volume per produced water spills is more than double that for spills of hydraulic fracturing fluids and chemicals – the cause of most of the produced water spills is failure of container integrity. As is the case with chemicals in hydraulic fracturing fluid, mobility, solubility and volatility are all properties affecting the likelihood of a spilled chemical reaching and impacting a water resource.[103] The volume, timing and composition of produced water are all factors that define the level of impact on water resources.[104] Produced water that is disposed of in unlined pits or impoundments can lead to chronic releases and have long-term impacts.[105] Provisions on flowback and produced water that

---

98 In most of the literature, flowback water refers to the water that is initially produced after the hydraulic fracturing operation. Produced water refers to hydraulic fracturing fluids and any formation water that returns to the surface. See EPA (n 45) ES-16.

99 Factors that influence the composition of produced water include the type of formation fractured, the composition of the injected fracturing fluids, subsurface processes and residence time (duration). EPA (n 45), ES-16.

100 Flow rates are high in volume initially but decrease over time. EPA (n 45), ES-16.

101 EPA (n 45), ES-16.

102 Ibid. ES-17.

103 Ibid. ES-18.

104 EPA (n 45), ES-19.

105 Environmental Protection Agency June 2015 report: 'Assessment of the Potential Impacts of Hydraulic Fracturing for Oil and Gas on Drinking Water Resources' External Review Draft EPA/600/R-15/047a, ES-19.

are indirectly applicable can be found in the MPRDA Regulations,[106] the NEMWA (defined as hazardous waste),[107] the NWA (under its pollution prevention section),[108] and the NEMA (also under its pollution prevention section).[109]

3.5 Waste Water Treatment and Waste Disposal

Water generated from the hydraulic fracturing process requires management and disposal.[110] There are a number of options available to manage contaminated waste water from SGD. These options include: the minimization of produced water generation, recycling and reuse within operations, treatment, disposal and beneficial reuse outside the operations.[111] From a practical point of view, the management of waste water translates into the following choices: via evaporation ponds, by disposal in underground disposal wells, treatment at waste treatment facilities followed by reuse or discharge to surface waters or treatment works, reuse with minimal or no treatment, and land application or road spreading.[112] The disposal and management of waste water could impact water resources in a number of ways, including: inadequate treatment of waste water prior to discharge to a receiving water source, accidental releases during transport or leakage from waste water storage pits, unpermitted discharge, migration of constituents in waste water following land application, inappropriate management of residual materials from treatment, and accumulation of waste water constituents in sediments near outfalls of waste treatment

---

106 Chapter 9 of the Regulations per MPRDA at 115 (Fracturing Fluids Management); 116 (Management of Flowback and Produced Fluids); 117 (Transportation of Fluids); 118 (Storage of Fluids) and 126 (Management of Spillage.

107 Schedule 3 of the Waste Amendment Act, 2014. NEMWA imposes controls on 'waste', 'hazardous waste', 'waste disposal facility' and requires the licensing of any 'waste management activity' in accordance with Chapter 5 of the Act. Also see Oelofse, Schoonraad and Baldwin (n 88), 7.

108 Section 19 of NWA.

109 Section 28 of NEMA.

110 For purposes of this research the emphasis is to focus on the water aspect of SGD. There are, however, numerous other possible waste streams from SGD, for example construction and demolition waste, drill cuttings and drilling muds. Oelofse, Schoonraad and Baldwin (n 88), 6.

111 Oelofse, S., Schoonraad, J. and Baldwin, D. (n 88), 8.

112 EPA (n 45), ES-19–20.

facilities.[113] Another possible impact on water resources could be by disposal of waste water into underground injection control wells.[114]

Decisions on choice of waste water management are generally based on the availability and costs of transport, disposal and treatment facilities.[115] Waste water treatment facilities or treatment works could discharge treated waste water to surface water, or send it back to well operators for reuse in other fracturing operations.[116] However, it is arguable that the waste water treatment works of municipalities in the Karoo, generally speaking, do not currently have the capacity to treat extra waste water and are ill-equipped to treat the specific type of waste water generated from SGD. The treatment and disposal of waste water have much in common with the flowback and produced water cycle discussed above. This is likewise a regulatory concern that is addressed to some extent in the provisions of the MPRDA Regulations,[117] the NEMWA (under its hazardous waste provisions),[118] the NWA (under its pollution prevention section),[119] and the NEMA (under its pollution prevention section).[120]

4 CONCLUSION

South Africa's law and governance framework related to water governance in the context of SGD is a mixed bag. What is clear is that the country does not have a single, integrated law that comprehensively provides regulation of all water aspects during the life cycle stages of SGD. Although a juridical mechanism exists to mitigate this fragmented approach to water-SGD governance in the form of cooperative governance, there is currently no intergovernmental agency or body to facilitate and enable cooperative governance and the alignment of different laws, agencies and governance efforts. We believe government would be well-advised to establish such an institution as soon as possible so that it can benefit from an integrated and holistic approach to water-SGD governance.

---

113 Ibid. ES-19.

114 Ibid. ES-20.

115 Ibid.

116 Ibid.

117 Chapter 9 of the Regulations at 116 (per MPRDA) refers to the management of flowback and produced fluids as per a waste management plan (at 125).

118 See n 107 above.

119 Section 19 of NWA.

120 Section 28 of NEMA.

Relatedly, it is clear that some progress has been made to enact laws and create governance structures to regulate the water aspects of SGD. This is evident from the various (albeit scattered) legal provisions we have briefly referred to above. Whereas the country had virtually no legal provisions dealing with water-SGD governance only a few years ago, at least it now has a range of provisions that could broadly assist in regulating many of the issues that arise in each stage of the SGD life cycle. A future option might be to consolidate all provisions that directly relate to SGD under a comprehensive and integrated SGD law or set of regulations. This would allow for more comprehensive and detailed regulation and simultaneously address the issue of fragmented governance noted above.

With the declaration of SGD as a controlled activity by the DWS, this lead water governance agency has asserted itself as a prominent and crucial regulatory authority in SGD governance, alongside other government departments. The involvement of DWS should considerably strengthen the regulatory framework in the protection of South Africa's water resources and it should ensure that water issues are more adequately represented in a predominantly economic-development oriented governance sector, notably one that could easily be dominated by the DMR and DE.

Finally, it is encouraging to see that South Africa is embarking on scientific research to inform its regulatory approach and regime of SGD. In the process it is also looking elsewhere to determine best water-SGD governance practices and relying on these to augment deficiencies and gaps in its own law and governance regime. But while there is the need to continuously look to adopt and apply best international practices to all water related aspects of SGD, there is the danger of not recognizing the unique local characteristics of South Africa's Karoo region. We would accordingly suggest a cautious and nuanced approach that is best suited to South Africa's unique natural and socio-economic conditions and circumstances.

Index